Chromosome Engineering in Plants

Chromosome Engineering in Plants

Contributors

Teresa Cuellar, Eric Belhassen et al.

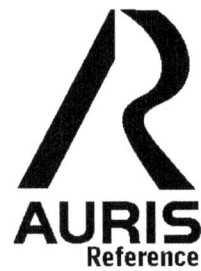

www.aurisreference.com

Chromosome Engineering in Plants

Contributors: Teresa Cuellar, Eric Belhassen et al.

Published by Auris Reference Limited

www.aurisreference.com

United Kingdom

Chromosome Engineering in Plants

ISBN: 978-1-78154-958-2

British Library Cataloguing in Publication Data
A CIP record for this book is available from the British Library

Printed in the United Kingdom

Exclusively distributed by CBS Publishers & Distributors Pvt. Ltd.

Sales & Distribution Rights only for India, Pakistan, Bangladesh, Sri Lanka, Nepal and Bhutan. This book is not to be sold outside these territories.

Contents

List of Abbreviations

AEs	Axial Elements
ALS	Acetolactate Synthase
BFB	Breakage-Fusion-Bridge
BIBAC	Binary Bacterial Artificial Chromosome
CCD	Chilled Charge-Coupled Device
ChIP	Chromatin Immunoprecipitation
CPC	Chromosome Passenger Complex
CS	Chinese Spring
CSC	Cellulose Synthase Complex
CSILs	Chromosome Segment Introgression Lines
DEGs	Differentially Expressed Genes
DSB	Double-Stranded Break
ESTs	Expressed Sequence Tags
FISH	Fluorescence In Situ Hybridization
HFDs	Histone Fold Domains
LEs	Lateral Elements
LINE	Long Interspersed Nuclear Element
LSC	Large Single Copy
MET	Multi-Environment Trials
MTOC	Microtubule-Organizing Centers
NCBI	National Center for Biotechnology Information
NORs	Nucleolar Organizer Regions
PEG	Polyethylene Glycol
RB	Right Border
RE	Relative Enrichment
SCMT	Silicon Carbide Mediated Transformation
SSR	Simple Sequence Repeats
ZFNs	Zinc-Finger Endonucleases

List of Contributors

Mansoor Zoveidavianpoor
Universiti Teknologi, Faculty of Petroleum & Renewable Energy Engineering, Malaysia

Ariffin Samsuri
Universiti Teknologi, Faculty of Petroleum & Renewable Energy Engineering, Malaysia

Seyed Reza Shadizadeh
Petroleum University of Technology, Abadan Faculty of Petroleum Engineering, Iran

Lixin Chen
Key Laboratory Soil and Water Conservation and Desertification Combating, Ministry of Education, College of Soil and Water Conservation, Beijing Forestry University, Beijing, People's Republic of China

Zhiqiang Zhang
Key Laboratory Soil and Water Conservation and Desertification Combating, Ministry of Education, College of Soil and Water Conservation, Beijing Forestry University, Beijing, People's Republic of China

Brent E. Ewers
Program in Ecology, Department of Botany, University of Wyoming, Laramie, Wyoming, United States of America

Stefan Wirtz
Department of Physical Geography, Trier University, Trier, Germany

Manuel Seeger
Department of Physical Geography, Trier University, Trier, Germany
Department of Land Degradation and Development, Wageningen University, Wageningen, The Netherlands

Andreas Zell
Department 7.3- Technical Physics, Saarland University, Saarbrü cken, Germany

Christian Wagner
Department 7.3- Technical Physics, Saarland University, Saarbru¨ cken, Germany

Jean-Frank Wagner
Department of Geology, Trier University, Trier, Germany

Johannes B. Ries
Department of Physical Geography, Trier University, Trier, Germany

Stanislaus J. Schymanski
Department of Environmental Systems Sciences, ETH Zurich, Zurich, Switzerland

Dani Or
Department of Environmental Systems Sciences, ETH Zurich, Zurich, Switzerland

Maciej Zwieniecki
Department of Plant Sciences, University of California Davis, Davis, California, United States of America

Rebecca T. Barnesa
Department of Earth Science, Rice University, Houston, Texas, United States of America

Morgan E. Gallagher
Department of Earth Science, Rice University, Houston, Texas, United States of America

Caroline A. Masiello
Department of Earth Science, Rice University, Houston, Texas, United States of America

Zuolin Liu
Department of Earth Science, Rice University, Houston, Texas, United States of America

Brandon Dugan
Department of Earth Science, Rice University, Houston, Texas, United States of America

Yifeng Chen
State Key Laboratory of Water Resources and Hydropower Engineering Science, Key Laboratory of Rock Mechanics in Hydraulic Structural Engineering, Wuhan University, P. R. China

Chuangbing Zhou
State Key Laboratory of Water Resources and Hydropower Engineering Science, Key Laboratory of Rock Mechanics in Hydraulic Structural Engineering, Wuhan University, P. R. China

Wenzel Kröber
Institute of Biology, Geobotany and Botanical Garden, Martin-Luther-University Halle-Wittenberg, Halle (Saale), Germany

Shouren Zhang
State Key Laboratory of Vegetation and Environmental Change, Institute of Botany, the Chinese Academy of Sciences, Beijing, China

Merten Ehmig
Institute of Biology, Geobotany and Botanical Garden, Martin-Luther-University Halle-Wittenberg, Halle (Saale), Germany

Helge Bruelheide
Institute of Biology, Geobotany and Botanical Garden, Martin-Luther-University Halle-Wittenberg, Halle (Saale), Germany
German Centre for Integrative Biodiversity Research (iDiv) Halle-Jena-Leipzig, Leipzig, Germany

Julia L. Barringer
U.S. Geological Survey, USA

Pamela A. Reilly
U.S. Geological Survey, USA

Arshad Ashraf
Water Resources Research Institute, National Agricultural Research Center, Islamabad, Pakistan

Fereshte Haghighi
Soil Conservation and Watershed Management Institute, Tehran,

Mirmasoud Kheirkhah
Soil Conservation and Watershed Management Research Institute, Tehran, Iran

Bahram Saghafian
Soil Conservation and Watershed Management Research Institute, Tehran, Iran

Preface

With an increasing human population and a decreasing amount of arable land, creative improvements in agriculture will be a necessity in the coming decades to maintain or improve the standard of living. The text *Chromosome Engineering in Plants* presents techniques for the modification of crops and other plant species in order to achieve the goal of developing the much needed novel approaches to the production of food, feed, fuel, fiber, and pharmaceuticals. First chapter focuses on chromosomal differentiation in *Helianthus annuus var.macrocarpus.* Recent advances in plant transformation have been presented in second chapter. In third chapter, we isolate a cDNA encoding $CENH_3$ from sunflowers and raise a peptide antibody against $CENH_3$. In fourth chapter, we review the current knowledge on the processes taking place during chromosome segregation in plant meiosis, focusing on the characterization of the molecular factors involved. Principles and applications of immunolocalization technique of chromosome-associated proteins in plants have been discussed in fifth chapter. Sixth chapter discusses about chromosomal location of *HWA1* and *HWA2*, complementary hybrid weakness genes in rice. Structure and stability of telocentric chromosomes in wheat have been described in seventh chapter. Major locus for chloride accumulation on chromosome 5A in bread wheat has been outlined in eighth chapter. In last chapter, we identify three CSILs with stronger fiber or high fiber strength that carried different *G. barbadense* chromosome segment(s) in the recurrent parent TM-1.

Chapter 1

CHROMOSOMAL DIFFERENTIATION IN HELIANTHUS ANNUUS VAR.MACROCARPUS: HETEROCHROMATIN CHARACTERIZATION AND RDNA LOCATION

Teresa Cuellar[1], Eric Belhassen[2], Begoña Fernández-Calvín[3], Juan Orellana[3], and Jose L Bella[1]

[1]Departemento de Biología, Unidad de Genética, Facultad de Ciencias, Universidad Autónomo de Madrid, E-28049 Madrid, Spain

[2]Génétique et Amélioration des Plantes, INRA, Place Pierre Viala 2, F-34060 Montpellier, France

[3]Departamento de Biotecnología, ETSI Agrónomos, Universidad Politécnica de Madrid, E-28040 Madrid, Spain

ABSTRACT

The $2n = 34$ chromosomes of the inbred line HA89, and the Flamme and Mirasol hybrids of *Helianthus annuus* var. *macrocarpus* possess centromeric heterochromatin as established by Giemsa C-banding. This heterochromatin can not be differentiated by fluorochromes such as DAPI or Chromomycin A3, with selective affinity for specific DNA base pairs. This situation probably results from either a balanced AT/GC composition of the involved repeat or the existence of alternating repetitive sequences of opposite base pair composition in these heterochromatic areas. However, there is also heterochromatin associated with the secondary constrictions on three pairs of chromosomes. This heterochromatin appears to be GC-rich according to its response to the fluorochrome treatments, thus indicating heterochromatin heterogeneity in *H. annuus*. Silver staining reveals the existence of active NORs associated with these secondary constrictions. *In situ* hybridization with an rDNA probe confirms these results and makes the existence of other inactive rDNA sites unlikely. These results are relevant to evolutionary and breeding studies on sunflowers.

INTRODUCTION

Although the genus Helianthus is of clear economic and evolutionary interest, the karyotypic status of the 50 or so species that make up the genus (Schilling & Heiser, 1981) has not been analysed to any great extent (Chandler, 1991). Sunflower chromo- somes have not been individually identified and numbered, so there are no cytological markers and no work on chromosome banding has been reported. Although some genetic linkage groups have been identified, they have yet to be localized to specific chromosomes (Rieseberg et al., 1993; Gentzbittel et at., 1995). Most of the karyological information about this taxon has come from the analysis of meiotic pairing in interspecific hybrids in the study of hybrid fertility, introgression and genome relationships (Chandler et al., 1986; Chandler, 1991). Our limited knowledge is a consequence of the *Correspondence. intrinsic difficulties of studying an organism which has a relatively large number of comparatively small chromosomes. Furthermore, their occurrence in an oily cytoplasm makes good quality preparations diffi- cult to obtain.

On the other hand, the use of banding techniques is a general procedure by which markers for individual chromosomes may be obtained, thereby permitting discrimination of species and varieties at the karyological level. In the sunflower the lack of such markers has made it difficult to analyse the introgression of economically interesting traits and the evolutionary relationships of the genus are not well understood. In fact, it has even been proposed that the recognized species of Helianthus are ecotypes of a single ecospecies (Kulshreshtha & Gupta, 1979). As we show in this study, the use of pectinase treatment improves the quality of sunflower chromo- somal preparations. This has allowed us to characterize the chromosomes of one of the commonest inbred lines and two varieties of the cultivated sunflower species Helianthus annuus L. We describe the distribution of the constitutive heterochromatin, its response to specific DNA base pair ligands such as 4'-6-diamidino-2-phenylindole (DAPI) or Chromomycin A3, and the number and position of the Nucleolar Organizer Regions (NORs), as demonstrated by silver staining and fluorescence in situ hybridization (FISH) with an rDNA probe.

MATERIALS AND METHODS

Seeds from the inbred line HA89 of H. annuus var. macrocarpus and the commercial hybrids Mirasol and Flamme were grown in the dark at 25°C. To obtain mitotic chromosomes arrested at metaphase, roots were collected after 3 days and placed in a 0.04 per cent solution of colchicine for 90 mi fixed in 3:1 absolute ethanol: glacial acetic acid and stored at 4°C. Roots were pretreated with a 0.5 per cent pectinase solution in standard pH 4.2 citrate buffer for 1 h at 37°C. The meristems were squashed in 45 per cent glacial acetic acid on microscope slides using a coverslip, the coverslip was removed with a razor blade after immersion in liquid nitrogen and the slides were air dried. Additionally, some root tips were stained employing the standard Feulgen technique and squashed on slides, according to Darlington & La Cour (1976, pp. 125—127). A minimum of 15 roots per line or variety was studied for each technique employed

C-banding and silver staining treatments were carried out according to Schwarzacher et al. (1980) and Lacadena et a!. (1984), respectively, with minor modifications. Air-dried slides were stained after a minimum of 3 days with 4'-6-diamidino-2-phenylindole (DAPI) or Chromomycin A3 (CMA3) and counterstained with Distamycin-A (DA) or Actinomycin-D (AD), as described by Schweizer (1976, 1980).

The 9 kbp EcoRI rDNA fragment of wheat containing the 18S, 5.8S and 26S rRNA coding regions (plus the spacers), obtained from the pTA71 clone (Gerlach & Bedbrook, 1979) was nick translated with digoxigenin-11-dUTP and in situ hybridized using the method of Pendás et a!. (1993). Hybridization sites were detected with antidigoxigenm antibody conjugated to fluorescein isothiocyanate (FISH). Some slides were counterstained with propidium iodide (2.5 jig mL1) and mounted in Vecta Shield antifade solution (Vector Laboratories).

Preparations were examined at 100 x magnification under a microscope equipped with a 100 W epifluorescence system and the appropriate filters for the observation of DAPI- and CMA3-stained chromosomes, and a double filter for the simultaneous observation of fluorescein and propidium iodide in the in situ hybridized chromosomes. Light field photographs were taken with Kodak Imagelink HO film and Kodak Plus-X film was used for fluorescence and in situ hybridization photographs.

RESULTS

The techniques used revealed no chromosomal differences among the line HA89 and the two hybrids of H. annuus studied.

Feulgen Staining

Helianthus annuus has 2n = 34 chromosomes of similar size which, following Al-Allaf & Godward (1979), can be grouped as four pairs of metacentric (M1—M4), eight pairs of submetacentric (SM5—SM12) and five pairs of subtelocentric (ST13—ST17) chromosomes (Fig. la). The similarity in size and morphology of the chromosomes within each group makes further differentiation difficult. However, in all the cases considered, two pairs of submetacentric (SM7 and SM1O) and one pair of subtelocentric (ST13) chromosomes show secondary constrictions (Figs la,b) which allow them to be distinguished.

C-banding

All the chromosomes possess small amounts of centromeric constitutive heterochromatin, and the three pairs of chromosomes bearing secondary constrictions (satellite chromosomes) also have associated heterochromatin (Fig. ib). Distal or interstitial C-heterochromatin has not been revealed in any member of the complement. Variation in size between different individuals of these heterochromatic regions has not been observed.

Fluorochrome Staining

DA-DAPI staining shows no positive response, but instead the chromosomes stain uniformly (Fig. 2a). Similarly, AD-DAPI does not reveal any clear differentiation, although some centromeric regions yield a diffuse, slightly positive response, although this cannot be reproduced photographically. However, in both cases the heterochromatin associated with the secondary constrictions appears negative, in contrast with the positive response it shows with DA-CMA3.

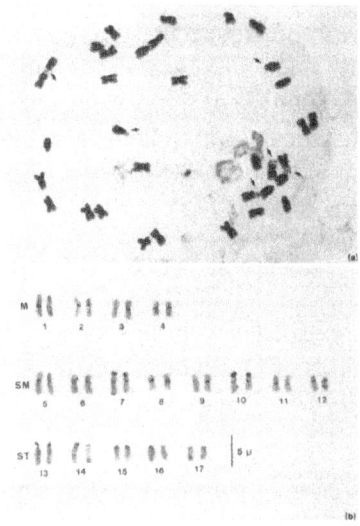

Figure 1. (a) Feulgen-stained mitotic metaphase chromosomes of the hybrid Flamme of Helianthus annuus. Note the existence of three pairs of chromosomes bearing sec-

ondary constrictions (arrows). (b) C-banded karyotype of the inbred line HA89. The heterochromatin is restricted to the centromeric areas and the secondary constrictions of the sate!- lite chromosomes (pairs SM7, SM1O and ST13).

No other differentiation is revealed with this latter treatment (Fig. 2b). None of these fluorescence techniques has differentiated non-C-heterochromatic regions.

Silver Staining

Silver precipitates attached to the secondary constrictions are shown by the three pairs of satellite chromosomes: SM7, SM1O and ST13 (Fig. 3a). Interphase nuclei show between one and six spots of silver. In a sample of 537 silver-stained interphase meristem cells, 39.8 per cent had one nucleolus, 25.6 per cent had two, 28.3 percent had three, 4.4 per cent four, 1.1 per cent five and 0.5 per cent six nucleoli.

FISH

The chromosomal in situ hybridization of the rDNA probe confirms the existence of ribosomal clusters associated with the secondary constrictions of the three pairs of chromosomes (Fig. 3b). Hybridization does not occur in any other location and interphase nuclei consistently show six hybridization sites (close to 99 per cent nuclei in a sample of 214 cells).

DISCUSSION

Despite the economic and evolutionary importance of the genus Helianthus its cytogenetics has hitherto not been investigated in any great detail. This is partially because of the intrinsic difficulty of obtain ing mitotic chromosome preparations of sufficient quality. The enzymatic pretreatment of the meristematic root tissues with pectinase helps to solve this problem, as is commonly observed in other plants. The chromosome number and karyomorphology of the line and varieties of H. annuus var. macrocarpus studied here agree with those described by other authors for other varieties (Al-Allaf & Godward, 1979; Chandler, 1991). However, there have been no previous studies of heterochromatin distribution, chromosomal response to specific DNA sequence fluorochromes, or rDNA activity and location.

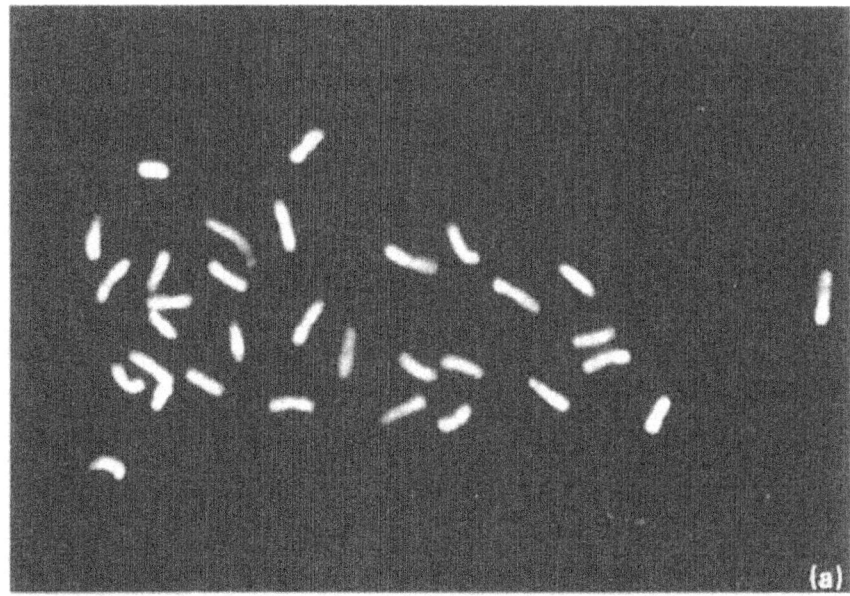

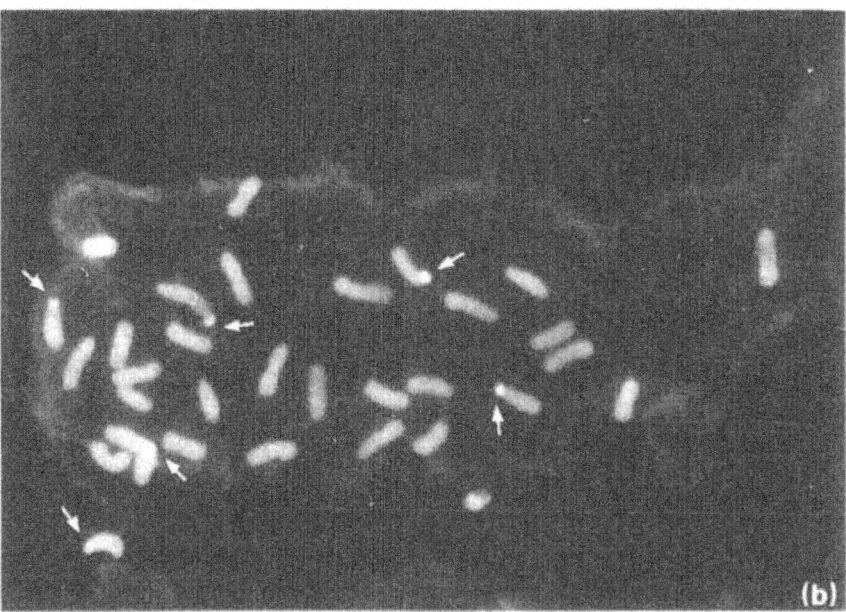

Figure 2. Fluorescent tn-staining of the mitotic metaphase chromosomes of the commercial hybrid Flamme of Helianthus annuus. The DA-DAPI treatment (a) does not show clear differentiation whereas there is positive response to the DA-CMA3 staining of the heterochromatin associated with the secondary constrictions (b).

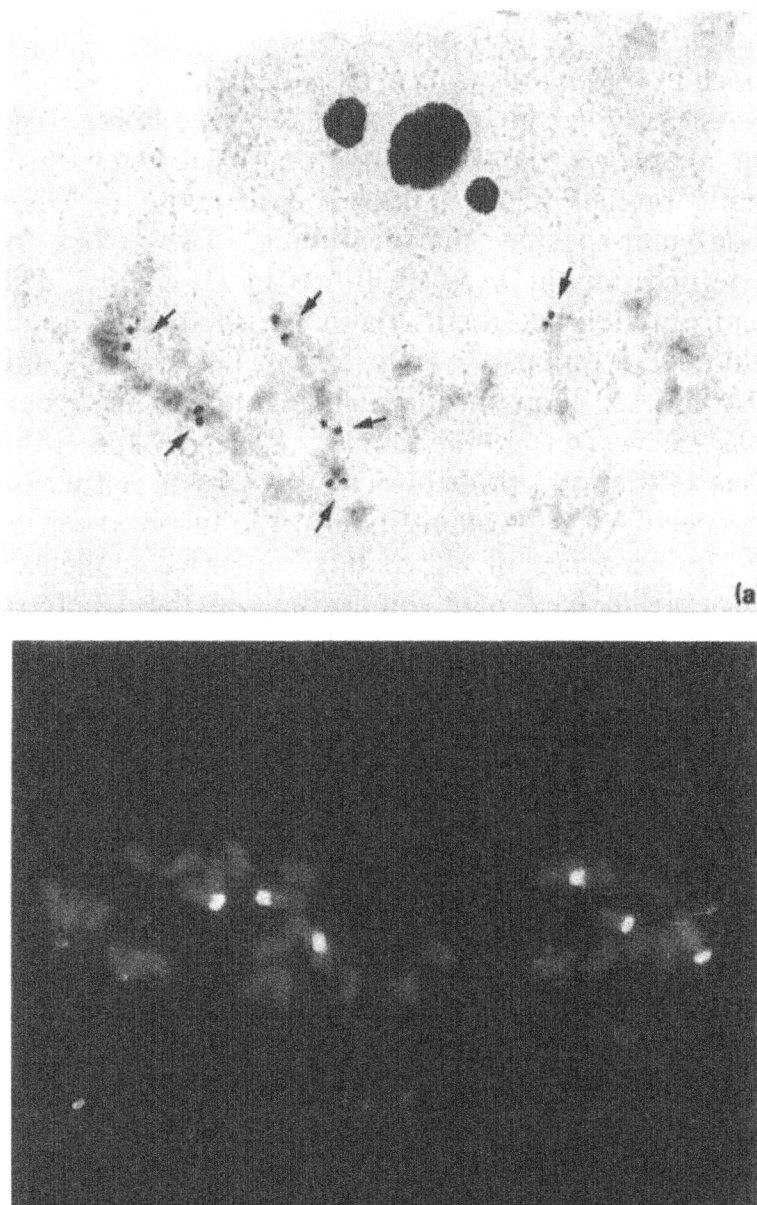

Figure 3. In the inbred line HA89 of Helianthus annuus the silver staining (a) shows six silver spots (NOR5) attached to the secondary constrictions of pairs SM7, SM1O and ST13 (arrows) in the mitotic metaphase chromosomes. These results are confirmed by FISH with an rDNA probe (b). Note also the three nucleoli in the interphase nucleus in (a).

The absence of variation in the C-heterochromatin distribution within and between the inbred line and varieties studied here suggests that this is very probably the standard heterochromatin pattern of the macro carpus variety, which is the most commonly cultivated sunflower. However, this does not imply that this pattern is invariable within H. annuus. It is reasonable to expect that in natural populations of H. annuus there exist polymorphisms for the distribution, size or even composition of the heterochromatin. The lack of karyological information concerning this genus precludes comparison of the C-band distribution reported here with the banding pattern of other Helianthus species and varieties. C-band variation can provide valuable cytotaxonomic information as has been shown in work on the related Anthemidea (Schweizer & Ehrendorfer, 1983). It would therefore be useful to survey a larger range of germplasm to see if such variation exists in H. annuus.

As observed in other organisms (see John, 1988 for a review), the H. annuus equilocal heterochromatic regions have an identical response to DNA base pair-specific fluorochromes. Two types of heterochromatin were identified in H. annuus. The centromeric heterochromatin showed the same staining response with the DA-DAPI and DA-CMA3 fluorochromes, indicating no base pair bias (Schweizer, 1981; Sumner, 1990, pp. 155—185). This situation probably results from either a balanced ATIGC composition of the involved repeat or the existence of alternating repetitive sequences of opposite base pair composition in these heterochromatic areas. However, the heterochromatin in the secondary constrictions of the satellited chromosomes was DA-CMA3-positive, showing that it was GC-rich. In any case, the two types of heterochromatin revealed by these treatments confirm, at the cytological level, the coexistence of distinct families of DNA repetitive sequences in the genome of H annuus that have also been distinguished by molecular analysis (K. Sossey, personal communication). DNA content within the genus Helianthus exhibits a four-fold range of variation largely because of polyploidy; however, there is also considerable variation in DNA content and chromosome size among sunflower species with the same chromosome number. For example, H annuus (HA89) has the fourth lowest DNA content of the 19 species with $2n = 34$ chromosomes (Sims & Price, 1985). Intra- specific variation in DNA content amongst cultivated varieties and inbred lines of H. annuus has also been described (Cavallini et al., 1986; Chandler, 1991; Michaelson et a!., 1991). To what extent different amounts of heterochromatin may be involved in this DNA content variation remains to be elucidated, but from our results we know that HA89 has little constitutive heterochromatin, which is confined to small bands in the centromeric areas and secondary constrictions. Further molecular and cytological studies will investigate this relationship.

Silver staining reveals active ribosomal genes (Hubbell, 1985). In root meristem cells from HA89 and the two hybrids of H. annuus, the three pairs of satellite chromosomes show silver deposits attached to the secondary constrictions, revealing the position of active rDNA clusters. This coincidence with secondary constrictions as well as the GC richness of the associated heterochromatin are common features of NORs (Schweizer, 1980). The in situ hybridization with an rDNA probe confirms the number and location of the rDNA clusters and makes the existence of other inactive ribosomal sites unlikely. The existence of one to three silver spots in most of the interphase nuclei of root meristem cells (93.7 per cent) is probably the result of nucleolar fusion (Giménez-MartIn et a!., 1977) although FISH with the rDNA probe shows six clearly separated fluorescent dots in the majority of the interphase cells. The six chromosomes with rDNA clusters show silver deposits attached to them at metaphase, indicating that all of them were active during the preceding interphase (Miller et at., 1976).

The search for chromosomal markers in other species of the genus Helianthus is helpful in a taxon where (i) there are diploid, tetraploid and hexaploid species, all of which have a basic chromosome number of $x = 17$ and are probably of polyploid origin (Jackson & Murray, 1983; Chandler et a!., 1986); (ii) different races of the same Helianthus species may be chromosomally distinct, although successful crosses may be produced between perennial and annual species, and between diploid and polyploid species (Kulshreshtha & Gupta, 1979); (iii) the cytogenetical information comes almost exclusively from the analysis of interspecific hybrid meiotic configurations (Chandler, 1991). These chromosomal markers could be useful either from the taxonomic point of view to identify chromosomes from different genomes or to follow the introgression of alien chromosomes in cultivated sunflowers, as happens in other plant species (Gustafson & Dille, 1992; Werner et al., 1992). In any case, the existence of linkage maps in the sunflower (Rieseberg et al., 1993; Gentzbittel et a!., 1995) where ribosomal gene polymorphisms have been described (Choumane & Heizmann, 1988) might permit the chromosomes with NORs to be assigned to linkage groups. Current work on sunflower RFLP maps may also yield in situ hybridization probes of use in assigning linkage groups to chromosomes.

The study of the constitutive heterochromatin distribution and fluorochrome response, as well as the number and location of the rDNA sequences in other species and other varieties of Helianthus will provide complementary information on their evolutionary status and, hopefully, suitable markers for breeding studies. Combined research on breeding, cytogenetics and molecular techniques are required in the sunflower to understand natural and artificial interspecific gene transfer (introgression) and the selective forces involved (Belhassen et at., 1994), as well as to provide new tools to evaluate variability in this taxon.

ACKNOWLEDGEMENTS

We are grateful to Drs J. Gosálvez, P. L. Mason and M. Navarrete for their very valuable help. This work has been partially supported by the Spanish-French programme of 'Acciones Integradas', and grants AGF93—0869 and SAF93—0093 from the Spanish CICYT.

REFERENCES

1. Al-Allaf, S, and Godward, M B E. 1979. Karyotype analysis of four varieties of*Helianthus annuus* L. *Cytologia*, **44**, 319–323.

2. Belhassen, E, Auge, G, Ji, J, Billot, C, Frenández-Martínez, J, Ruso, J, and Vares, D. 1994. Dynamic management of genetic resources: first generation analysis of sunflower artificial populations. *Génét Sél Évol*, **26**, 241–253.

3. Cavallini, A, Zolfino, C, Cionini, G, Cremonini, R, Natali, L, Sassoli, O, and Cionini, P G. 1986. Nuclear DNA changes within *Helianthus annuus* L.: cytophotometric, karyological and biochemical analyses. *Theor Appl Genet*,**73**, 20–26.

4. Chandler, J M. 1991. Chromosome evolution in sunflower. In: Tsuchiya, P. and Gupta, P. K (eds) *Chromosome Engineering in Plants: Genetics, Breeding, Evolution*, part B, pp. 229–249. Elsevier, Amsterdam.

5. Chandler, J M, Jan, C, and Beard, B H. 1986. Chromosomal differentiation among the annual *Helianthus* species. *Syst Bot*, **11**, 354–371.

6. Choumane, W, and Heizmann, P. 1988. Structure and variability of nuclear ribosomal genes in the genus *Helianthus*. *Theor Appl Genet*, **76**, 481–489.

7. Darlington, C D, and La Cour, L F. 1976. *The Handling of Chromosomes*, 6th edn. George Allen & Unwin Ltd, London.

8. Gentzbittel, L, Vear, F, Zhang, Y X, Berville, A, and Nicolas, P. 1995. Development of a consensus linkage map of cultivated sunflower (*Helianthus annuus* L.). *Theor Appl Genet*, **90**, 1079–1086.

9. Gerlach, W L, and Bedbrook, J R. 1979. Cloning and characterization of ribosomal RNA genes from wheat and barley. *Nucl Acids Res*, **7**, 1869–1885.

10. Giménez-Martín, G, De La Torre, C, López-Sáez, J F, and Esponda, P. 1977. Plant nucleolus: structure and physiology. *Cytobiologie*, **14**, 421–462.

11. Gustafson, J P, and Dille, J E. 1992. Chromosome location of *Oryza sativa*recombination linkage groups. *Proc Natl Acad Sci USA*, **89**, 8646–8650.

12. Hubbell, H R. 1985. Silver staining as an indicator of active ribosomal genes.*Stain Technol*, **60**, 285–294.

13. Jackson, R C, and Murray, B G. 1983. Colchicine induced quadrivalent formation in *Helianthus*: evidence for ancient polyploidy. *Theor Appl Genet*,**64**, 219–222.

14. John, B. 1988. The biology of heterochromatin. In: Verma, R. S. (ed) *Molecular and Structural Aspects of Heterochromatin*, pp. 1–147. Cambridge University Press, Cambridge.

15. Kulshreshtha, V B, and Gupta, P K. 1979. Cytogenetic studies in the genus*Helianthus* L. *Cytologia*, **44**, 325–334.

16. Lacadena, J R, Cermeño, M C, Orellana, J, and Santos, J L. 1984. Evidence for wheat-rye nucleolar competition (amphiplasty) in triticale by silver staining procedure. *Theor Appl Genet*, **67**, 207–213.

17. Michaelson, M J, Price, H J, Spencer-Johnston, J, and Ellison, J R. 1991. Variation of nuclear DNA content in *Helianthus annuus* (Asteraceae). *Am J Bot*, **78**, 1238–1243.

18. Miller, D A, Dev, V G, Tantravahi, R, and Miller, O J. 1976. Suppression of human nucleolus organizer activity in mouse-human somatic hybrid cells. *Exp Cell Res*, **101**, 135–243.

19. Pendás, A M, Morán, P, and García-Vázquez, E. 1993. Ribosomal genes are interspersed throughout a heterochromatic arm in Atlantic salmon. *Cytogenet Cell Genet*, **63**, 128–130.

20. Rieseberg, L H, Choi, H, Chan, R, and Spore, C. 1993. Genomic map of a diploid hybrid species. *Heredity*, **70**, 285–293.

21. Schilling, E E, and Heiser, C B, Jr, 1981. Infrageneric classification of*Helianthus* compositae. *Taxon*, **30**, 393–403.

22. Schwarzacher, T, Ambros, P, and Schweizer, D. 1980. Application of Giemsa banding to orchid karyotype analysis. *Plant Syst Evol*, **134**, 293–297.

23. Schweizer, D. 1976. Reverse fluorescent chromosome banding with Chromomycin and DAPI. *Chromosoma*, **58**, 307–324.

24. Schweizer, D. 1980. Fluorescent chromosome bands in plants: applications, mechanisms and implications for chromosome structure. In: Davies, D. R. and Hopwood, R. A (eds) *The Plant Genome: Proceedings of the 4th John Innes Symposium*, pp. 61–72. John Innes Charity, Norwich.

25. Schweizer, D. 1981. Counterstaining-enhanced chromosome banding. *Hum Genet*, **57**, 1–14.

26. Schweizer, D, and Ehrendorfer, F. 1983. Evolution of C-band patterns in Asteraceae-Anthemideae. *Biol Zbl*, **102**, 637–655.

27. Sims, L E, and Price, H J. 1985. Nuclear DNA content variation in *Helianthus*(Asteraceae). *Am J Bot*, **72**, 1213–1219.

28. Sumner, A T. 1990. *Chromosome Banding*. Unwin Hyman Ltd., London.

29. Werner, J E, Endo, T R, and Gill, B S. 1992. Towards a cytogenetically based physical map of the wheat genome. *Proc Natl Acad Sci USA*, **89**, 11307–11311.

Chapter 2

RECENT ADVANCES IN PLANT TRANSFORMATION

INTRODUCTION

The world's agriculture and farming are heavily dependent on crops that provide food and fibers for human use, either directly or through livestock. For the past two centuries, modern technology has improved agricultural practices, thereby augmenting conventional plant breeding methods to achieve improved yield and quality of crops. However, multiple factors such as population growth, environmental stress, ecological considerations, and demand for renewable energy have led to the demand for further improvements in the quality and quantity of crops. Plant genetic engineering offers new avenues in this regard and has become one of the most important molecular tools in the modern molecular breeding of crops.

Advancements in plant genetic engineering have made it possible to transfer genes into crop plants from unrelated plants and even from nonplant organisms; as a result, many crop species are being genetically modified for better agronomical traits, including disease resistance, insect tolerance, better nutritional values, and other desirable qualities.. Presently, foreign genes from various origins and production of products in transgenic plants represent a new aspect of the molecular agriculture revolution. In addition, transgenic plants have great impact on nonagricultural applications and represent an alternative for the production of medicinally useful and recombinant proteins and vaccines. However, in some crop plants, the lack of efficient transformation methods to introduce foreign DNA remains an obstacle to the application of plant genetic engineering.

Over the last decade, some significant achievements have been made in the development of new and efficient transformation methods in plants. Methods for delivering exogenous DNA to plant cells and gene transformation in general can be divided into two major categories: indirect and direct DNA deliveries. In the former approach, genes of interest are introduced into the target cell via bacteria, e.g., Agrobacterium tumefaciens or Agrobacterium rhizogenes. In contrast, the latter approach does not employ bacterial cells as mediators to transfer DNA to plant cells. Although various delivery methods have been reported, including the use of other bacterial strains (i.e., TransBacter™ Technology), Agrobacterium-mediated transformation remains the method of choice for plant transformation. The enduring success of the Agrobacterium-mediated approach is primarily attributable to the continuous improvements in plant tissue culture and T-DNA transfer processes. As a result, reproducible and efficient Agrobacterium-mediated protocols have been developed for many dicot and some monocot crops. Agrobacterium-mediated transformation possesses intrinsic advantages over direct DNA delivery systems. These advantages include the ability to transfer large intact segments of DNA, simple transgene insertions with defined ends and low copy number, stable integration and inheritance, and consistent gene expression over the generations.

The lower rates of success achieved with Agrobacterium in monocots and recalcitrant plant species have led to the development of specific direct DNA transfer methods, one of which is microparticle bombardment. This technology, first developed by Sanford and coworkers, is often termed biolistics or gene gun. Particle bombardment has a high success rate in monocot species in which agroinfection is limited. However, biolistic technology possesses several intrinsic disadvantages, including a low transformation efficiency as compared with Agrobacterium (when agroinfection works), a high frequency integration of the vector backbone and a loss of transgene cassette integrity, and transgene silencing due to multicopy insertions. Alternative delivery systems have also been used for gene transfer in plants, including electroporation, microinjection, silicon carbide, and chloroplast transformation. Of these, silicon carbide-mediated transformation represents one of the least complicated methods of plant transformation. Yet, each of these methods has their own limitations in the successful production of transgenic plants.

To date, Agrobacterium-mediated and particle bombardment transformation are the most commonly used methods for plant transformation. Nevertheless, the resulting transgenic plants are of course subject to biosafety issues related to the presence of vector backbone sequences and/or selectable marker genes, irrespective of the delivery method used. Transgenic plants produced by Agrobacterium-mediated transformation are likely subject to integration of vector backbone sequences. This leads to multiple transgene copies in the transgenic plants, complicating the regulatory process of genetically engineered plants. In recent years, there has been an important advancement in generation of vector-backbone and selectable-marker-free transgenic plants while still enabling the use of marker genes to select and identify transgenic plants.

Several strategies have been proposed and are in use for the production of marker-free transgenic plants. The methods that have been developed include simultaneous transformation of two marker genes (cotransformation), the movement of a transgene segment within the genome (transposition), and recombination between two specific sequences that are not necessarily homologous (site-specific recombination). Currently, there is a great demand for simultaneous expression of multiple genes for expression of complex traits in plants. With recent advancement in molecular biology and vector construction technology, it is possible to achieve stable expression of multiple transgenes in a single genome. The recent development of minichromosome technology might represent a strategy for gene stacking in plants. The minichromosome technique can be used to incorporate desirable traits such as insect, bacterial, or fungal resistance, herbicide tolerance, and increased crop quality. Hence, new techniques are in demand to boost yield or to improve crop traits. Finally, the genetic engineering of plants has already begun to play a crucial role in the production of biofuels. This chapter discusses various methods and recent advances in plant transformation technology.

PLANT TRANSFORMATION METHODS

Recent advances in genetic transformation have made it possible to transfer genes of both academic and agronomic importance into various crop species. A prerequisite for successful transformation system is an efficient regeneration protocol when tissue culture-based transformation process is employed. The very basis of plant regeneration relies on the realization that plant somatic cells are totipotent and can be stimulated to regenerate into whole plants. However, this insight is limited because, in reality, only a limited number of plant species and certain types of explant tissues have been found to be capable of regenerating whole plants under appropriate culture conditions. Therefore, much effort has been aimed at establishing and improving plant regeneration systems. Yet, efficient regeneration alone does not necessarily lead to efficient transformation.

There is a need to develop advanced transformation methods that would not only incorporate the required characteristics (stable and desirable transgene integration and expression) into plants but also enable generation of transgenic events in a high-throughput manner. These requirements are particularly relevant now in the crop post-genome era in which ever-increasing amounts of genome sequence information, BAC clones, ESTs, and full-length cDNAs are available. This situation presents both new challenges and opportunities for plant transformation research. At present, Agrobacterium and microprojectile are the commonly used methods for this purpose; other methods, such as electroporation and microinjection, are still used only rarely. The following sections discuss in detail the recent advances in each of the plant transformation methods.

Agrobacterium Mediated Transformation

In this method, A. tumefaciens or A. rhizogenes is employed to introduce foreign genes into plant cells. A. tumefaciens is a soilborne gram-negative bacterium that causes crown-gall, a plant tumor. The tumor-inducing capability of this bacterium is due to the presence of a large Ti (tumor-inducing) plasmid in its virulent strains. Similarly, Ri (root-inducing) megaplasmids are found in virulent strains of A. rhizogenes, the causative agent of "hairy root" disease. Both Ti- and Ri-plamids contain a form of "T-DNA" (transferred DNA). The T-DNA contains two types of genes: oncogenic genes, encoding enzymes involved in the synthesis of auxins and cytokinins (causing tumor formation), and genes involved in opine production. The T-DNA element is flanked by two 25-bp direct repeats called the left border (LB) and right border (RB), respectively, which act as a cis element signal for the T-DNA transfer. Both oncogenic and opine catabolism genes are located inside the T-DNA of the Ti plasmid whereas the virulence (vir) genes are situated outside the T-DNA on the Ti plasmid and bacterial chromosome. These vir genes are organized into several operons (virA, virB, virC, virD, virE, virF, virG, and virH) on the Ti-plasmid and other operons (chvA, chvB, and chvF) that are chromosomal and are essential for T-DNA transfer.

The mechanism of gene transfer from A. tumefaciens to plant cells involves several steps, which include bacterial colonization, induction of the bacterial virulence system, generation of the T-DNA transfer complex, T-DNA transfer, and integration of the T-DNA into the plant genome. The process of T-DNA transfer is initiated upon receipt of specific signals (e.g., phenolic compounds) received from host cells. Previous observations suggested that wounding or vigorous cell division also promotes T-DNA transfer, presumably due to induction by phenolic compounds produced during cell repair or during the formation of new cells. In response, a signal received by virA activates a cascade of other vir protein machinery genes. However, very little is known about the nature and function of the factors that Agrobacterium utilizes, for instance, specific receptors on the host cell surface and/or cell wall. Subsequently, virD1 and virD2 proteins nick both the left and right borders on the bottom strand of the T-DNA. The resulting single-stranded T-DNA molecule (T-strand), together with several vir proteins, is then exported into the host cell cytoplasm through a channel formed by the Agrobacterium VirD4 and VirB protein complex. Before its entry into the host cell cytoplasm, the VirD2–T-strand conjugate is most likely coated by VirE2, forming the T-complex. VirE2 is a single-stranded DNA-binding Agrobacterium protein that is transported into the plant cell, where it presumably functions to protect the T-DNA from degradation.

The Agrobacterium T-complex is likely transported through the host cell cytoplasm by a cellular-motor-assisted mechanism. In a recent report, a dynein-like Arabidopsis protein (DLC3), coupled with another protein (VIP1), has been proposed to function in the intracellular transport of the Agrobacterium T-complex. Recently, an additional Arabidopsis protein, VIP2 (VirE2 interacting protein2), has been demonstrated to play a major role in T-DNA integration into the plant genome. The T-complex then enters the cell nucleus by an active mechanism mediated by the nuclear import machinery of the host cell. This facilitates integration of the T-strand into the host genome at random positions by a process of nonhomologous, or more precisely, illegitimate recombination

Many recent reviews have addressed mechanisms related to T-DNA transfer.Characterization of the mechanisms governing the T-DNA transfer process is very important for plant transformation studies and should facilitate the identification of conditions to maximize T-DNA transfer. The best example of this is the use of a phenolic compound (e.g., acetosyringone) as well as a low-pH media and temperature to induce T-DNA transfer during the Agrobacterium infection stage.

Advancement in molecular biology techniques have enabled the development of binary Ti vectors that are compatible with utilization of both Agrobacterium strains and Escherichia coli. Development of the binary vector and bacterial strain systems for plant transformation is achieved by placing virulence genes on a separate plasmid (the large Ti-plasmid) and the gene to be transferred on separate vector (the small binary vector). Since most gene manipulations are carried out in E. coli, the binary plasmids are designed to replicate in both E. coli and Agrobacterium. Recent advancements in vector-cloning techniques have led to the development of binary bacterial artificial chromosome (BIBAC) vectors that enhance the frequency of T-DNA transfer of large-sized DNA fragments. The key features that make BIBAC vectors useful include its extremely low copy number and high stability when they are replicated in either Agrobacterium or E. coli cells.

The development of superbinary vectors made it possible to transform monocot plants for the first time. A superbinary vector represents an improved version of a binary vector; the vector carries a 14.8 kb KpnI fragment containing the virB, virG, vir C genes derived from pTiBo542. These genes are responsible for the supervirulence phenotype of A. tumefaciens strain A281.

The superbinary vector has been highly efficient in transforming various plants, particularly recalcitrant species, such as important cereal crops.

The integration and enhancement of gene expression in the plant genome greatly depends on the promoter that is fused at the 5¢ end of the gene of interest. The most widely used foreign regulatory elements include the 35S promoter of the cauliflower mosaic virus and the transcriptional terminator of the Agrobacterium nopaline synthase gene (nos), which together promote high-level gene expression in transgenic plants. The 35S promoter is a constitutive promoter that is used in vector constructs to drive target gene expression in many plant species. Recently, a new, stronger promoter has been developed. This "super promoter" is a hybrid construct combining a triple repeat of the octopine synthase (ocs) activator sequence plus the mannose synthase (mas) activator elements fused to the mas promoter. An initial study performed with this construct in maize (Zea mays) and tobacco (Nicotiana tabacum) confirmed the stable expression of superpromoter – GUS fusion gene in both the plant species. In tobacco, activity of the superpromoter is higher in mature leaves than young leaves, whereas in maize, the activity differed little among the tested aerial portions of the plant.

In order to achieve efficient Agrobacterium-mediated T-DNA transfer, several factors must be taken into consideration, including the plant genotypes, sources of explants, Agrobacterium strains, medium salt strength and pH, duration and temperature of Agrobacterium–explant interactions (inoculation and cocultivation), and use of T-DNA-inducing compounds. Agrobacterium-mediated transformation of higher plants is now well established for dicotyledonous species. In recent years, the frequency of gene transfer to monocotyledonous species has also been greatly improved (Table 1). A variety of explants can be used as target material for Agrobacterium-mediated transformation, including embryonic cultures, immature or mature zygotic embryos, mature seed-derived calli, meristems, shoot apices, primary leaf nodes, excised leaf blades, roots, cotyledons (including or excluding nodal areas), stem segments, and callus suspension cultures. These explant tissues have been able to regenerate through either a somatic embryogenesis or organogenesis regime. As an alternative to the organogenesis regeneration regime, somatic embryogenesis offers the advantage of single cell regeneration. However, the types and physiological conditions of explants used are critical to successful regeneration and subsequent recovery of whole transgenic plants. For example, in sorghum, a higher transformation efficiency was achieved in immature embryos taken from field grown plants than in the immature embryos from greenhouse-grown plants (34).

Table 1: Major gene transfer methods used for transformation of monocot and dicotyledonous plant species

Plant species	Types of explants	Transfer method	Gene transferred	Transformation efficiency (%)	Reference
Monocots					
Oryza sativa	C	Ag	*cry1Ac*	2	(152)
	C	MpB	*shGH*	79.5	(153)
	SA	EP	*npt* II, *ppt*	13.8	(154)
Hordeum vulgare	ImE	Ag	*hpt* II, *luc*	25	(155)
	EC	MpB	*AtNDPK2*	0.15	(156)
Saccharum sp.	AxB	Ag	*npt* II, *bar*, *gusA*	50	(157)
	C	EP	*gusA*	80	(158)
Sorghum bicolor	C	Ag	*Man*, *gfp*	8.3	(159)
Triticum sp.	ImE	Ag	*bar*, *gusA*	9.7	(160)
Zea mays	ImE	Ag	*gusA*, *bar*	12.2	(41)
	ImE	MPB	*gusA*, *hpt* II	31	(73)
Dicots					
Arabidopsis thaliana	S	Ag	T-DNA	26	(161)
	L	MPB	*gusA*	–	(162)
Arachis hypogea	CN	Ag	*gusA*	38	(163)
	SE	MPB	*VP2*, *gusA*	12.3	(164)
	EL	EP	*gusA*	3	(165)
Brassica oleracea	ML	Cl-MPB	*cry1Ab*	11.1	(166)
Cajanus cajan	CN	Ag	*npt* II, *H*	51	(167)
Eucalyptus sp.	ApS	Ag	*gusA*	9	(168)
Glycine max	CN	Ag	*bar*, *gusA*	5.5	(169)
	SE	MPB	*Os-mALS*	60	(170)
	Fl	PTP	*phyA*	13	(171)
Gossypium hirsutum	EC	Ag	*cry1Ia5*	83	(7)
	EC	SCW	*AVP1*, *npt* II	64	(172)
Malus domestica	IVS	Ag	*Lc*	50	(173)
Pinus sp.	EC	Ag	*npt* II, *bar*, *gusA*	65–98	(174)

Ag Agrobacterium-mediated transformation, AtNDPK2 Arabidopsis nucleoside diphosphate kinase gene, ApS apical shoot, AVP1 Arabidopsis vacuolar pyrophosphatase, AxB axilllary bud, bar bialaphos-resistance gene, C callus, Cl-MPB chloroplast-mediated MPB, CN cotyledonary node, cry1Ia5 Bacillus thuringiensis toxins, EC embryogenic callus, EL embryonic leaflets, EP electroporation, Fl flower, gusA b-glucuronidase, H hemagglutinin protein, hGH human growth hormone, hpt II hygromycin phosphotransferase II, ImE immature embryo, IVS in vitro shoot, L leaf, Lc maize leaf color regulatory gene, ML mature leaf, MPB microprojectile bombardment-mediated transformation, npt II neomycin phosphotransferase II, Os-mALS acetolactate synthase derived from rice, PTP pollen tube pathway transformation, phyA phytase A, S seed, SA shoot apex, SCW silicon-carbide-whiskers-mediated transformation, SE somatic embryo, T-DNA transfer DNA

Competence for transformation can be enhanced in recalcitrant explants by phytohormone treatments. The maximal percent of calli showed higher transient GUS activity when picloram was used in Typha latifolia. Similarly, in Hibiscus cannabinus, preculturing of explants for 2 days in benzyladenine (BA) containing medium enhanced transient GUS expression . Phytohormone treatment activates cell division and dedifferentiation in many tissues. The stimulation of cell division by phytohormones suggests that efficient Agrobacterium transformation may occur at a particular stage of the plant cell cycle.

The procedures that promote Agrobacterium cells to come into close contact with the plant cells around wounded tissue sites have been found to enhance T-DNA transfer. For example, "dip-wounding," which is prewounding of the explants prior to cocultivation with blade dipped in Agrobacterium suspension, increases transformation frequency as high as 10-fold (Table 2, Xinlu Chen, Xiujuan Su, and Zhanyuan J. Zhang, unpublished data). When "dip-wounding" is combined with the use of phenolic compounds in inoculation and cocultivation media, the attraction of Agrobacterium is presumably enhanced at wounded sites, which facilitates increased access of bacteria to plant cells.

Table 2: Impact of dip-wounding on soybean regeneration and transformation

Inoculation procedure	Number of explants			Percentage of explants		Number of events	Percentage of events
	To start	Good	Green	Good	Green		
Standard	350	172	129	49.0	37.0a	1	0.3a
Dip-wounding	350	198	236	57.0	67.0b	12	3.4b

In a number of plant species, explants are hypersensitive to Agrobacterium infection, forming necrotic barriers; this can be overcome by the use of antioxidants to reduce the oxidative burst. Tissue browning/necrosis associated with Agrobacteriummediated transformation has been reported in various types of explants of both dicotyledonous and monocotyledonous species. Antioxidants such as polyvinylpyrrolidone (PVPP), dithiothreitol (DTT), cysteine, glutathione, lipoic acid, ascorbic acid, and citric acid are now commonly used to reduce tissue browning/necrosis of explants during plant transformation. In sugarcane, pretreatment of explants on media containing ascorbic acid and cysteine prior to transformation results in higher transformation efficiency. In grapes and rice, the addition of antioxidants, such as polyvinylpyrrolidone and dithiothreitol also increases the transformation efficiency. In soybean, which is difficult to transform, a higher transformation rate was achieved by including l-cysteine, DTT, and sodium thiosulphate in cocultivation media. These antioxidants also enhanced the transformation in maize Hi-II. Such increased T-DNA transfer enabled the use of standard binary vectors to routinely achieve efficient transformation without the use of a superbinary vector. In contrast, lipoic acid was found to enhance GUS transient expression and transformation efficiency in tomato.

In addition to the strategies discussed above, other treatment conditions have been devised more recently to promote Agrobacterium-mediated transformation. Desiccation of explants prior to Agrobacterium infection enhances transformation efficiency in sugarcane, whereas addition of surfactants such as Silwet-L77 or pluronic acid F68 enhances transformation in wheat. Such desiccation helps to reduce cell damage due to the reduction of cell turgidity, whereas the use of surfactants may induce wounds and thinner cell walls in the explant tissues, thereby promoting Agrobacterium attachment to explants and ultimately T-DNA transfer.Another method is treatment of plant tissues and Agrobacterium to brief sonication, which allows Agrobacterium and T-DNA entry into the tissues. In loblolly pine, sonication was found to enhance not only transient transformation but also the recovery of hygromycin-resistant lines.

Plant transformation frequency is also associated with cell division or dedifferentiation of the host explants. Recent studies have revealed the phase of the plant cell cycle at the time of transformation to be a major determinant of transformation and regeneration efficiency. To achieve a stable transgenic event, the differentiated cell for regeneration should adopt a "stem-cell-like" state for pluripotentiality to renter the S phase of the cell cycle. The transformation competence of the cells is high in S and G2 phase/M phase, and lower in G0 and G1 phases. A cell cycle study identified the RepA, HP1, E2Fa, CycD3, and CycD1 genes to be involved in the pluripotency of cells in G1-S phase and further progression through S and G2 phases . Gordon-Kamm et al. demonstrated substantial improvement in maize transformation by overexpressing the RepA gene. Arabidopsis histone H2A, which is expressed at higher levels in tissues that are more susceptible to Agrobacterium infection, is essential for T-DNA integration in somatic cells. Co-overexpressing E2Fa together with its dimerization partner, DPa, resulted in increased cell proliferation in cotyledons, leading to approximately twice the number of cells as wild type.

Transcriptome analysis of E2Fa–DPa-overexpressing plants showed upregulation of 14 genes that are involved in DNA replication and S phase onset (54). VIP1 (VirE2 interacting protein 1) in Arabidopsis; those genes were also found to be upregulated during the pluripotent stage. Overexpression of VIP1 increased the rate of transient and stable plant transformation and predicted its role in interacting with histone proteins. A recent study in Phaseolous coccineus has identified the proteins PIN and CUC to be involved in shoot apical meristem formation. The use of transgenic marker genotypes, such as WUSCHEL (WUS)- reporter or STM-reporter (SHOOTMERISTEMLESS) construct should be useful in identifying meristemoids in early stages in development.

Agrobacterium rhizogenes strains contain a T-DNA region located on the Ri plasmid that carries genes involved in root initiation, which are essential for production of hairy roots. Studies on the function of Ri T-DNA-encoded genes performed using the agropine-type Ri plasmids led to the identification of 18 ORF, including rolA, rolB, rolC, and rolD genes. It was evident that these genes also participate in the production of hairy roots. Several of the experiments in this study were carried out to inactivate or overexpress various rol genes, generating stable transgenic lines with various alterations in plant phenotypes and root morphology.

In general, however, A. rhizogenes-mediated root transformation has received considerably less attention than A. tumefaciens transformation. The main reason for this is the difficulty in regenerating plants from hairy roots transformed by A. rhizogenes. Therefore, this delivery system has been predominantly used to generate transgenic roots for transient assays. One of the most advanced systems in such type of assay is the production of "composite plants". The important characteristic feature of hairy roots is their ability to grow in plant hormone-free media. These growth characteristics have made hairy roots a useful tool for secondary metabolite production, use in metabolic engineering, and studies of root biology in general. Recently, it has been demonstrated that hairy-root cultures can be adapted for T-DNA activation tagging studies. Hairy roots are also used to genetically investigate root–bacterial interactions in soybean. For example, various studies have revealed that hairy roots derived from various soybean cultivars maintain their cognate nematode resistance or susceptible phenotypes. Recently, hairy roots have been used to produce recombinant proteins. This system is also suitable for high-throughput analysis of root-related transgene expression in Medicago or soybean root tissues and is now expected to be applied to root transformation for highthroughput functional analysis of certain gene expressions in other plant species.

In summary, Agrobacterium-mediated transformation has been very successfully employed recently in transformation of both dicots and particularly monocots, the latter of which had long been thought to be unable to host Agrobacterium. These successes are attributable to the development of the superbinary vector, the use of antioxidants, and optimization of the composition of inoculation and cocultivation media. This trend of success in transformation of various plant species will continue not just because of new ideas and approaches in improving Agrobacterium transformation but also because of the obvious advantages of such a natural gene delivery system.

Microprojectile (Particle) Bombardment Transformation

Microprojectile bombardment is one of the direct gene transfer methods for development of transgenics. This method was developed in 1980s to genetically engineer plants that were recalcitrant to transformation with Agrobacterium. Subsequently, the technique has been widely used to produce transgenic plants in a wide range of plant species. The first particle delivery method was developed by Sanford and coworkers. The Sanford device was extensively modified to produce the PDS-1000/He machine, which was licensed to DuPont. The technique involves coating microcarriers (gold or tungsten particles approx 0.6–1.0 mm in diameter) with the DNA of interest and then accelerating them at high velocities, to penetrate into the cell of essentially any organism.

Briefly, the microcarriers are spread evenly on circular plastic film (macrocarrier). The entire unit is then placed below the rupture disk in the main vacuum chamber of the biolistic device. A variety of rupture disks are available that burst at pressures ranging from 450 to 2,200 psi. Below the macrocarrier is a stopping screen, in which a wire-mesh is designed to retain the macrocarrier, while allowing the microcarriers to pass through. The target tissue is placed below the launch assembly unit. Under a partial vacuum, the microprojectile is fired, and helium is then allowed to fill the gas-acceleration tube. The helium pressure builds up behind a rupture disk, which bursts at a specific pressure, thus releasing a shock wave of helium that forces the macrocarriers down onto the stopping screen. The microcarriers leave the circular plastic film and continue flying down the chamber to hit and penetrate the target tissue, thus delivering the DNA.

Several factors must be considered for successful gene transfer using particle bombardment technology. These factors include the design of a suitable vector with a small size and high copy number, as well as the quantity and quality of the delivered DNA. The entire process must be performed under sterile conditions to prevent contaminations of target tissue during subsequent tissue culture. The types and sizes of microcarriers are important choices because they affect the depth of penetration of the accelerated microcarrier as well as the degree of damage to the target cells. Gold particles ranging from 0.6 to 3.0 μm in diameter are commercially available. The degree of penetration required will depend on the thickness of the cell wall, the type of tissue being transformed, and the depth of the target cell layers. Variation in the helium pressures, the level of vacuum generated, the size of the particles, and the position of target tissues will dictate the momentum and penetrating power at which the microprojectiles strike the tissue. All of these parameters are under the experimenter's control and must be optimized for a given target tissue.

Treatment of the target tissues prior to and after particle bombardment has a significant effect on the frequency of recoverable transgenic cell lines and plants. An attractive feature of particle bombardment is its ability to transfer foreign DNA into any cell or tissue type whose cell wall and plasma membrane can be penetrated. Embryogenic and meristematic tissues are the most commonly employed target tissues for the production of genetically transformed plants. Particle bombardment of embryogenic tissues has been successfully exploited to produce transgenic plants in a wide range of agronomically important plants, including legumes, tuber crops, starchy staples, trees, commodity crops, and all of the major cereals. In the case of bombarding apical meristems, the treatment, physiological status, and age of the mother plants prior to excision of the explants must be taken into consideration. Use of an osmotic pretreatment or partial drying of the target cells prior to bombardment is a commonly used strategy to increase the frequency of successful transformation.

One of the advantages of particle bombardment is the possible expression of multiple transgenes in the target tissue, which can be achieved by fusion of genes within the same plasmid that is then bombarded into the target tissues. In recent years, multiple independent gene expression cassettes have been successfully transferred using particle bombardment in wheat, rice, and soybean. The use of microprojectile bombardment has made it easy to transfer large DNA fragments into the plant genome, although the integrity of the DNA is a concern. Integration of yeast artificial chromosomes (YACs) into the plant genome by particle bombardment has been successful with inserts of up to 150 kb. Although requiring further development, integration of large DNA fragments promises to be an important tool in future plant research and crop biotechnology.

In summary, over the last two decades, many published studies have utilized microprojectile technology in both monocot and dicot plant species (Table 1). In the process, a diverse range of agronomic traits have been transferred and imparted to crop plants via particle bombardment, which remains the most promising technique to genetically engineer plastids. However, this technology is limited due to several drawbacks, such as the integration of multiple copies of the desired transgene, in addition to superfluous DNA sequences that are associated with the plasmid vector. Multicopy integrations and superfluous DNA can lead to silencing of the gene of interest in the transformed plant. This problem was overcome by transferring the desired coding region only with its control elements into the target cells of plant genome.

Electroporation Mediated Transformation

Electroporation-mediated transformation requires the application of strong electric field pulses to cells and tissues and is known to cause some type of structural rearrangement of the cell membrane. In vitro introduction of DNA into cells is now the most common application of electroporation. The technique was originally developed for protoplast transformation but has subsequently been shown to work with intact plant cells as well. A voltage of 25 mV and an amperage of 0.5 mA for 15 min are the most often used parameters. However, factors such as surface concentration of DNA and tolerance of cells to membrane permeation may affect electroporation efficiency. Using the electroporation method, successful transformation has been achieved with protoplasts of both monocot and dicot plants. The first report of fertile transgenic rice utilized electroporation of DNA into embryogenic protoplasts. However, using protoplasts as explants for regeneration of transformants limits the use of electroporation for stable transformation because the protoplastto-plant regeneration system has not been developed in most plant species. The electroporation of plant cells and tissues is very similar in its principles to the electroporation of protoplasts. This approach enabled the recovery of the first transgenic plants in barley. In sugarcane, a gene was transferred into intact meristem tissue using electroporation-mediated transformation (Table 1). While electroporation was proposed as an alternative to biolistics, it is not nearly as efficient. Compared to biolistics, it is inexpensive and simple, but the technique has only been successful in a few plant species. The thick cell walls of intact tissues represent key physical barriers to electroporation.

PEG/Liposome Mediated Transformation

Polyethylene glycol (PEG)-mediated transformation is a method used to deliver DNA using protoplasts as explants. The method is similar to electroporation in that the DNA to be introduced is simply mixed with the protoplast, and uptake of DNA is then stimulated by the addition of PEG, rather than an electrical pulse. PEG-mediated transformation has several advantages because it is easy to handle and no specialized equipment is required. However, the technique is rarely used due to the low frequency of transformation and because many species cannot be regenerated into whole plants from protoplasts. In addition, fertility may be a concern because of the somaclonal variation of the transgenic plants derived from protoplast cultures. Nonetheless, using this method, transgenic maize and barley have been produced. Thus, protoplast transformation is feasible in cereals, even though fertility problems in the regenerants are often encountered. In cotton, transformation was achieved using combined polybrene–spermidine-based callus treatment.

Related to PEG-mediated transformation is the liposomemediated transformation technique. In this method, DNA enters protoplasts via endocytosis of liposomes. Generally, this process involves three steps: adhesion of liposomes to the protoplast surface, fusion of liposomes at the site of adhesion, and release of the plasmid inside the cell. Liposomes are microscopic spherical vesicles that form when phospholipids are hydrated. Liposomes being positively charged tend to attract negatively charged DNA and cell membrane. In this process, the engulfed DNA is free to integrate into the host genome. However, there have been very few successful reports on the application of this technique in plant species because the technique is very laborious and is associated with low efficiency. In tobacco, intact YACs were transformed via lipofection-PEG technique.

Silicon Carbide Mediated Transformation (SCMT)

Kaeppier et al. first reported the use of silicon-mediated transformation, which is one of the least complicated methods. In this method, small needle-type silicon carbide whiskers are mixed with plant cells and the gene of interest, and the mixture is then vortexed. In the process, the whiskers pierce the cells, permitting DNA entry into the cells. The fibers most often used in this procedure have an elongated shape, a length of 10–80 mm and a diameter of 0.6 mm and show high resistance to expandability. The method is simple, inexpensive, and effective on a variety of cell types. The efficiency of SCMT depends on fiber size, vortexing, the shape of vortexing vessel, as well as the plant material used for transformation. The SCMT technique has been used in a variety of plants, including maize, rice, wheat, tobacco, etc. Furthermore, silicon carbide fibers have been found to improve the efficiency of Agrobacterium-mediated transformation. The main disadvantages of SCMT include low transformation efficiency and damage to cells, thereby negatively affecting their regeneration capacity. Furthermore, this method imposes health hazards due to fiber inhalation if not performed properly.

More recently, two related technologies have been developed: silicon fibers have been reported to increase callus transformation by 30–50% in rice, and mesoporous silica nanoparticles have been used to deliver DNA and chemicals into both plant cells and intact leaves. Mesoporous silica nanoparticles are synthesized from a reaction between tetraethyl orthosilicate and a template made of micellar rods.

Microinjection

Microinjection is the direct mechanical introduction of DNA into the nucleus or cytoplasm using a glass microcapillary injection pipette. Using a microscope, cells (protoplasts) are immobilized in low-melting-point agar with a holding pipette and gentle suction; DNA is then injected into the cytoplasm or nucleus. The microinjection technique requires relatively expensive technical equipment for the micromanipulation of single cells under a microscope and involves precise injection of small amounts of DNA solution; the procedure is also very time-consuming. The injected cells or clumps of cells are subsequently cultured in tissue culture systems and regenerated into plants. Successful regeneration of microinjected rapeseed, tobacco, and barley has produced genome-integrated, stable transformants. However, microinjection has achieved only limited success in plant transformation due to the thick cell walls of plants and, more challengingly, to a lack of availability of a single-cell-to-plant regeneration system in most plant species.

Chloroplast Mediated Transformation

In genetically modified plants, the gene of interest usually integrates into the nucleus; however, it is also possible to transfer the gene into the plastid. The chloroplast genome is highly conserved among plant species and typically consists of double-stranded DNA of 120–220 kb, arranged in monomeric circles or in linear molecules. In most higher plant species, the chloroplast genome has two similar inverted repeat (IR_A and IR_B) regions of 20–30 kb, that separate a large single copy (LSC) region and small single copy (SSC) region. Both the microprojectile or protoplastmediated transformation methods are capable of delivering DNA to plastids, but to achieve successful transformation, chloroplast-specific vectors are required. The basic plastid transformation vector is comprised of chloroplast-specific expression cassettes and target-specific flanking sequences. Integration of the transgene into the chloroplast occurs via homologous recombination of the flanking sequences used in the chloroplast vectors. The first successful chloroplast transformation (of the aada gene, which confers spectinomycin resistance) was reported in Chlamydomonas. In higher plants, plastid transformation has been accomplished in tobacco with various foreign genes. In recent years, several crop chloroplast genomes have been transformed through organogenesis, and maternal inheritance has been observed. In economically important crops such as cotton, efficient plastid transformation has been achieved through somatic embryogenesis by bombarding embryogenic cell cultures.

Several transgenes engineered through chloroplast transformation have conferred valuable agronomic traits in plants, including insect and pathogen resistance, and both drought and salt resistance. In soybean expressing the Cry1Ab gene, insecticidal activity against velvet bean caterpillar was conferred to the transformed plant (Table 1). Advancement in chloroplast engineering has made it possible to use chloroplasts as bioreactors for the production of recombinant proteins and biopharmaceuticals. Because plastid genes are maternally inherited, transgenes inserted into these plastids are not disseminated by pollen. Additional advantages of this transformation system include the ability to express several genes as a polycistronic unit, thereby potentially eliminating position effects and gene silencing in chloroplast genetic engineering.

Native Gene Transfer

Transformation of native genes (including regulatory elements) into plants without a selectable marker is highly desirable to overcome consumers' concerns about GM crops. Historically, the development of transgenics with important agronomic traits depended on the use of genes derived from other organisms. Over the past decade, however, rapid advances in plant molecular biology have resulted in a major shift from bacteria and viruses to plants as important gene sources. A broad variety of plant genes associated with agronomically important traits have now been identified. For example, plants containing modified acetolactate synthase (ALS) genes displayed the same high levels of sulfonylurea tolerance as transgenic plants that expressed bacterial ALS tolerance genes. Likewise, the occurrence of glyphosate tolerance in a goosegrass (Eleusine indica) biotype has been associated with a mutated 5-enolpyruvylshikimate-3-phosphate synthase (EPSPS) gene. With respect to native gene transfer research, the application of transposon tagging and map-based cloning methods have resulted in the identification of more than 50 functionally active resistance (R) genes, several of which are currently being used as viable alternatives to foreign antimicrobial genes in crop improvement programs. One of the most agronomically important R-genes isolated is the Solanum bulbocastanum RB gene, which provides resistance to the potato late blight fungus Phytophthora infestans.

Rapid progress has also been made in the development of plant-based gene alternatives and recovery of various insecticidal proteins that are involved in insect resistance. It has been suggested that a 30-kDa maize cysteine protease can be used to enhance maize tolerance to caterpillars and armyworms. Alternatively, silencing or overexpressing key biosynthetic genes can enhance a plant's ability to produce insecticidal secondary metabolites. In tobacco, suppression of a P450 hydroxylase gene resulted in a 19-fold increase in cembratrieneol levels in trichomes, dramatically enhancing aphid resistance.

About 40 diverse plant genes have been used to enhance tolerance to abiotic stresses. In Arabidopsis, overexpression of C-repeat-binding transcription factor (CBF) increased survival resistance to freezing, drought, and salt.

Various methods such as promoter trapping and RNA fingerprinting have facilitated the identification of hundreds of plant promoters (e.g., ubiquitin and actin gene promoters), many of which contain regulatory elements that support high-level gene expression in most tissues of transgenic plants. Use of the same genetic material available in the plant benefits genetic engineering approaches into existing plant breeding programs.

MARKER GENES AND METHODS TO REMOVE SELECTABLE MARKERS FROM TRANSGENICS

After explants are transformed with the requisite vector having the gene of interest, the transformed cells/tissues need to be selected, which requires a selectable marker gene in the vector. In current transformation systems, a selectable marker gene is codelivered with the gene of interest to identify and resolve rare transgenic cells from nontransgenic cells. However, during transformations, only a few plant cells accept the integration of foreign DNA; most cells remain nontransgenic. Several selectable marker genes are currently in use, and most utilize antibiotic or herbicide selection (Table 3). Of these, the npt II gene (encoding neomycin phosphotransferase II), which imparts kanamycin resistance, is commonly used in the development of transgenics. Visual observation of gene expression is achieved using reporter genes that can represent important components for transient and stable expression studies. The most commonly used reporter genes in plant transformation are the ß-glucuronidase (GUS) gene and the green fluorescent protein gfp. Selectable marker genes are required to recover stably transformed plants. However, due to environmental concerns, as well as human health risks, the use of nonselectable markers needs to be promoted. In recent years, the following techniques have been used for the production of selectable-marker-free plants

Cotransformation

Cotransformation is a strategy that utilizes two plasmid vectors: one containing the gene of interest and the other containing a selectable marker gene. With this approach, integration of genes may be either within a single locus or at unlinked loci. The method achieves a high cotransformation efficiency and is usually carried out using biolistic or Agrobacterium-mediated transformation. The approach may permit the simultaneous introduction of many genes, independent of gene sequence, with a limited number of selectable marker genes. For example, in one report, nine genes were transferred into the rice genome by biolistics. Because the cotransformed genes are integrated at a single locus, they cosegregate. However, Agrobacterium-mediated cotransformation has the advantage over biolistic transformation that cotransformed genes often integrate into different loci, which results in the segregation of unlinked selectable marker genes from the gene of interest, thereby permitting production of marker-free transgenic plants. Cotransformation via Agrobacterium uses either two mixed Agrobacterium populations (each carrying a different binary vector) or a single Agrobacterium population (carrying two different binary vectors) (Fig. 1).

Table 3: Various selectable markers and reporter genes commonly used in transgenic plants

Gene	Enzymes encoded	Substrate	Gene source	Reference
Selectable markers				
bar	Phosphinothricin acetyl transferase	Phosphinothricin	*Streptomyces hygroscopicus*	(175)
BADH	Betaine aldehyde dehydrogenase	Betaine aldehyde	*Spinacia oleracea*	(176)
bxn	Bromoxynil nitrilase	Oxynils	*Klebsiella pneumonia*	(177)
cat	Chloramphenicol acetyl transferase	Chloramphenicol	*Escherichia coli* Tn5	(178)
dhfr	Dihydrofolate reductase	Methotrexate	*Candida albicans*	(179)
EPSPS	5-Enolpyruvyl shikimate-3-phosphate synthase	Glyphosate	*Petunia ×hybrida*	(180)
gox	Glyphosate oxidoreductase	Glyphosate	*Ochrobactrum anthropi*	(181)
hpt II	Hygromycin phospho-transferase II	Hygromycin B	*E. coli*	(182)
ManA	Phosphomannose isomerase	D-Mannose	*E. coli*	(183)
npt II	Neomycin phosphotrans-ferase II	Kanamycin	*E. coli* Tn5	(184)
xylA	Xylose isomerase	D-Xylose	*Streptomyces rubignosus*	(185)
Reporter genes				
uidA/GUS	β-Glucuronidase	X-gluc	*E. coli*	(186)
gfp	Geen fluorescent protein		*Aequorea victoria*	(187)
lacZ	Galactosidase	X-gal	*E. coli*	(188)
luc	Luciferase	Luciferin	*Photinus pyralis*	(189)
	Oxalate oxidase	Oxalic acid	*Triticum aestivum*	(190)

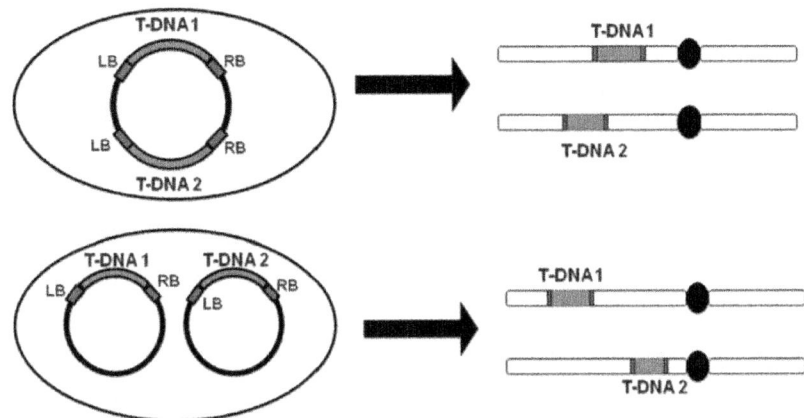

Figure 1. Schematic representation of two T-DNA on single or two separate vectors in Agrobacterium. LB and RB: T-DNA left and right borders, respectively. After transformation, the cotransformed T-DNA integrates, along with its flanking sequences (the LB and RB), into the plant genome at different loci.

Transposon Mediated Transformation

Recent developments in genetic engineering have made it possible to use transposon-based transformation to eliminate selectable marker genes in transgenics; for example, the maize Ac/Ds transposable element system has been used to excise selectable marker genes from plants. In this approach, the marker gene is flanked by the inverted repeat sequences of the Ds element. Subsequent to the transformation and T-DNA integration, expression of the Ac transposase from within the T-DNA results in excision of the gene of interest from the T-DNA insert containing the selectable marker gene. As a result, the gene of interest is transferred from the T-DNA site to another chromosomal location. Successful application of the system therefore requires the activity of the Ac transposase for the development of markerfree transgenic plants.

Repositioning within the genome can also enhance the expression profile of the gene of interest. In tomato, transposition of the GUS reporter gene and subsequent generation of npt II-free plants has been accomplished with both single and multiple T-DNA insertions. For example, hpt II-free rice plants were created that expressed Bt endotoxin encoded by the cry1B gene excision; reinsertion of the transgene occurred at very high frequencies (25–37%), preserving high levels of resistance to striped stem borer. However, because this technology relies on crossing plants to segregate the gene of interest from the marker gene and transposase, it is of limited use in plants that are vegetatively propagated or which have a long reproductive cycle.

Multiautosystem

To improve the selectable-marker-free plant technology, an IPT-type MAT (multiautotransformation) vector system was developed as an alternative to the Ac/Ds transposable element system. A MAT vector system uses the IPT (isopentenyltransferase) gene and Ac element under the control of the CaMV35S promoter. However, IPT expression in the transformed plant generates an abnormal phenotype called the "extreme shooty phenotype." Thus, subsequent to transformation, the IPT gene is removed using the Ac transposable element from the T-DNA, leaving only the gene of interest in the inserted copy of the T-DNA. This results in marker-free transgenic plants with a normal phenotype. However, there are several drawbacks of using a MAT system for marker gene removal. Specifically, there is the possibility of variable rates of transposition efficiency andalso of reinsertion of the transposable element.

Because of potential biosafety concerns, attempts have been made to generate transgenic plants devoid of selectable marker genes and vector backbone sequences from the T0 generation. Thus, methods have been devised to reduce the frequency of vector backbone sequence integration during plant transformation. In maize, the ovary-drip method was used to increase transformation frequency using linear green fluorescent protein (GFP) cassettes (Ubi-GFP-nos) flanked by 25-bp T-DNA borders as the transfer gene. In another approach, multiple tandem LB repeats were used to suppress the transfer of vector backbone in rice transformation. And more recently, soybean transformation experiments have been conducted to reduce vector backbone sequences by using nonlethal genes that interfere with plant development (113), resulting in a high frequency of vector-backbone-free transgenics. Multiple border sequences can also be used to generate vector-backbonefree transgenics. And finally, in maize, backbone-free transgenics were developed using vectors having a selectable marker gene in the vector backbone and the gene of interest in the T-DNA.

Site-Specific Recombination

The most widely used site-specific recombination system is the Cre–lox system from bacteriophage P1, which is very effective in the generation of marker-free plants. In this system, the plant is transformed with a T-DNA vector carrying the gene of interest with lox sites (34 bp repeats in direct orientation) flanking the selectable marker. In the second round of transformation, Cre recombinase is introduced to achieve precise excision of the marker gene. Specifically, Cre catalyzes the recombination between the lox repeat sequences, thereby eliminating the marker gene in the progeny. This system has been used in various plant species to generate marker-free transgenics. Using the Cre– lox system, it is also possible to resolve multiple transgene copies into only a single copy per recipient genome. Recent advances in this technology have led to the use of different inducible promoter systems; for example, a strategy has been developed using the B estradiol-inducible promoter system in which an artificial transcription factor, XVE, was constructed for use in plants with its cognate promoter. Using the Cre/lox system under the control of an inducible promoter, successful marker-free transgenic plants were developed. Recently, successful excision of loxP-flanked selectable marker has been achieved using a flowerspecific promoter in rice. However, to achieve a high transformation efficiency, more refinement is needed to improve this technology.

Two T-DNA Mediated Transformation

Using two T-DNAs, one bearing a selectable marker and the other containing the gene of interest, in a single vector, markerfree transgenic plants have been produced. This approach yields higher frequencies of cotransformation than a strategy using a mixture of A. tumefaciens strains carrying separate vectors. In one study, GUS and hpt II genes cotransformed into tobacco showed segregation of both the genes at unlinked loci, resulting in hpt marker-free plants. In another report, a 100% cotransformation frequency was achieved in tobacco when the selected T-DNA was twice as large as the nonselected T-DNA. In maize, cotransformation with an octopine strain carrying a binary vector with two T-DNAs yielded cotransformation frequencies of 93% for the bar and GUS genes in the T0 generation. In barley, a similar approach using smaller vectors yielded a cotransformation frequency of 66%, but only 24% of these segregated as marker-free plants. These studies clearly demonstrate that marker-free plants can be generated with variable efficiencies using Agrobacterium-mediated cotransformation followed by segregation of the genes in the subsequent sexual generations.

GATEWAY PLANT TRANSFORMATION VECTORS

Gateway cloning technology offers a fast and reliable highthroughput, restriction-enzyme-free cloning strategy for plasmid construction. The Gateway technology is based on the sitespecific recombination reaction mediated by bacteriophage 1 DNA fragments flanked by recombination sites (att). These sites can be transferred into vectors containing compatible recombination sites (attL attR or attB, attP) in a reaction mixture mediated by the Gateway clonase mix. The backbone of all described Gateway-compatible plant transformation vectors is the plasmid pPZP200. Two recombination reactions, catalyzed by LR and BP recombinases (clonases), respectively, are used in Gateway cloning. The first step, catalyzed by LR, inserts the gene of interest into the

Gateway vector at the attL and attR sites. The resulting construct is called the entry clone. All entry clones have attLs flanking the gene of interest. These are necessary in the Gateway system because these attL sites are cut to form sticky ends by the Gateway clonase. These sticky ends match with the sticky ends of the destination vector, which contains attR restriction sites. This process is called a LR reaction and is mediated by LR clonase mix, which contains the recombination proteins necessary for excision and incision. The product formed in the LR reaction is called the expression clone, which represents a subclone of the starting DNA sequence, correctly positioned in a new vector backbone. The second Gateway step is the BP reaction, which is the reverse of the LR reaction. In the BP reaction, the DNA insert flanked by 25 bp attB sites is transferred from the expression clone into a vector donated by a plasmid containing the attP sites. The final product is termed the destination clone and contains the transferred DNA sequence. Alternatively, these two sequential steps can be reversed to meet specific cloning needs (Fig. 2). The BP reaction thus allows rapid, efficient, directional PCR Cloning.

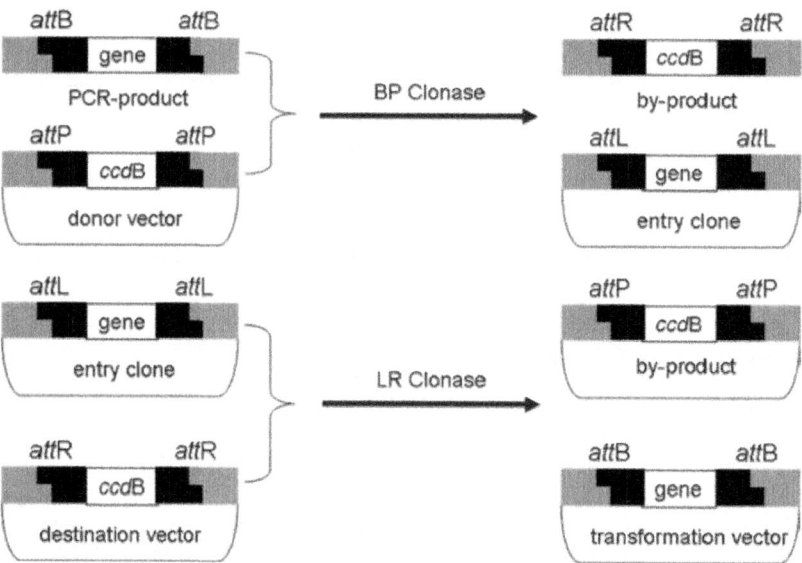

Figure 2. Diagram of gateway cloning technology to clone a PCR-product containing the gene of interest into plant transformation vector, without the use of restriction or ligation enzymes. ccdB, a bacterial suicide gene whose replacement by the incoming transgene-containing PCR product, will allow the transformed bacterial cells to survive, ensuring the presence of the transgene in either the entry or destination clone.

With advances in Gateway technology, it is possible to create many vectors with fused or directly linked multiple transgenes. Recently, for RNA interference, a high throughput cloning system has been designed using pHELLSGATE vectors. In these vectors, the recombination sites do not affect gene silencing performance, in contrast to the conventional restriction enzyme cloning vectors. In Arabidopsis, overexpression of large DNA fragments was found to be effective using Gateway technology. Molecular analysis of T0 and T1 generations confirmed that Gateway-compatible constructs were active and also that the att recombination sites did not inhibit transgene activity or interfere with enhancer activity in vivo.

However, the use of Gateway-RNAi transformation vectors can be a challenge due to the requirement for generating doublestranded RNA via inverted repeats. Thus, because the att recombination sites need to be duplicated, there is an increased chance of altering the orientation of the intron spacer whose splicing efficiency may be adversely impacted, causing reduced RNAimediated silencing efficacy. One solution to this problem is to use a double-intron spacer in the opposite orientation. However, the effectiveness of such a design needs to be validated in more experiments with different plant species.

Recently, the "pEarleyGate" vectors have been designed for Agrobacterium-mediated plant transformations; these vectors translationally fuse FLAG, HA, cMyc, AcV5, or tandem affinity purification epitope tags onto target proteins, with or without an adjacent fluorescent protein. A high-throughput protocol has recently been developed using the Gateway binary vector R4pGWBs for transformation of Arabidopsis thaliana. This vector is designed for the one-step construction of chimeric genes between any promoter and any cDNA. Also, autofluorescent protein tags are useful due to their ability to visualize cellular processes in vivo. For example, the pSAT vectors are useful for both autofluorescent protein tagging and multiple gene transfer. In many plant laboratories, Gateway binary vector technology has enabled in planta expression of recombinant proteins fused to fluorescent tags. Efficient expression of fluorescent protein has been achieved in Nicotiana benthamiana. For easy manipulation and efficient cloning of DNA fragments for gene expression studies, a new Gateway expression vector has been developed by combining the Gateway system and a recombineering system. The recombineering system uses bacteriophage-based homologous recombination in which genomic DNA in a bacterial artificial chromosome (BAC) is modified or subcloned without restriction enzymes or DNA ligase, thereby permitting the direct cloning of gDNA fragments from BACs to plant transformation vectors. The construct is converted into a novel Gateway Expression vector that incorporates cognate 5' and 3' regulatory regions, using recombineering, to replace the intervening coding region with the Gateway Destination cassette. Using this approach, efficient transformation has been achieved in Arabidopsis.

GENE TARGETING (ZINC-FINGER)

Zinc-finger endonucleases (ZFNs), or "molecular scissors," have recently been developed to target genes in plant systems. ZFNs are engineered proteins that have highly specific ZF domains fused to a sequence-independent nuclease domain. The current generation of ZFNs combines the nonspecific cleavage endonuclease domain of the Fok I restriction enzyme with several (usually three) zinc fingers domains that provide cleavage specificity. Subsequent to transformation, ZFNs introduce targeted double-stranded breaks in genomic DNA, thereby inducing recombination and repair processes at specific sites. In this process, homologous recombination will occur using the homologous sequence of the transferred gene to repair the double-strand DNA break (Fig. 3).

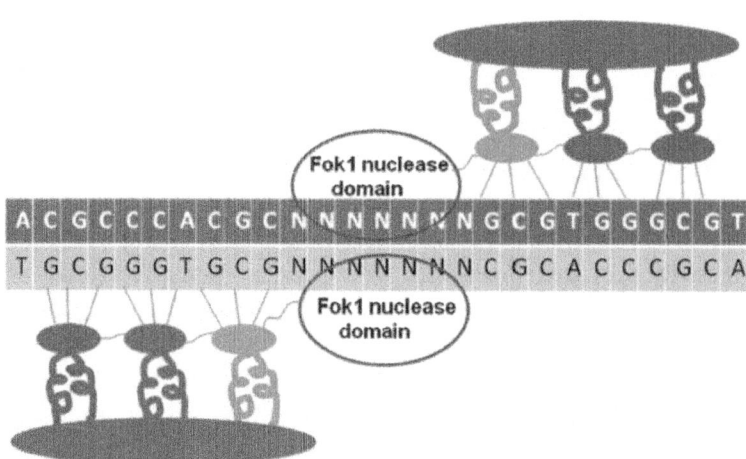

Figure 3. Schematic representation of a zinc finger nuclease (ZFN) binding to its target site. The three different zinc finger domain indicate different sequences. The zinc-finger domains are fused to the cleavage domain of endonuclease FokI to create the ZFN. The three-finger ZFN requires two copies of the 9 bp recognition site in an inverted orientation to produce a double-stranded break (DSB).

ZFN-mediated gene targeting promises to be a powerful tool in the development of novel crop species possessing beneficial agronomic traits. ZFNs have been utilized successfully in both tobacco and Arabidopsis. In Arabidopsis, the ZFN have been shown to cleave and stimulate mutation at specific genomic sites, resulting in a mutation frequency of 0.2 per target. Sitespecific cleavage and transgene integration has also been demonstrated in an engineered tobacco cell culture system. Such studies demonstrate the utility of zinc finger technology in plants and hold much promise for the development of novel transgenic plants.One major limitation in ZFN technology is the potential for cellular toxicity due to off-target DNA cleavage; thus, ZFNs are being modified to reduce this effect.

CHROMOSOME TRUNCATION

The recent development of plant artificial chromosome technology provides an opportunity for the stable expression and maintenance of multiple transgenes in a single genome. Plant artificial chromosomes are produced through telomere-mediated truncation of endogenous chromosomes. Telomeres are comprised of arrays of tandemly repeat sequences present at each end of a plant chromosome. To facilitate in vivo telomere-mediated truncation, an array of telomere sequences is cloned into a vector and then delivered into the genome via Agrobacterium-mediated T-DNA transformation or particle bombardment. Upon integration of the construct into the genome, the telomeric sequences can at some frequency be recognized as a bona fide chromosome terminus in the recipient chromosome, thereby truncating the chromosome at the insertion site. Recently, a 2.6-kb telomere repeat array isolated from Arabidopsis has been successfully transformed in maize to generate telomere-truncated chromosomes whose structure was verified by marker gene expression and FISH (fluorescent in situ hybridization)-based karyotyping. Plant minichromosome technology can also be combined with site-specific recombination systems to facilitate the stacking of multiple transgenes. This strategy for construction of engineered chromosomes should easily be extended to other plant species because it does not rely on species-specific cloned centromere sequences.

FUTURE PROSPECTS

Over the last two decades numerous transformation techniques have been developed for plants. Both Agrobacterium-mediated and microprojectile bombardment-based transformations are now standard laboratory techniques in plant labs. Recent advancements in genetic engineering techniques have led to the identification of genes and proteins involved in the DNA transfer mechanism accomplished by Agrobacterium transformation. Yet, despite this progress, many economically important crops and tree species remain highly recalcitrant to Agrobacterium infection. Thus, much effort is being made to develop regeneration protocols that can efficiently integrate exogenous genes to develop stable transgenics using both Agrobacterium- and biolisticmediated transformation. Novel techniques are also under development to develop genetically transformed plants with desired characteristics.

The generation of transgenic plants requires the use of various selectable marker genes that are introduced together with the gene of interest. Yet, it is abundantly clear that marker-free transgenic plants will be required in the future, thereby requiring more progress in genomics, cloning technology, and vector design, so as to eliminate the need for residual bacterial selectable marker genes in the future. Genetically engineered plants also play an important role in the ongoing study of gene function and metabolic pathways. Promising research has led to the identification of genes that control organogenesis or somatic embryogenesis which may function as selectable marker genes. The reintroduction of native genes and regulatory elements into plants therefore represents a viable means to achieve exogenous marker-free plants. A wide variety of plant genes associated with agronomically important traits have now been identified.

Chloroplast transformation technology is gaining importance due to its unique advantage of gene stacking without concern about gene silencing and of creating an opportunity to produce multivalent vaccines in a single transformation step. The recently developed zinc finger nuclease technology holds much promise in both basic and applied agricultural biotechnology. ZFN-assisted gene targeting and chromatin remodeling studies should also aid in characterizing gene function in plants. Resources are now available for engineering ZFNs in numerous plant species. Advancements in genetic engineering have also led to the development of minichrosome technology, which may provide a solution to gene stacking. Large DNA sequences, including multiple genes, could be introduced into the genome using this technology. Therefore, advances in transgenic technology would provide a solution for production of improved crop species to meet the world's demands for food, feed, fiber, and fuel. Hence, it is hoped that in the near future, GM plants with minimal genomic modifications can be developed.

REFERENCES

1. Job, D. (2002) Plant biotechnology in agriculture. Biochimie 84, 1105–1110.

2. Vain, P. (2007) Thirty years of plant transformation technology development. Plant Biotechnol. J. 5, 221–229.

3. Fischer, R., Stoger, E., Schillberg, S., Christou, P., and Twyman, R.M. (2004) Plant based production of biopharmaceuticals. Curr. Opin. Plant Biol. 7, 152–158.

4. Tzfira, T., and Citovsky, V. (2006) Agrobacterium-mediated genetic transformation of plants: biology and biotechnology. Curr. Opin. Biotechnol. 17, 147–154.

5.   Broothaerts, W., Mitchell, H.J., Weir, B., Kaines, S., Smith, L.M., Yang, W., Mayer, J.E., Roa-Rodríguez, C., and Jefferson, R.A. (2005) Gene transfer to plants by diverse species of bacteria. Nature 433, 629–633.

6.   Hiei, Y., Ohta, S., Komari, T., and Kumashiro, T. (1994) Efficient transformation of rice (Oryza sativa) mediated by Agrobacterium and sequence analysis of the boundaries of the T-DNA. Plant J. 6, 271–282.

7.   Leelavathi, S., Sunnichan, V.G., Kumria, R., Vijaykanth, G.P., Bhatnagar, R.K., and Reddy, V.S. (2004) A simple and rapid Agrobacterium-mediated transformation protocol for cotton (Gossypium hirsutum L.): embryogenic calli as a source to generate large numbers of transgenic plants. Plant Cell Rep. 22, 465–470.

8.   Zhang-Hua, H., Jin-Qing, C., Guan-Ting, W., Wei, J., Chun-Xiu, L., Rui-Zhi, H., Fu-Lin, W., Zhi-Hong, L., and Xiao-Yun, C. (2005) Highly efficient transformation and plant regeneration of tall fescue mediated by Agrobacterium tumefaciens. J. Plant. Physiol. Mol. Biol. 31, 149–159.

9.   Sanford, J.C. (1990) Biolistic plant transformation. Physiol. Plant. 79, 206–209.

10. Taylor, N.J., and Fauquet, C.M. (2002) Microparticle bombardment as a tool in plant science and agricultural biotechnology. DNA Cell Biol. 21, 963–977.

11. Rakoczy-Trojanowska, M. (2002) Alternative methods of plant transformation – a short review. Cell Mol. Biol. Lett. 7, 849–858.

12. Scutta, C.P., Zubko, E., and Meyer, P. (2002) Techniques for the removal of marker genes from transgenic plants. Biochimie 84, 1119–1126.

13. Darbani, B., Eimanifar, A., Stewart, C.N., Jr., and Camargo, W.N. (2007) Methods to produce marker-free transgenic plants. Biotechnol. J. 2, 83–90.

14. Yu, W., Lamb, J.C., Han, F., and Birchler, J.A. (2006) Telomere-mediated chromosomal truncation in maize. Proc. Natl. Acad. Sci. USA 103, 17331–17336.

15. Zupan, J., Muth, T.R., Draper, O., and Zambryski, P. (2000) The transfer of DNA from Agrobacterium tumefaciens into plants: a feast of fundamental insights. Plant J. 23, 11–28.

16. Christie, P.J. (1997) Agrobacterium tumefaciens T-complex transport apparatus: a paradigm for a new family of multifunctional transporters in Eubacteria. J. Bacteriol. 179, 3085–3094.

17. Tzfira, T., Vaidya, M., and Citovsky, V. (2002) Increasing plant susceptibility to Agrobacterium infection by overexpression of the

Arabidopsis nuclear protein VIP1 Proc. Natl. Acad. Sci. USA 99, 10435–10440.

18. Anand, A., Krichevsky, A., Schornack, S., Lahaye, T., Tzfira, T., Tang, Y., Citovsky, V., and Kirankumar, S.M. (2007) Arabidopsis VIRE2 INTERACTING PROTEIN2 is required for Agrobacterium T-DNA Integration in plants. Plant Cell 19, 1695–1708.

19. Tzfira, T., and Citovsky, V. (2002) Partnersin-infection: host proteins involved in the transformation of plant cells by Agrobacterium. Trends Cell Biol. 12, 121–129.

20. Bako, L., Umeda, M., Tiburcio, A.F., Schell, J., and Koncz, C. (2003) The VirD2 pilot protein of Agrobacterium-transferred DNA interacts with the TATA box-binding protein and a nuclear protein kinase in plants. Proc. Natl. Acad. Sci. USA 100, 10108–10113.

21. Gelvin, S.B. (2000) Agrobacterium and plant genes involved in T-DNA transfer and integration. Annu. Rev. Plant. Physiol. 51, 223–256.

22. Veena, J.H., Doerge, R.W., and Gelvin, S.B. (2003) Transfer of T-DNA and vir proteins to plant cells by Agrobacterium tumefaciens induces expression of host genes involved in mediating transformation and suppresses host defense gene expression. Plant J. 35, 219–236.

23. Lacroix, B., Kozlovosky, S.V., and Citovsky, V. (2008) Recent patents on Agrobacteriummediated gene and protein transfer for research and biotechnology. Recent Pat. DNA Gene Seq. 2, 69–81.

24. Hoekema, A., Hirsch, P.R., Hooykaas, P.J.J., and Schilperoort, R.A. (1983) A binary plant vector strategy based on separation of vir and T-region of the Agrobacterium tumefaciens Ti plasmid. Nature 303, 179–180.

25. Hamilton, C.M. (1997) A binary-BAC system for plant transformation with high-molecular-weight DNA. Gene 200, 107–116.

26. Rui-Feng, H., Yuan-Yuan, W., Bo, D., Ming, T., Ai-Qing, Y., Li-Li, Z., and Guang-Cun, H. (2006) Development of transformation system of rice based on binary bacterial artificial chromosome (BIBAC) vector. Acta Gen. Sin. 33, 269–276.

27. Komari, T. (1990) Transformation of cultured cells of Chenopodium quinoa by binary vectors that carry a fragment of DNA from the virulence region of pTiBo542. Plant Cell Rep. 9, 303–306.

28. Lee, L., and Gelvin, S.B. (2008) T-DNA binary vectors and systems. Plant Physiol. 146, 325–332.

29. Ishida, Y., Saito, H., Ohta, S., Hiei, Y., Komari, T., and Kumashiro, T. (1996) High efficiency transformation of maize (Zea mays) mediated by Agrobacterium tumefaciens. Nat. Biotechnol. 14, 745–750.

30. Khanna, H.K., and Daggard, G.E. (2003) Agrobacterium tumefaciens-mediated transformation of wheat using a super binary vector and a polyamine supplemented regeneration medium. Plant Cell Rep. 21, 429–436.

31. Lee, L.-Y., Kononov, M.E., Bassuner, B., Frame, B.R., Wang, K., and Gelvin, S.B. (2007) Novel pant transformation vectors containing the superpromoter. Plant Physiol. 145, 1294–1300.

32. Cheng, M., Lowe, B.A., Spencer, T.M., Ye, X., and Armstrong, C.L. (2004) Factors influencing Agrobacterium-mediated transformation of monocotyledonous species. In Vitro Cell Dev. Biol. Plant 40, 31–45.

33. Opabode, J.T. (2006) Agrobacterium mediated transformation of plants: emerging factors that influence efficiency. Biotech. Mol. Biol. Rev. 1, 12–20.

34. Zhao, Z., Cai, T., Tagliani, L., Miller, M., Wang, N., Pang, H., Rudert, M., Schroeder, S., Hondred, D., Seltzer, J., and Pierce, D. (2000) Agrobacteriummediated sorghum transformation. Plant Mol. Biol. 44, 789–798.

35. Nandakumar, R., Chen, L., and Rogers, S.M.D. (2004) Factors affecting the Agrobacterium-mediated transient transformation of the wetland monocot, Typha latifolia. Plant Cell Tiss. Organ Cult. 79, 31–38.

36. Herath, S.P., Suzuki, T., and Hattori, K. (2005) Factors influencing Agrobacteriummediated genetic transformation of kenaf. Plant Cell Tiss. Organ Cult. 82, 201–206.

37. Chateau, S., Sangwa, R.S., and SangwanNorreel, B.S. (2000) Competence of Arabidopsis thaliana genotypes and mutants for Agrobacterium tumefaciens-mediated gene transfer: role of phytohormones. J. Exp. Bot. 51, 1961–1968.

38. Dan, Y. (2008) Biological functions of antioxidants in plant transformation. In Vitro Cell Dev. Biol. Plant 44, 149–161.

39. Enriquez-Obregon, G.A., Vazquez-Padron, R.I., Prieto-Samsonov, D.L., de la Riva, G.A., and Selman-Housein, G. (1998) Herbicide-resistant sugarcane (Saccharum officinarum L.) plants by Agrobacteriummediated transformation. Planta 205, 20–27.

40. Olhoft, P.M., and Somers, D.A. (2001) l-Cysteine increases Agrobacterium mediated T-DNA delivery into soybean cotyledonary-node cells. Plant Cell Rep. 20, 706–711.

41. Vega, J., Yu, W., Kennon, A.M., Chen, X., and Zhang, Z.J. (2008) Improvement of Agrobacterium-mediated transformation in Hi-II maize (Zea mays L.) using standard binary vectors. Plant Cell Rep. 27, 297–305.

42. Frame, B.R., Shou, H., Chikwamba, R.K., Zhang, Z., Xiang, C.I., Fonger, T.M., Pegg, S.E.K., Li, B., Nettleton, D.S., Pei, D., and Wang, K. (2002) Agrobacterium tumefaciens-mediated transformation of maize embryos using a standard binary vector system. Plant Physiol. 129, 13–22.

43. Dan, Y.A. (2004) A novel plant transformation technology-Lipoic acid. In vitro Cell Dev. Biol. Plant 42, 18.

44. Dan, Y., Armstrong, C.L., Dong, J., Feng, X., Fry, J.E., Keithly, G.E., Martinell, B.J., Roberts, G.A., Smith, L.A., Tan, L.J., and Duncan, D.R. (2009) Lipoic acid – a unique plant transformation enhancer. In Vitro Cell Dev. Biol. Plant (DOI: 10.1007/s11627- 009-9227-5).

45. Cheng, M., Fry, J.E., Pang, S., Zhou, I., Hironaka, C., Duncan, D.R.I., Conner, T.W.L., and Wang, Y. (1997) Genetic transformation of wheat mediated by Agrobacterium tumefacien. Plant Physiol. 115, 971–980.

46. Tang, W. (2003) Additional virulence genes and sonication enhance Agrobacterium tumefaciens-mediated loblolly pine transformation. Plant Cell Rep. 21, 555–562.

47. Pena, L., Perez, R.M., Cervera, M., Juarez, J.A., and Navarro, L. (2004) Early events in Agrobacterium-mediated genetic transformation of citrus explants. Ann. Bot. (Lond.) 94, 67–74.

48. Arias, R.S., Filichkin, S.A., and Strauss, S.H. (2006) Arias divide and conquer: development and cell cycle genes in plant transformation. Trends Biotechnol. 24, 267–273.

49. Grafi, G. (2004) How cells dedifferentiate: a lesson from plants. Dev. Biol. 268, 1–6.

50. Riou-Khamlichi, C., Huntley, R., Jacqmard, A., and Murray, J.A. (1999) Cytokinin activation of Arabidopsis cell division through a D-type cyclin. Science 283, 1541–1544.

51. De Veylder, L., Beeckman, T., Beemster, G.T.S., Engler, J.D.A., Ormenese, S., Maes, S., Naudts, M., Der Schueren, E.V., Jacqmard, A., Engler, G., and Inze, D. (2002) Control of proliferation, endoreduplication and differentiation by the Arabidopsis E2Fa–DPa transcription factor. EMBO J. 21, 1360–1368.

52. Gordon-Kamm, W., Dilkes, B.P., Lowe, K., Hoerster, G., Sun, X., Ross, M., Church, L., Bunde, C., Farrell, J., Hill, P., Maddock, S., Snyder, J., Sykes, L., Li, Z., Woo, Y.-M., Bidney, D., and Larkins, B.A. (2002)

Stimulation of the cell cycle and maize transformation by disruption of the plant retinoblastoma pathway. Proc. Natl. Acad. Sci. USA 99, 11975–11980.

53. Mysore, K.S., Nam, J., and Gelvin, S. (2000) An Arabidopsis histone H2A mutant is deficient in Agrobacterium T-DNA integration. Proc. Natl. Acad. Sci. USA 97, 948–953.

54. Vlieghe, K., Vuylsteke, M., Florquin, K., Rombauts, S., Maes, S., Ormenese, S., Hummelen, P.V., de Peer, Y.V., Inze, D., and De Veylder, L. (2003) Microarray analysis of E2Fa–DPa-overexpressing plants uncovers a cross-talking genetic network between DNA replication and nitrogen assimilation. J. Cell Sci. 116, 4249–4259.

55. Zambre, M., Geerts, P., Maquet, A., Montagu, M.V., Dillen, W., and Angenon, G. (2001) Regeneration of fertile plants from callus in Phaseolus polyanthus Greenman (year bean). Ann. Bot. (Lond.) 88, 371–377.

56. Taoka, K.-I., Yanagimoto, Y., Daimon, Y., Hibara, K.-I., Aida, M., and Tasaka, M. (2004) The NAC domain mediates functional specificity of CUP-SHAPED COTYLEDON proteins. Plant J. 40, 462–473.

57. Slightom, J.L., Durand-Tardif, M., Jouanin, L., and Tepfer, D. (1986) Nucleotide sequence analysis of TL-DNA of Agrobacterium rhizogenes agropine type plasmid. Identification of open reading frames. J. Biol. Chem. 261, 108–121.

58. Collier, R., Fuchs, B., Walter, N., Kevin, L.W., and Taylor, C.G. (2005) Ex vitro composite plants: an inexpensive, rapid method for root biology Plant J. 43, 449–457.

59. Giri, A., and Narasu, M.L. (2000) Transgenic hairy roots: recent trends and applications. Biotechnol. Adv. 18, 1–22.

60. Georgiev, M.I., Pavlov, A.I., and Bley, T. (2007) Hairy root type plant in vitro systems as sources of bioactive substances. Appl. Microbiol. Biotechnol. 74, 1175–1185.

61. Seki, H., Nishizawa, T., Tanaka, N., Niwa, Y., Yoshida, S., and Muranaka, T. (2005) Hairy root-activation tagging: a highthroughput system for activation tagging in transformed hairy roots. Plant Mol. Biol. 59, 793–807.

62. Narayanan, R.A., Atz, R., Denny, R., Young, N.D., and Somers, D.A. (1999) Expression of soybean cyst nematode resistance in transgenic hairy roots of soybean. Crop Sci. 39, 1680–1686.

63. Skarjinskaia, M., Karl, J., Araujo, A., Ruby, K., Rabindran, S., Streatfield, S.J., and Yusibov, V. (2008) Production of recombinant proteins in clonal

root cultures using episomal expression vectors. Biotechnol. Bioeng. 100, 814–819.

64. Kuster, H., Vieweg, M.F., Manthey, K., Baier, M.C., Hohnjec, N., and Perlick, A.M. (2007) Identification and expression regulation of symbiotically activated legume genes. Phytochemistry 68, 8–18.

65. Breitler, J.C., Labeyrie, A., Meynard, D., Legavre, T., and Guiderdoni, E. (2002) Efficient microprojectile bombardmentmediated transformation of rice using gene cassettes. Theor. Appl. Genet. 104, 709–719.

66. Christou, P. (1995) Strategies for varietyindependent genetic transformation of important cereals, legumes and woody species utilising particle bombardment. Euphyica 85, 13–27.

67. Campbell, B.T., Baeziger, P.S., Mitra, A., Sato, S., and Clemente, T. (2000) Inheritance of multiple genes in wheat. Crop Sci. 40, 1133–1141.

68. Schmidt, M.A., Lafayette, P.R., Artelt, B.A., and Parrott, W.A. (2008) A comparison of strategies for transformation with multiple genes via microprojectile-mediated bombardment. In Vitro Cell Dev. Biol. Plant 44, 162–168.

69. Agrawal, P.K., Kohli, A., Twyman, R.M., and Christou, P. (2005) Transformation of plants with multiple cassettes generates simple transgene integration patterns and high expression levels. Mol. Breed. 16, 247–260.

70. Schmidt, M.A., Tucker, D.M., Cahoon, E.B., and Parrott, W.A. (2005) Towards normalization of soybean somatic embryo maturation. Plant Cell Rep. 24, 383–391.

71. Pawlowski, W.P., and Somers, D.A. (1996) Transgene inheritance in plants genetically engineered by microprojectile bombardment. Mol. Biotechnol. 6, 17–30.

72. Kohli, A., Gahakwa, D., Vain, P., Laurie, D.A., and Christou, P. (1999) Transgene expression in rice engineered through particle bombardment: molecular factors controlling stable expression and transgene silencing. Planta 208, 88–97.

73. Lowe, B.A., Prakash, N.S., Melissa, W., Mann, M.T., Spencer, T.M., and Boddupalli, R.S. (2009) Enhanced single copy integration events in corn via particle bombardment using low quantities of DNA. Transgenic Res. 18, 831–840 (DOI: 10.1007/s11248-009-9265-0).

74. Shimamoto, K., Teralda, R., Izawa, T., and Fujimoto, H. (1989) Fertile transgenic rice plants regenerated from transformed protoplasts. Nature 338, 274–276.

75. Salmenkallio-Marttila, M., Aspegren, K., Kerman, S., Kurt, U., Mannonen, L., Ritala, A., Teeriz, T.H., and Kauppinen, V. (1995) Transgenic barley (Hordeum vulgare L.) by electroporation of protoplasts Plant Cell Rep. 15, 301–304.

76. Daveya, M.R., Anthonya, P., Powera, J.B., and Loweb, K.C. (2005) Plant protoplasts: status and biotechnological perspectives. Biotechnol. Adv. 23, 131–171.

77. Sawahel, W.A. (2001) Stable genetic transformation of cotton plants using polybrene spermidine treatment. Plant Mol. Biol. Rep. 19, 377a–377f.

78. Gad, A.E., Rosenberg, N., and Altman, A. (1990) Liposome-mediated gene delivery into plant cells. Physiol. Plant. 79, 177–183.

79. Wordragen, M.V., Roshani, S., Ruud, V., Regis, P., Abvan, K., and Pim, Z. (1997) Liposome-mediated transfer of YAC DNA to tobacco cells. Plant Mol. Biol. Rep. 15, 170–178.

80. Kaeppler, H., Somers, D.A., Rines, H.W., and Cockburn, A.F. (1992) Silicon carbide fiber-mediated stable transformation of plant cells. Theor. Appl. Genet. 84, 560–566.

81. Frame, B.R., Drayton, P.R., Bagnall, S.V., Lewnau, C.J., Bullock, W.P., Wilson, H.M., Dunwell, J.M., Thompson, J.A., and Wang, K. (1994) Production of fertile transgenic maize plants by silicon carbide whisker-mediated transformation. Plant J. 6, 941–948.

82. Singh, N., and Chawla, H.S. (1999) Use of silicon carbide fibers for Agrobacteriummediated transformation in wheat. Curr. Sci. 76, 1483–1485.

83. Nagatani, N., Honda, H., Shimada, T., and Kobayashi, T. (1997) DNA delivery into rice cells and transformation using silicon carbide whiskers. Biotechnol. Tech. 11, 781–786.

84. Torney, F., Trewyn, B.G., Lin, V.S.Y., and Wang, K. (2007) Mesoporous silica nanoparticles deliver DNA and chemicals into plants. Nature Nanotech. 2, 295–300.

85. Nandiyanto, A.B.D., Kim, S.G., Iskandar, F., and Okuyama, K. (2009) Synthesis of silica nanoparticles with nanometer-size controllable mesopores and outer diameters. Microporous Mesoporous Mater. 120, 447–453.

86. Crossway, A., Oakes, J.W., Irvine, J.M., Ward, B., Knauf, V.C., and Shewmaker, C.K. (1986) Integration of foreign DNA following microinjection of tobacco mesophyll protoplasts. Mol. Gen. Genet. 202, 179–185.

87. Jones-Villeneuve, E., Huang, B., Prudhome, I., Bird, S., Kemble, R., Hattori, J., and Miki, B. (1995) Assessment of microinjection for introducing DNA into uninuclear microspores of rapeseed. Plant Cell Tiss. Organ Cult. 40, 97–100.

88. Holm, P.B., Olsen, O., Schnorf, M., BrinchPederse, H., and Knudsen, S. (2000) Transformation of barley by microinjection into isolated zygote protoplasts. Transgenic Res. 9, 21–32.

89. Lilly, J.W., Havey, M.J., Jackson, S.A., and Jiang, J. (2001) Cytogenomic analysis reveal the structural plasticity of the chloroplast genome in higher plants. Plant Cell 13, 245–254.

90. Maliga, P. (2004) Plastid transformation in higher plants. Annu. Rev. Plant Biol. 55, 289–313.

91. Verma, D., and Daniell, H. (2007) Chloroplast vector systems for biotechnology applications. Plant Physiol. 145, 1129–1143.

92. Boynton, J.E., Gillham, N.W., Harris, E.H., Hosler, J.P., Johnson, A.M., and Jones, A.R. (1988) Chloroplast transformation in Chlamydomonas with high velocity microprojectiles. Science 240, 1534–1538.

93. Lee, S.M., Kang, K., Chung, H., Yoo, S.H., Xu, X.M., Lee, S.B., Cheong, J.J., Daniell, H., and Kim, M. (2006) Plastid transformation in the monocotyledonous cereal crop, rice (Oryza sativa) and transmission of transgenes to their progeny. Mol. Cells 21, 401–410.

94. Kumar, S., Dhingra, A., and Daniell, H. (2004) Stable transformation of the cotton plastid genome and maternal inheritance of transgenes. Plant Mol. Biol. 56, 203–216.

95. Daniell, H., Chebolu, S., Kumar, S., Singleton, M., and Falconer, R. (2005) Chloroplast derived vaccine antigens and other therapeutic proteins. Vaccine 23, 1779–1783.

96. Rommens, C.M. (2004) All-native DNA transformation: a new approach to plant genetic engineering. Trends Plant Sci. 9, 457–464.

97. Bernasconi, P. et al. (1995) A naturally occurring point mutation confers broad range tolerance to herbicides that target acetolactate synthase. J. Biol. Chem. 270, 17381–17385.

98. Baerson, S.R., Rodriguez, D.J., Tran, M., Feng, M., Biest, N.A., and Dill, G.M. (2002) Glyphosate-resistant goosegrass identification of a mutation in the target enzyme 5-enolpyruvylshikimate-3-phosphate synthase. Plant Physiol. 129, 1265–1275.

99. Song, J., Bradeen, J.M., Naess, S.K., Raasch, J.A., Wielgus, S.M., Haberlach, G.T., Liu, J., Kuang, H., Austin-Phillips, S., Buell, C.R.,

Helgeson, J.P., and Jiang, J. (2003) Gene RB cloned from Solanum bulbocastanum confers broad spectrum resistance to potato late blight. Proc. Natl. Acad. Sci. USA 100, 9128–9133.

100. Pechan, T., Cohen, A., Williams, W.P., and Luthe, D.S. (2002) Insect feeding mobilizes a unique plant defense protease that disrupts the peritrophic matrix of caterpillars. Proc. Natl. Acad. Sci. USA 99, 13319–13323.

101. Wang, E., Wang, R., DeParasis, J., Loughrin, J.H., Gan, S., and Wagner, G.J. (2001) Suppression of a P450 hydroxylase gene in plant trichome glands enhances naturalproduct-based aphid resistance. Nat. Biotechnol. 19, 371–374.

102. Kasuga, M., Liu, Q., Miura, S., YamaguchiShinozaki, K., and Shinozaki, K. (1999) Improving plant drought, salt, and freezing tolerance by gene transfer of a single stressinducible transcription factor. Nat. Biotechnol. 17, 287–291.

103. Miki, B., and Mc Hugh, S. (2004) Selectable marker genes in transgenic plants: applications, alternatives and biosafety. J. Biotechnol. 107, 193–232.

104. Daley, M., Knauf, V.C., Summerfelt, K.R., and Turner, J.C. (1998) Co-transformation with one Agrobacterium tumefaciens strain containing two binary plasmids as a method for producing marker-free transgenic plants. Plant Cell Rep. 17, 489–496.

105. Komari, T., Hiei, Y., Saito, Y., Murai, N., and Kumashiro, T. (1996) Vectors carrying two separate T-DNAs for co-transformation of higher plants mediated by Agrobacterium tumefaciens and segregation of transformants free from selection markers. Plant J. 10, 165–174.

106. Hua, Y., and Rommens, C.M. (2007) Transposition-based plant transformation. Plant Physiol. 143, 570–578.

107. Cotsaftis, O., Sallaud, C., Breitler, J.C., Meynard, D., Greco, R., Pereira, A., and Guiderdoni, E. (2002) Transposon-mediated generation of T-DNA- and marker-free rice plants expressing a Bt endotoxin gene. Mol. Breed. 10, 165–180.

108. Saelim, L., Phansiri, S., Suksangpanomrung, M., Netrphan, S., and Narangajavana, J. (2009) Evaluation of a morphological marker selection and excision system to generate marker-free transgenic cassava plants. Plant Cell Rep. 28, 445–455.

109. Ebinuma, H., and Komamine, A. (2001) MAT (Multi-Auto-Transformation) vector system. The oncogenes of Agrobacterium as

positive markers for regeneration and selection of marker-free transgenic plants. In Vitro Cell Dev. Biol. Plant 37, 103–113.

110. Kuraya, Y., Ohta, S., Fukuda, M., Hiei, Y., Murai, N., Hamada, K., Ueki, T., Imaseki, H., and Komari, T. (2004) Suppression of transfer of non-T-DNA "vector backbone" sequences by multiple left border repeats in vectors for transformation of higher plants mediated by Agrobacterium tumefaciens. Mol. Breed. 14, 309–320.

111. Yang, A., Su, Q., and An, L. (2009) Ovarydrip transformation: a simple method for directly generating vector- and marker-free transgenic maize (Zea mays L.) with a linear GFP cassette transformation. Planta 229, 793–801.

112. Yang, A., Su, Q., An, L., Liu, J., Wu, W., and Qiu, Z. (2009) Detection of vector- and selectable marker-free transgenic maize with a linear GFP cassette transformation via the pollen-tube pathway. J. Biotechnol. 139, 1–5.

113. Ye, X., Williams, E.J., Shen, J., Esser, J.A., Nichols, A.M., Petersen, M.W., and Gilbertson, L.A. (2008) Plant development inhibitory genes in binary vector backbone improve quality event efficiency in soybean transformation. Transgenic Res. 17, 827–838.

114. Lu, H.J., Zhou, X.R., Gong, Z.X., and Upadhyaya, N.M. (2001) Generation of selectable marker-free transgenic rice using double right-border (DRB) binary vectors. Aust. J. Plant Physiol. 28, 241–248.

115. Hajdukiewicz, P., Svab, Z., and Maliga, P. (1994) The small, versatile pPZP family of Agrobacterium binary vectors for plant transformation. Plant Mol. Biol. 25, 989–994.

116. Fu, X., Duc, L.T., Fontana, S., Bong, B.B., Tinjuangjun, P., Sudhakar, D., Twyman, R.M., Christou, P., and Kohli, A. (2000) Linear transgene constructs lacking vector backbone sequences generate low-copynumber transgenic plants with simple integration patterns. Transgenic Res. 9, 11–19.

117. Dale, E., and Ow, D. (1991) Gene transfer with subsequent removal of the selection gene from the host genome. Proc. Natl. Acad. Sci. USA 88, 10558–10562.

118. Wang, Y., Chen, B., Hu, Y., Li, J., and Lin, Z. (2005) Inducible excision of selectable marker gene from transgenic plants by the Cre/lox site-specific recombination system. Transgenic Res. 14, 605–614.

119. Srivastava, V., and Ow, D.W. (2001) Singlecopy primary transformants of maize obtained through the co-introduction of a recombinase-expressing construct. Plant Mol. Biol. 46, 561–566.

120. Zuo, J., Niu, Q.-W., Moller, S.G., and Chua, N.-H. (2001) Chemical-regulated, site-specific DNA excision in transgenic plants. Nat. Biotechnol. 19, 157–161.

121. Bai, X., Qiuyun, W., and Chu, C. (2008) Excision of a selective marker in transgenic rice using a novel Cre/loxP system controlled by a floral specific promoter. Transgenic Res. 17, 1035–1043.

122. McCormac, A.C., Fowler, M.R., Chen, D.F., and Elliott, M.C. (2001) Efficient co-transformation of Nicotiana tabacum by two independent T-DNAs, the effect of T-DNA size and implications for genetic separation. Transgenic Res. 10, 143–155.

123. Miller, M., Tagliani, L., Wang, N., Berka, B., Bidney, D., and Zhao, Z.Y. (2002) Highefficiency transgene segregation in co-transformed maize plants using an Agrobacterium tumefaciens 2T-DNA binary system. Transgenic Res. 11, 381–396.

124. Matthews, P.R., Waterhouse, P.M., Thornton, S., Fieg, S.J., Gubler, F., and Jacobsen, J.V. (2001) Marker gene elimination from transgenic barley, using co-transformation with adjacent "twin T-DNAs" on a standard Agrobacterium transformation vector. Mol. Breed. 7, 195–202.

125. Karimi, M., Inze, D., and Depicker, A. (2002) GATEWAY vectors for Agrobacteriummediated plant transformation. Trends Plant Sci. 7, 193–195.

126. Karimi, M., Bjorn, D.M., and Hilson, P. (2005) Modular cloning in plant cells. Trends Plant Sci. 10, 103–105.

127. Karimi, M., Depicker, A., and Hilson, P. (2007) Recombinational cloning with plant gateway vectors. Plant Physiol. 145, 1144–1154.

128. Chen, Q.-J., Zhou, H.-M., Chen, J., and Wang, X.C. (2006) A gateway-based platform for multigene plant transformation. Plant Mol. Biol. 62, 927–936.

129. Helliwell, C., and Waterhouse, P. (2003) Constructs and methods for high-throughput gene silencing in plants. Methods 30, 289–295.

130. Daxinger, L., Hunter, B., Sheikh, M., Jauvion, V., Gasciolli, V., Vaucheret, H., Matzke, M., and Furner, I. (2008) Unexpected silencing effects from T-DNA tags in Arabidopsis. Trends Plant Sci. 13, 4–6.

131. Heilersig, B.H.J.B., Loonen, A.E.H.M., Wolters, A.-M.A., and Visser, R.G.F. (2006) Presence of an intron in inverted repeat constructs does not necessarily have an effect on efficiency of post-transcriptional gene silencing. Mol. Breed. 17, 307–316.

132. Earley, K.W., Haag, J.R., Pontes, O., Opper, K., Juehne, T., Song, K., and Pikaard, C.S. (2006) Gateway-compatible vectors for plant functional genomics and proteomics. Plant J. 45, 616–629.

133. Nakagawa, T., Kurose, T., Hino, T., Tanaka, K., Kawamukai, M., Niwa, Y., Toyooka, K., Matsuoka, K., Jinbo, T., and Kimura, T. (2007) Development of series of gateway binary vectors, pGWBs, for realizing efficient construction of fusion genes for plant transformation. J. Biosci. Bioeng. 104, 34–41.

134. Nakagawa, T., Nakamura, S., Tanaka, K., Kawamukai, M., Suzuki, T., Nakamura, K., Kimura, T., and Ishiguro, S. (2008) Development of R4 gateway binary vectors (R4pGWB) enabling high-throughput promoter swapping for plant research. Biosci. Biotechnol. Biochem. 72, 624–629.

135. Curtis, M.D., and Grossniklaus, U. (2003) A gateway cloning vector set for high-throughput functional analysis of genes in planta. Plant Physiol. 133, 462–469.

136. Tzfira, T., Tian, G.-W., Lacroix, B., Vyas, S., Li, J., Leitner-Dagan, Y., Krichevsky, A., Taylor, T., Vainstein, A., and Citovsky, V. (2005) pSAT vectors: a modular series of plasmids for autofluorescent protein tagging and expression of multiple genes in plants. Plant Mol. Biol. 57, 503–516.

137. Zhong, S., Lin, Z., Fray, R.G., and Grierson, D. (2008) Improved plant transformation vectors for fluorescent protein tagging. Transgenic Res. 17, 985–989.

138. Martin, K., Kopperud, K., Chakrabarty, R., Banerjee, R., Brooks, R., and Goodin, M.M. (2009) Transient expression in Nicotiana benthamiana fluorescent marker lines provides enhanced definition of protein localization, movement and interactions in planta. Plant J. 59, 150–162.

139. Copeland, N.G., Jenkins, N.A., and Court, D.L. (2001) Recombineering: a powerful new tool for mouse functional genomics. Nat. Rev. Genet. 2, 769–779.

140. Rozwadowski, K., Yang, W., and Kagale, S. (2008) Homologous recombination-mediated cloning and manipulation of genomic DNA regions using gateway and recombineering systems. BMC Biotechnol. 8, 88.

141. Cai, C.Q., Doyon, Y.W., Ainley, M., Miller, J.C., et al. (2009) Targeted transgene integration in plant cells using designed zinc finger nucleases. Plant Mol. Biol. 69, 699–709.

142. Durai, S., Mani, M., Kandavelou, K., Wu, J., Porteus, M.H., and Chandrasegaran, S. (2005) Zinc finger nucleases: customdesigned

molecular scissors for genome engineering of plant and mammalian cells. Nucleic Acids Res. 33, 5978–5990.

143. Lloyd, A., Plaisier, C.L., Carroll, D., and Drews, G.N. (2005) Targeted mutagenesis using zinc-finger nucleases in Arabidopsis. Proc. Natl. Acad. Sci. USA 102, 2232–2237.

144. Ordiz, M.I., Barbas, C.F., and Beachy, R.N. (2002) Regulation of transgene expression in plants with polydactyl zinc finger transcription factors Proc. Natl. Acad. Sci. USA 99, 13290–13295.

145. Szxzepek, M., Brondani, V., Buchd, I., Serrano, L., Segal, D., and Cathomen, T. (2007) Structure-based redesign of the dimerization interface reduces the toxicity of zinc-finger nucleases. Nat. Biotechnol. 25, 786–793.

146. Weichang, Y., Fangpu, H., Gao, Z., Vega, J.M., and Birchler, J.A. (2007) Construction and behavior of engineered minichromosomes in maize. Proc. Natl. Acad. Sci. USA 104, 8924–8929.

147. Yu, W., Han, F., and Birchler, J.A. (2007) Engineered minichromosomes in plants. Curr. Opin. Biotechnol. 18, 425–431.

148. Kato, A., Zheng, Y.Z., Auger, D.L., PhelpsDurr, T., Bauer, M.J., Lamb, J.C., and Birchler, J.A. (2005) Minichromosomes derived from the B chromosome of maize. Cytogenet. Genome. Res. 109, 156–165.

149. Veena (2008) Engineering plants for future: tools and options Physiol. Mol. Biol. Plants 14, 131–135.

150. Vain, P. (2005) Plant transgenic science knowledge. Nat. Biotechnol. 23, 1348–1349.

151. Chapotin, S.M., and Wolt, J.D. (2007) Genetically modified crops for the bioeconomy: meeting public and regulatory expectations. Transgenic Res. 16, 675–688.

152. Kim, E.H., Suh, S.C., Park, B.S., Shin, K.S., Kweon, S.J., Han, E.J., Park, S.-H., Kim, Y.-S., and Kim, J.-K. (2009) Chloroplasttargeted expression of synthetic cry1Ac in transgenic rice as an alternative strategy for increased pest protection. Planta 230, 397–405.

153. Kim, T.-G., Baek, M.-Y., Lee, E.-K., Kwon, T.-H., and Yang, M.-S. (2008) Expression of human growth hormone in transgenic rice cell suspension culture. Plant Cell Rep. 27, 885–891.

154. De Padua, V.L.M., Ferreira, R.P., Meneses, L., Uchoa, L., Marcia, M.-P., and Mansur, E. (2001) Transformation of Brazilian elite Indica-Type Rice (Oryza Sativa L.) by electroporation of shoot apex explants. Plant Mol. Biol. Rep. 19, 55–64.

155. Cho, M.-J., Choi, H.W., Okamoto, D., Zhang, S., and Lemaux, P.G. (2003) Expression of green fluorescent protein and its inheritance in transgenic oat plants generated from shoot meristematic cultures. Plant Cell Rep. 21, 467–474.

Chapter 3

SUNFLOWER CENTROMERES CONSIST OF A CENTROMERE-SPECIFIC LINE AND A CHROMOSOME-SPECIFIC TANDEM REPEAT

Kiyotaka Nagaki[1], Keisuke Tanaka[2], Naoki Yamaji[1], Hisato Kobayashi[2], and Minoru Murata[1]

[1]Applied Genomics Unit, Institute of Plant Science and Resources, Okayama University, Kurashiki, Japan

[2]NODAI Genome Research Center, Tokyo University of Agriculture, Setagaya, Japan

ABSTRACT

The kinetochore is a protein complex including kinetochore-specific proteins that plays a role in chromatid segregation during mitosis and meiosis. The complex associates with centromeric DNA sequences that are usually species-specific. In plant species, tandem repeats including satellite DNA sequences and retrotransposons have been reported as centromeric DNA sequences. In this study on sunflowers, a cDNA-encoding centromere-specific histone H3 (CENH3) was isolated from a cDNA pool from a seedling, and an antibody was raised against a peptide synthesized from the deduced cDNA. The antibody specifically recognized the sunflower CENH3 (HaCENH3) and showed centromeric signals by immunostaining and immunohistochemical staining analysis. The antibody was also applied in chromatin immunoprecipitation (ChIP)-Seq to isolate centromeric DNA sequences and two different types of repetitive DNA sequences were identified. One was a long interspersed nuclear element (LINE)-like sequence, which showed centromere-specific signals on almost all chromosomes in sunflowers. This is the first report of a centromeric LINE sequence, suggesting possible centromere targeting ability. Another type of identified repetitive DNA was a tandem repeat sequence with a 187-bp unit that was found only on a pair of chromosomes. The HaCENH3 content of the tandem repeats was estimated to be much higher than that of the LINE, which implies centromere evolution from LINE-based centromeres to more stable tandem-repeat-based centromeres. In addition, the epigenetic status of the sunflower centromeres was investigated by immunohistochemical staining and ChIP, and it was found that centromeres were heterochromatic.

INTRODUCTION

Asterales is the most diverged order of dicots and includes 11 families and 27,000 species. Among the 11 families, Asteraceae is the largest and includes the sunflower and daisy. Sunflowers (*Helianthus annuus* L., $2n = 2x = 34$, genome size $= 2.43$ Gb/haploid) are one of the most important crops in Asterales because their seeds can be used for oil production (Bennett et al., 1982). Sunflowers have been genetically and cytogenetically investigated (Feng et al., 2013), and a genome sequencing project is now in progress (http://sunflowergenome.org/). For karyotypic analyses, repetitive DNA sequences including rDNA, sunflower-specific tandem repeats and bacterial artificial chromosome (BAC) clones have been used as probes, but none of these probes showed centromeric localization (Ceccarelli et al., 2007; Talia et al., 2010; Feng et al., 2013). In other words, at present, there is no genetic and cytogenetic marker for sunflower centromeres.

The kinetochore is a special protein complex formed on centromeric regions that ensures equal and accurate distribution of chromatids to daughter cells during mitosis and meiosis (Amor et al., 2004). Among the constitutive proteins of kinetochores, centromere-specific histone H3 (CENH3) acts as a base for assembling other kinetochore proteins, and its presence epigenetically determines the kinetochore position (Perpelescu and Fukagawa, 2011). The first identified CENH3 was CENP-A in humans (Earnshaw and Rothfield, 1985), and its orthologs have been isolated from 11 of 63 APG III orders, including Poales, Asparagales, Rosales, Fabales, Malpighiales, Malvales, Brassicales, Myrtales, Solanales, Asterales, and Apiales, in this decade (Talbert et al., 2002; Zhong et al., 2002; Nagaki et al., 2004,2005, 2012b; Nagaki and Murata, 2005; Sanei et al., 2011; Wang et al., 2011; Neumann et al., 2012, 2015; Dunemann et al., 2014; Masonbrink et al., 2014; He et al., 2015; Maheshawari, 2015). However, no CENH3 has been isolated from Asterales species.

The CENH3s possess relatively conserved histone fold domains (HFDs) and highly variable N-terminal tails, and the HFD has been shown to be important for centromere localization (Vermaak et al., 2002; Black et al., 2004; Lermontova et al., 2006). The Loop 1 region on the HFD is longer than that of canonical histone H3, which is common among CENH3s. This region allows more compact packing of nucleosomes with CENH3 compared to those with canonical histone H3 (Black et al., 2004; Tek et al., 2010, 2011; Nagaki et al., 2012a; Maheshawari, 2015).

CENH3 is a component of core histones in the centromeric regions, and it directly binds to DNA. Additionally, CENH3 localizes to functional centromeres (Warburton et al., 1997; Nasuda et al., 2005; Han et al., 2006). Therefore, centromeric DNA sequences have been identified in plant species by chromatin immunoprecipitation (ChIP) using anti-CENH3 antibodies (Zhong et al., 2002; Nagaki et al., 2003, 2004, 2009, 2011, 2012a,b; Nagaki and Murata, 2005;Houben et al., 2007; Tek et al., 2010, 2011; Wang et al., 2011; Gong et al., 2012; Neumann et al., 2012; He et al., 2015). In most cases, species-specific microsatellites, minisatellites, macrosatellites, and retrotransposons have been identified as the centromeric sequences, and these sequences are located on all centromeric regions in the species (Zhong et al., 2002; Nagaki et al., 2003, 2009, 2011; Nagaki and Murata, 2005; Houben et al., 2007; Tek et al., 2010,2011; Wang et al., 2011; Neumann et al., 2012; He et al., 2015).

Histone modifications are the key components in epigenetics, and they have been much investigated in this decade (Desvoyes et al., 2014; Sharma et al., 2015). In plant species, two different types of heterochromatin distribution are reported to be related to the genome size of the species (Houben et al., 2003). In plant species with a genome size smaller than 510 Mb, chromocenters appear on interphase nuclei, and heterochromatic modifications occur specifically on the chromocenters. *A. thaliana* is a good example for small-genome species, as it shows heterochromatic modifications on centromeric and pericentromeric regions not only at interphase but also at metaphase. Euchromatic modifications of the species were observed in the regions with no heterochromatic modifications (Houben et al., 2003;Jasencakova et al., 2003). In contrast, no chromocenters appear on interphase nuclei in plant species with genomes larger than 530 Mb. However, in these species, heterochromatic and euchromatic modifications are observed to disperse on interphase nuclei (Houben et al., 2003), and neither modification appears on centromeric and pericentromeric regions during metaphase.

The epigenetic modifications around centromeric regions have also been analyzed by ChIP using centromeric DNA sequences. In human cells, the centromere-specific histone H3 variants (CENP-A) coexist with heterochromatic modified histones at almost all stages during the cell cycle, but the histones are instantaneously modified with a euchromatic modification from anaphase to early G1 (Ohzeki et al., 2012). In rice, a CENH3-binding region in a centromere, Cen8, was revealed by ChIP to be heterochromatic (Nagaki et al., 2004). However, euchromatic markers were also detected in genic regions and 167-bp CentO variants in rice centromeres (Yan et al., 2006; Zhang et al., 2013).

In this study, we isolated a cDNA encoding CENH3 from sunflowers and raised a peptide antibody against CENH3. The antibody showed centromere-specific signals in immunostaining and immunohistochemical staining experiments. A ChIP-Seq experiment was also conducted to isolate DNA sequences that coexist with CENH3. Additionally, the epigenetic status of the centromeres was successfully revealed by immunohistochemical staining using anti-CENH3 and anti-modified histone antibodies as well as by ChIP-qPCR using these antibodies and isolated DNA sequences from ChIP-Seq.

MATERIALS AND METHODS

Plant Material

Sunflower seeds (*H. annuus* L., $2n = 2x = 34$, number 1802-065684) were obtained from a commercial source (LIC, Okayama, Japan).

Identification of a Sunflower Expressed Sequence Tag (EST) Encoding CENH3

An EST sequence encoding sunflower CENH3 (HaCENH3) was identified from the gene indices using the tblastn program (http://compbio.dfci.harvard.edu/tgi/) and the amino acid sequence of NtCENH3-1 (GenBank accession number: BAH03514, Nagaki et al., 2009) as a query.

RNA Isolation and PCR

Total RNA was isolated from a 3-day-old sunflower seedling using the RNeasy Plant Mini kit (Qiagen, Hilden, Germany). To determine the full-length cDNA sequence of a sunflower CENH3 gene, rapid amplification of cDNA ends (RACE) was conducted. For 3'RACE, the primer HaCENH3-3RACE was designed from an EST encoding a putative sunflower CENH3 sequence found during the BLAST search; this primer was used with the SMARTer RACE cDNA Amplification Kit (Clontech, CA, USA). Another primer, HaCENH3-5RACE, was designed from the sequences determined by 3'RACE and was used to determine the 5' end.

Sequencing and Sequence Analyses

The 3'- and 5'-RACE products were cloned into a pGEM-T easy vector (Promega, WI, USA) and sequenced from both ends using a BigDye Terminator v1.1 cycle sequencing kit and an ABI PRISM 3130xl genetic analyzer (Applied Biosystems, CA, USA). A putative amino acid sequence was deduced from the DNA sequences and used as a query sequence for a protein

BLAST search on the NCBI website (http://blast.ncbi.nlm.nih.gov/Blast. cgi?CMD=Web&PAGE_TYPE=BlastHome). The deduced HaCENH3 amino acid sequence was aligned with orthologs identified by the protein BLAST search and canonical histone H3 of rice using the Clustal X software program (Thompson et al., 1997). Phylogenetic relationships among the CENH3s were analyzed by the neighbor-joining method (Saitou and Nei, 1987).

Immunostaining

Based on the deduced HaCENH3 amino acid sequence, a peptide corresponding to the N-terminus of HaCENH3 (H_2N-ARTKHPAKRSSGIPADGRSS-COOH) was synthesized and injected into two rabbits. The raised antisera were purified using an affinity Sepharose column consisting of the aforementioned peptide.

Immunostaining was conducted as previously described (Nagaki et al., 2012b). In brief, root tips of 3-day-old sunflowers were fixed in microtubule stabilizing buffer (50 mM PIPES, pH 6.9, 5 mM MgSO4, and 5 mM EGTA) containing 3% (w/v) paraformaldehyde and 0.2% (v/v) Triton X-100. The fixed tips were washed and digested with a mixture of 1% (w/v) cellulase Onozuka RS (Yakult Pharmaceutical Industry, Tokyo, Japan) and 0.5% (w/v) pectolyase Y-23 (Seishin Pharmaceuticals, Tokyo, Japan) and then compressed onto slides coated with poly-L-lysine (Matsunami, Osaka, Japan). A 1:100 dilution of the purified anti-HaCENH3 antibody and monoclonal anti-modified-histone antibodies produced in mice (anti-histone H3 dimethyl K4 (H3K4me2): MBL (Nagoya, Japan) MABI0303 and anti-histone H3 dimethyl K9 (H3K9me2): MBL MABI0317) were applied to the slides. To detect acetylations of histone H4, an anti-histone H4 acetyl (H4Ac) antibody raised in rabbits (Millipore, MA, USA: #06-598) was used. The antibodies were detected using 1:1000 dilutions of Alexa Fluor 555-labeled anti-rabbit antibodies (Molecular Probes, OR, USA) and Alexa Fluor 488-labeled anti-mouse antibodies (Molecular Probes), respectively. Chromosomes were counterstained with 0.1 µg/ml 4,6-diamino-2-phenylindole (DAPI). Immunosignals and stained chromosomes were captured using a chilled charge-coupled device (CCD) camera, AxioCam HR (Carl Zeiss, Oberkochen, Germany), and images were pseudo-colored and processed using AxioVision software (Carl Zeiss).

Immunohistochemical Staining

Immunohistochemical staining was conducted as described with minor modifications (Yamaji and Ma, 2007; Nagaki et al., 2012b). Three-day-old sunflower roots were fixed as described above for immunostaining, and the fixed roots were sectioned at 100-µm thickness using a microslicer (LinearSlicer PRO10; Dosaka EM). These sections were transferred onto slides and then macerated. For three-color detection, 1:100 dilutions of the purified anti-HaCENH3 rabbit antibody and anti-α-tubulin mouse antibody (Sigma, MO, USA: T6199) were applied to the slides. For four-color detection, 1:100 dilutions of the purified anti-HaCENH3 rabbit antibody, anti-α-tubulin rat antibody (Abcam, Cambridge, UK: ab64332) and monoclonal anti-H3K9me2 mouse antibody (MBL MABI0317) were applied to the slides. After washing in PBS, the primary antibodies were detected using 1:1000 diluted secondary antibodies, the Alexa Fluor 555-labeled anti-rabbit antibodies and the Alexa Fluor 488-labeled anti-mouse antibodies for the three-color detection and Alexa Fluor 647-labeled anti-rabbit antibodies (Molecular Probes), Alexa Fluor 488-labeled anti-rat antibodies (Molecular Probes), and Alexa Fluor 546-labeled anti-mouse antibodies (Molecular Probes) for the four-color detection. Then, nuclei and chromosomes were counterstained with DAPI. Immunosignals and stained chromosomes were observed with a laser-scanning confocal microscope (LSM700; Carl Zeiss). The obtained data were analyzed using AxioVision software.

Chromatin Immunoprecipitation (ChIP)

ChIP was performed as previously described (Nagaki et al., 2012b) with minor modifications using the anti-HaCENH3 antibody and the anti-modified histone antibody (anti-H3K9me2: MBL #MABI0317, anti-H3K4me2: MBL #MABI0303 and anti-H4Ac: Millipore #06-598). Nuclei were isolated from the leaves of 1-month-old sunflowers and then digested with micrococcal nuclease (Sigma) to produce chromatin. Following overnight incubation of the chromatin with the antibodies at 4°C, the antibodies were captured using Dynabeads Protein G (Invitrogen, CA, USA). For mock experiments, a normal rabbit serum was used instead of the antibodies. DNA was purified from the chromatin with the captured antibodies by phenol/chloroform extraction followed by ethanol precipitation.

ChIP-Seq and Repeatexplorer Analysis

ChIP-Seq was conducted using precipitated DNA from the input and HaCENH3 fractions in the ChIP. Libraries were constructed using the NEBNext ChIP-Seq Library Prep Reagent Set for Illumina (New England Biolabs, MA, USA), and the libraries were read by MiSeq (Illumina, CA, USA) with the paired-end 2 × 300 bp protocol. Conversion of raw base-call data to sequence data in the fastq format, identification of reads derived from each sample by index sequences, and adapter trimming were performed using MiSeq reporter 2.3.32. The sequence data were analyzed by a similarity-based clustering program,

RepeatExplorer (http://www.repeatexplorer.org) (Novák et al., 2010) with default parameters.

qPCR

qPCR was conducted using SYBR Premix Ex Taq II (Tli RNaseH Plus) (Takara, Shiga, Japan) and primers for qPCR with a StepOne instrument (Applied Biosystems). The primers were designed based on the sequences in the clusters of the RepeatExplorer analysis. The precipitated DNA in the ChIP experiment was used as a template, and the mock was used as a negative control. Relative enrichment (RE) was calculated by the following formula: RE = amount of the sequence in the antibody fraction/amount of the sequence in the mock. The qPCR results were assessed by Student's t-test.

Fluorescence *in Situ* Hybridization (FISH)

Probes were amplified using primer sets designed based on the sequences in clusters of the RepeatExplorer analysis using sunflower genomic DNA as a template. The amplified DNA was cloned into pGEM T-easy, and sequences were confirmed using a BigDye Terminator v1.1 cycle sequencing kit and an ABI PRISM 3130xl genetic analyzer. To detect nucleolar organizing regions, an 18S-5.8S-28SrDNA clone from wheat (pTa71) was used (Gerlach and Bedbrook, 1979). To characterize sunflower chromosomes, a reported 386-bp tandem repetitive sequence, HAG004N15 (Ceccarelli et al., 2007), was amplified using specific primers and cloned. The sequence was confirmed as described above.

FISH analysis of mitotic chromosomes was performed as previously described (Nagaki et al., 2011). Chromosomes were prepared from the root tips of 3-day-old sunflowers. The plasmid DNA was labeled by nick translation using a DIG-Nick Translation Mix (Roche, Basel, Switzerland) or a Biotin-Nick Translation Mix (Roche). The digoxigenin- and biotin-labeled probes were visualized using rhodamine-conjugated anti-digoxigenin antibody (Roche) and Alexa Fluor 488-conjugated streptavidin (Molecular Probes), respectively.

RESULTS

Isolation and Sequence Analyses of CENH3 in Sunflowers

To identify the HaCENH3 gene, a BLAST search was conducted in a sunflower EST database of the gene index project using the amino acid sequence of NtCENH3-1 as a query. One EST group (TC48348) containing a 675-bp sequence showed 68% identity to the query sequence. However, because of low quality in the first 150 bp of the EST, no start codon was found in the sequence. To obtain a full-length cDNA of the gene, 5'- and 3'-RACE experiments were performed using sunflower seedling cDNA as a template. As a result, a putative full-length cDNA containing a 429-bp ORF that encodes 143 amino acids (GenBank accession number LC075743) was obtained.

The amino acid sequence deduced from the ORF showed similarity to the sequences of some other plants, with CENH3 from *Daucus muricatus* showing the highest similarity (70%). The amino acid sequence of HaCENH3 was aligned with those of CENH3s from other plant species and rice canonical histone H3. The alignment indicated that the N-terminal amino acid at position 14–44 of HaCENH3 did not show any similarity to those of the other CENH3s or the canonical histone H3, and it also indicated that HaCENH3 possessed a longer loop 1 domain than canonical histone H3. The longer loop 1 is a feature of CENH3s.

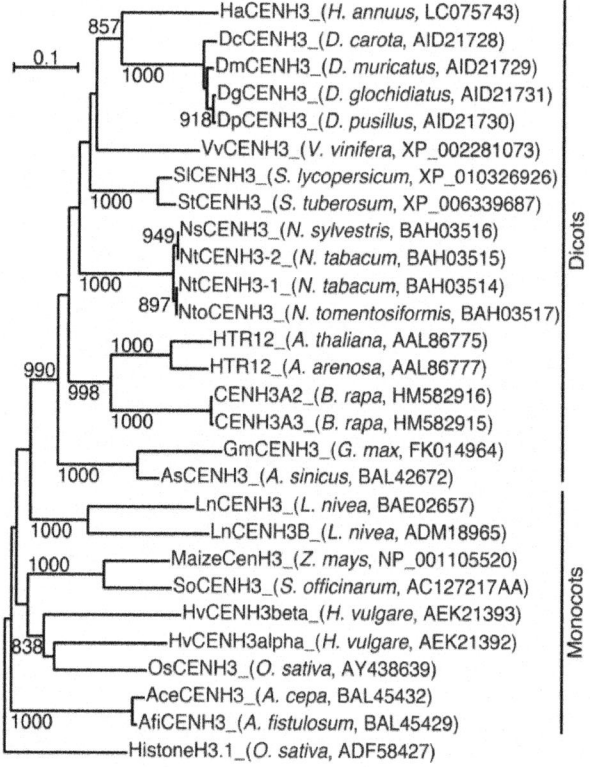

Figure 1. Phylogenetic tree based on the amino acid sequences of plant CENH3. The species name and GenBank accession number are indicated in parentheses. Rice ca-

nonical histone H3 was used as an outgroup. Bootstrap values greater than 800 in 1000 tests are indicated on the branches.

In a phylogenetic analysis using the alignment, HaCENH3 was classified into a dicot clade (Figure 1). As expected, HaCENH3 was found to be most closely related to the CENH3s of *Daucus* species, but the sequence was placed outside of a clade containing the *Daucus* CENH3s.

Centromere Localization of HaCENH3

An anti-HaCENH3 antibody was raised against a synthetic peptide comprising N-terminal amino acid residues 2–21 of the deduced HaCENH3 amino acid sequence. To confirm its specificity to the centromeres, immunostaining using the antibody was conducted, and centromere-specific immunosignals appeared on all sunflower chromosomes (Figure 2). Additionally, in immunohistochemical staining, microtubule signals on all chromosomes were associated with all of the anti-HaCENH3 immunosignals at metaphase.

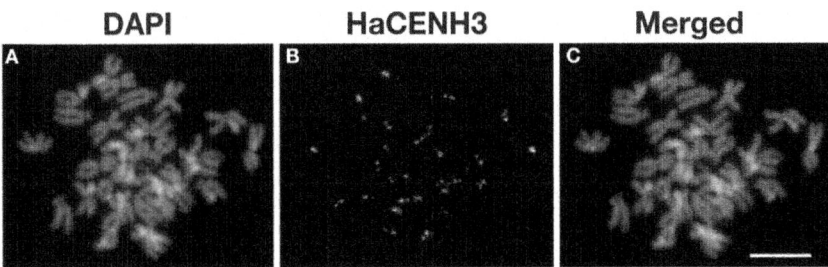

Figure 2. Immunostaining of sunflower metaphase chromosomes using an anti-Ha-CENH3 antibody.(A) DAPI-stained chromosomes. (B) Immunosignals of an anti-Ha-CENH3 antibody. (C) Merged image of (A,B). Scale bar, 10 μm.

DNA Sequences that Interact with Centromeric Nucleosomes

To investigate DNA sequences that coexist with HaCENH3, a ChIP-Seq experiment was conducted using the anti-HaCENH3 antibody and chromatin extracted from sunflower leaves. DNA fragments from the input and the HaCENH3 fraction in the ChIP were sequenced using MiSeq with the paired-end 2 × 300 bp protocol and deposited in DDBJ (Accession number: DRA003719). After the initial quality checks, 1,795,002 paired reads from the input and 2,012,000 paired reads from the HaCENH3 fraction were analyzed using the RepeatExplorer program. By this analysis, a total of 372 clusters containing at least 0.01% of the used sequences were generated. Then, enrichment ratios (ERs) were calculated by the following formula: ER = HaCENH3 ChIP reads/

input reads in the each cluster, and the clusters were sorted by ER. Out of the 372 clusters formed, 41 clusters showed an ER higher than 2.0.

To confirm the ChIP-Seq results, ChIP-qPCR was conducted using the DNA sequences in five clusters (HaCENH3CL1, 20, 22, 124, and 289) selected from the 41 clusters having an ER higher than 2.0 (Figure 3). In the ChIP-qPCR, a sunflower ubiquitin gene (GenBank accession number: X14333) was used as a negative (non-centromeric) control. The DNA sequences from all five clusters were significantly increased (27-fold for HaCENH3CL1, 154-fold for HaCL289, 289-fold for HaCENH3CL22, 344-fold for HaCENH3CL22, and 370-fold for HaCENH3CL124) compared with the negative control in the CENH3 fractions ($P < 0.01$ via Student's t-test, $n = 4$), whereas a non-centromeric control sequence, HAG004N15, did not increase ($P = 0.14$ via Student's t-test, $n = 4$). Since the ER in the ChIP-Seq and the RE in the ChIP-qPCR showed a high correlation coefficient ($r = 0.78$) according to the Pearson product-moment correlation, the ChIP-qPCR data support the ChIP-Seq results. The significant increase in HaCENH3CL1 showing the minimum ER (2.0) among the five clusters in the ChIP-qPCR suggested that sequences in the 41 clusters having an ER higher than 2.0 in the ChIP-Seq coexisted with HaCENH3 in the sunflower genome.

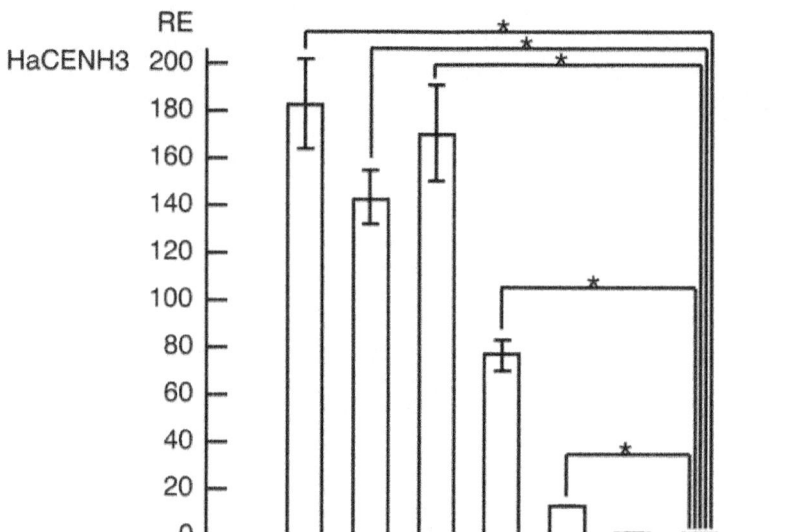

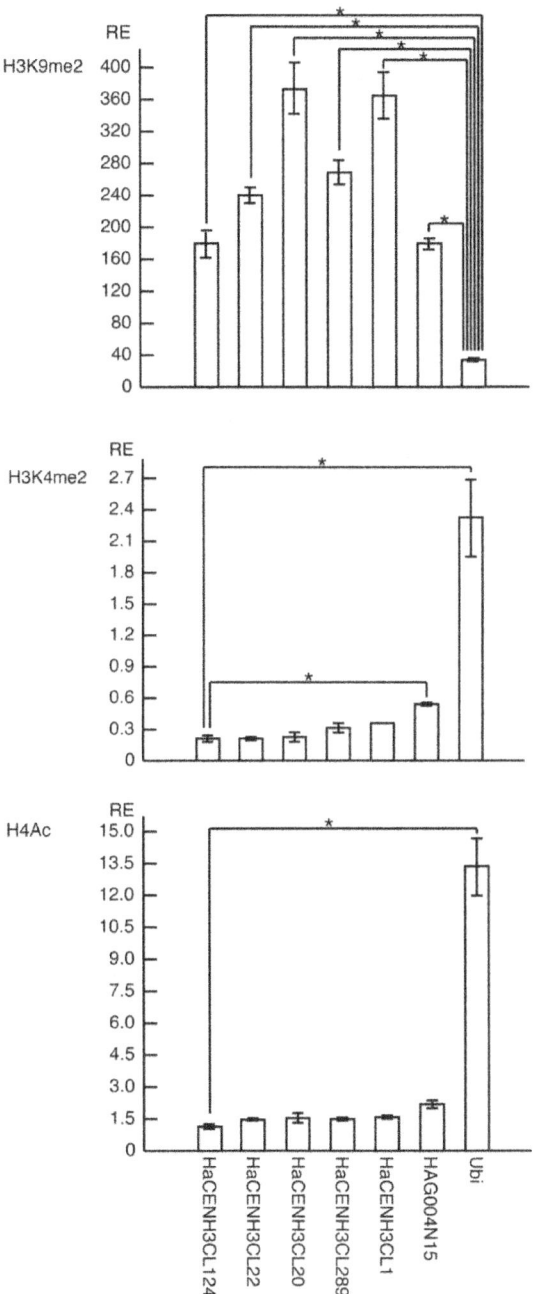

Figure 3. ChIP-qPCR analysis of centromeric DNA sequences of sunflowers. The columns and error bars represent the average relative enrichments (REs) and the standard errors from four independent ChIP reactions, respectively. The coding region of the

sunflower ubiquitin gene (X14333) was used as a non-centromeric (negative) control in the anti-HaCENH3 ChIP and as a negative control in the anti-H3K9me2 ChIP. Since HaCENH3CL124 showed the lowest REs among the sequences in the anti-H3K4me2 and H4Ac ChIP, HaCENH3CL124 was used as a negative control. The statistical significance of differences between the negative controls and other sequences was determined using Student's t-test ($*P < 0.01$).

The HaCENH3CL124 had the highest ER (52.0) and contained 187-bp tandem repeat sequences. The consensus DNA sequence of the tandem repeat was relatively AT-rich (58.3%), and no sequence similarity to DNA sequences in the GenBank database was found.

Of the 41 clusters, the program suggested that 25 clusters contained long interspersed nuclear element (LINE)-like sequences, and the sequences involve an ORF encoding an endonuclease (ENDO) and reverse transcriptase (RT) complex of LINE. These sequences show approximately 60% similarity to each other. Almost all RepeatExplorer cluster graphs of the LINE showed typical patterns of retrotransposons (line shape) rather than typical patterns of tandem repeats (star or ring shape). As exceptional cases, two clusters, HaCENH3CL78 and HaCENH3CL115, involved 12 and four ENDO/RT-related reads, and showed star shape. Since RepeatExplorer splits some satellite repeats with long monomers, e.g., rDNA, into multiple clusters, connections of the LINE-like sequences clusters were investigated. In the investigation, rDNA was used as a positive control of a satellite repeat with long monomers, and 17 clusters related to rDNA were found in the investigation. Ends of the rDNA clusters showed similarity to other clusters. On the other hand, almost all of the LINE clusters did not show frequent similarity hits observed among the rDNA clusters. As exceptional cases, two clusters, HaCENH3CL78 and HaCENH3CL115, showed frequent similarity hits to HaCENH3CL8, suggesting these are not LINE-like elements. Twenty-two of the remaining 23 LINE-like clusters showed ERs higher than 10.0. Additionally, junction of the elements and insertion sites were surveyed. If clusters include ends of mobile elements, junctions should be visible as sites with heterologous sequences at ends of contigs in the cluster. For example, a clear junction was observed in an end of HaCENH3CL189 showing a typical cluster graph of LTR-retrotransposon, and similar junctions were also observed in some of the LINE-like clusters. However, sunflower DNA sequences that were similar to the LINE-like sequences were not found in the GenBank DNA database. These results suggested that most centromeric regions of sunflowers are not involved in the current sunflower genome sequencing project because the project has not involved repetitive DNA sequence-rich regions (http://sunflowergenome.org/). An additional 12 of 41 clusters contained transposable element-like sequences, whereas DNA sequences in the remaining four clusters showed no homology to those registered in the GenBank DNA database.

To confirm the centromeric localization of the DNA sequences immunoprecipitated with anti-HaCENH3 in the ChIP-Seq and ChIP-qPCR, these sequences were used as probes in FISH analysis (Figure 4). A probe containing the HaCENH3CL124 sequence (pHaCENH3CL124-1, GenBank accession number: LC075744) showed centromeric signals on a pair of chromosomes with a secondary constriction (Figures 4A–D). To identify the chromosomes showing the HaCENH3CL124 signal, another FISH experiment in which the repeat and a reported FISH marker, HAG004N15, were used as probes was conducted; it revealed that the chromosome in question was chromosome 8 in Ceccarelli's report (Ceccarelli et al., 2007) (Figures 4E–M). A cloned probe containing the LINE-like sequence from HaCENH3CL20 (pHaCENH3CL20-1, GenBank accession number:

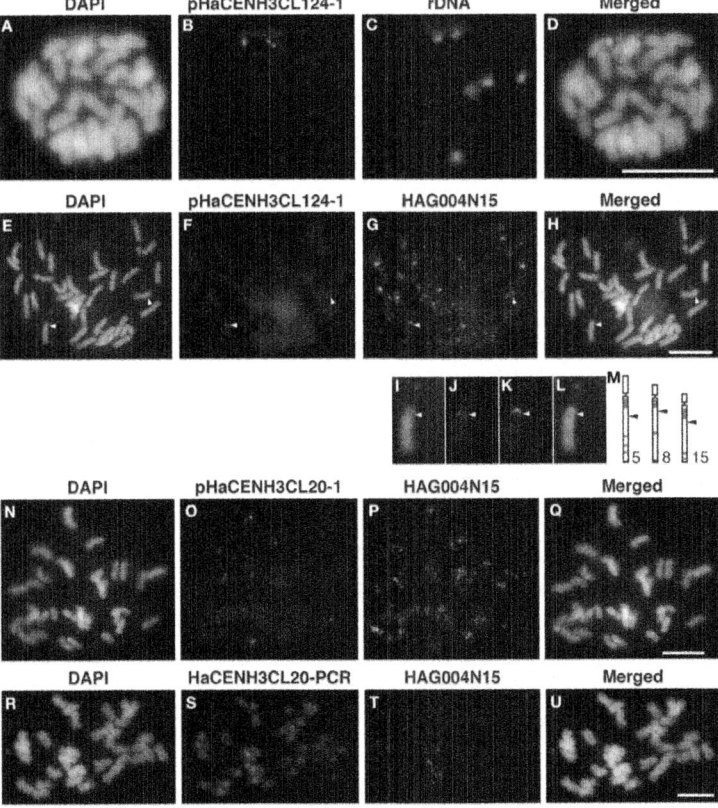

Figure 4. FISH using the enriched sequence from ChIP-Seq. (A,E,I,N,R) DAPI-stained sunflower chromosomes. (B,F,J) FISH signals of pHaCENH3CL124-1. (C)

FISH signals of rDNA. (G,K,P,T) FISH signals of HAG004N15. (O) FISH signals of pHaCENH3CL20-1. (S) FISH signals of the PCR products of HaCENH3CL20. (D) A merged image of (A–C). (H) A merged image of (E–G). (L) A merged image of (I–K). (Q) A merged image of (N–P). (U) A merged image of (R–T). (I–L) Enlarged images of (E–H). (M) Karyograms from Ceccarelli's report (2007). White arrowheads in (E–H) and black arrowheads in (M) indicate centromeres on the HaCENH3CL124-positive chromosomes and the reported chromosomes, respectively. Scale bar, 10 μm.

LC075745) showed centromeric signals on almost all of the chromosomes (Figures 4N–Q). Since the LINE-like sequence showed centromeric signals, we named it HaCEN-LINE. In FISH using HaCENH3CL20-PCR products as probes, much stronger signals on all the centromeric and pericentromeric regions were found, but faint signals were also observed on the arm regions (Figures 4R–U). These results imply that the PCR products contain not only some HaCEN-LINE variants but also some non-centromeric LINE variants; centromeric variants are on all of the centromeres. Two Ty3-related DNA sequences from HaCENH3CL1 (pHaCENH3CL1-1, GenBank accession number: LC075747) and HaCENH3CL189 (pHaCENH3CL189-1, GenBank accession number: LC075746), when used as FISH probes, showed dispersed patterns with some strong spots on centromeres. These partial centromere localizations coincided with the partial enrichment of the sequences in the ChIP-Seq and ChIP-qPCR (Figure 3), suggesting that these localize on both centromeric and non-centromeric arm regions.

Epigenetic Status of Sunflower Centromeres

Post translational histone modifications of sunflower centromeres were investigated by immunostaining with four different antibodies against H3K4me2, H3K9me2, H3K9Ac, and H4Ac (Figures 5 and 6). In the interphase cells, HaCENH3 signals scattered as dots on nuclei, and polar organization, Rabl orientation, and chromocenters were not observed (Figures 5A–E and 6A–E). H3K4me2, one of the representative euchromatic modifications, was detected on almost all regions of nuclei with many small dot signals, but the dot signals did not overlap with the HaCENH3 signals (Figures 5A–E). Similarly, the H3K4me2 signals did not co-localize with the HaCENH3 signals at prophase or metaphase (Figures 5F–O). Another euchromatic modification, H3K9Ac, showed a similar tendency at interphase and prophase, but almost all H3K9Ac signals disappeared at metaphase. Additionally, another euchromatic modification, H4Ac, showed stronger signals on the arms of metaphase chromosomes, but the signals on the centromeric and pericentromeric regions were weaker than those on the arms.

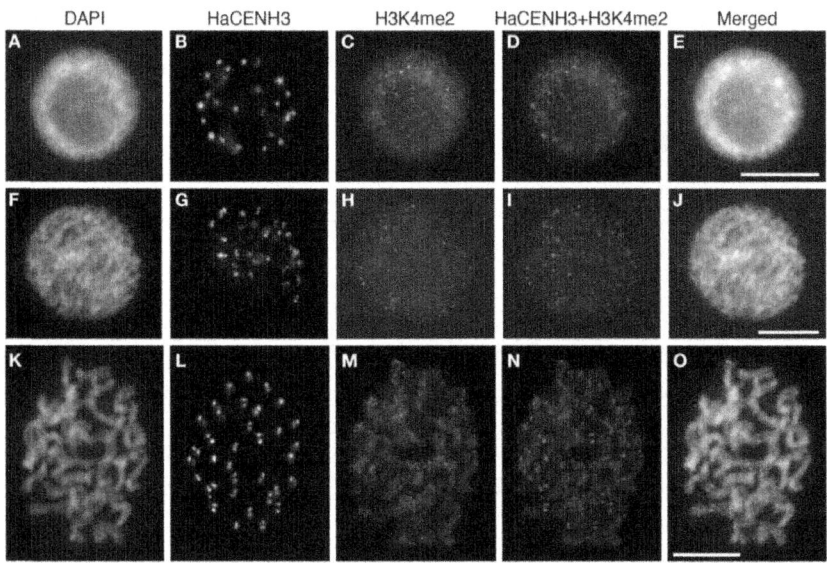

Figure 5. Immunostaining using the anti-HaCENH3 and anti-H3K4me2 antibodies. (A–E) An interphase nucleus. (F–J) Prophase chromosomes. (K–O) Metaphase chromosomes. Scale bar, 10 μm.

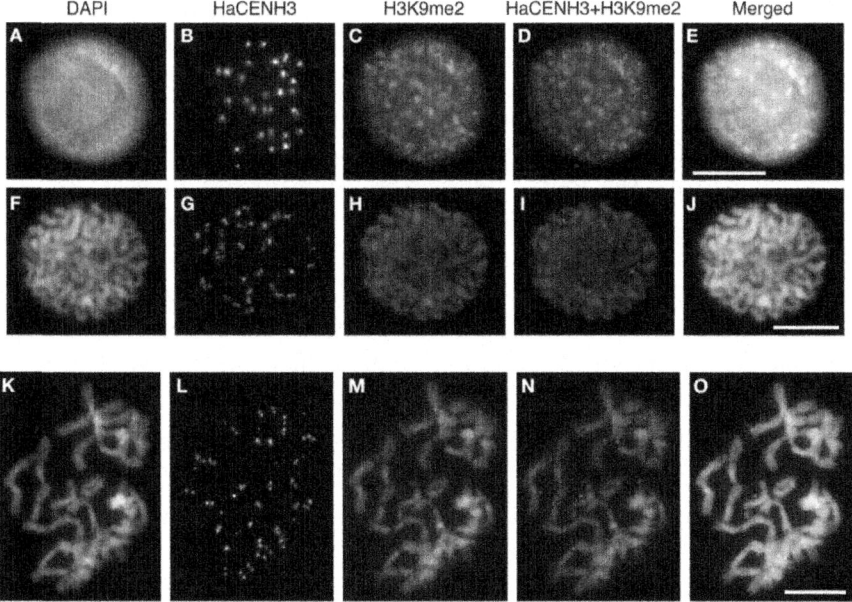

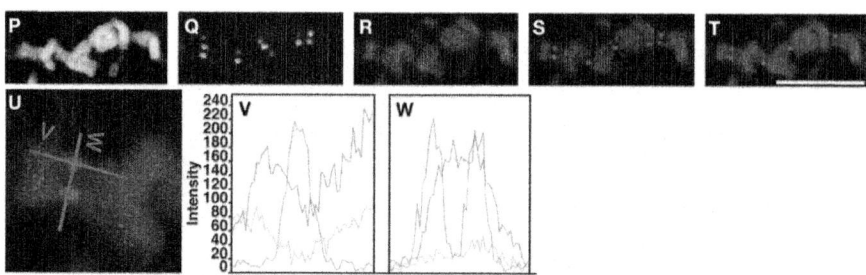

Figure 6. Immunostaining and immunohistochemical staining using the anti-Ha-CENH3 and anti-H3K9me2 antibodies. (A–O) Immunostaining images. (A–E) An interphase nucleus. (F–J) Prophase chromosomes. (K–O) Metaphase chromosomes. (P–T) Immunohistochemical staining images of metaphase chromosomes. Scale bar, 10 μm. (U) Scanning of metaphase chromosomes. Scanned positions are indicated as red lines. (V) A vertical scan of the chromosome. (W) A horizontal scan of the chromosome.

Immunosignals from H3K9me2 that represented heterochromatin status were distributed on an entire region of nuclei with dots in the interphase cells, but the dot sizes were larger than these of H3K4me2 (Figures 6A–E). Although the larger dot signals did not overlap with the HaCENH3 signals, the HaCENH3 signals were colocalized with faint H3K9me2 signals. In the prophase and metaphase cells, the H3K9me2 signals appeared mainly on the chromosome arms, whereas centromeres and pericentric regions of chromosomes showed faint signals for histone modification (Figures 6F–O). Co-localization of HaCENH3 and H3K9me2 was confirmed by immunohistochemical staining using a laser-scanning confocal microscope (Figures 6P–W). On metaphase chromosomes, the HaCENH3 signals were colocalized with weak H3K9me2 signals, and both arms showed strong H3K9me2 (Figures 6U,V). In a horizontal-scanning of centromeres on a pair of chromatids (Figures 6U,W), a DAPI-staining peak appeared at the middle of the paired chromatids, and HaCENH3 signals were detected on both sides of the paired chromatids. On the other hand, H3K9me2 showed no peaks in the scan, and weak H3K9me2 signals appeared constantly on the centromeric region.

Histone modifications on centromeric DNA sequences were also investigated by ChIP-qPCR (Figure 3). In the analyses, the ubiquitin gene was used as a negative control for a heterochromatin marker, H3K9me2, and as a positive control for two euchromatin markers, H3K4me2 and H4Ac. In ChIP-qPCR with the anti-H3K9me2 antibody, all five HaCENH3-positive sequences and a non-centromeric repetitive sequence (HAG004N15) significantly accumulated compared with the negative control ($P < 0.01$ via Student's t-test, $n = 4$, Figure 3). For modification of H3K4me2, HaCENH3CL124 showed the lowest RE

(0.21), and the REs of the positive control (ubiquitin) and HAG004N15 were 2.32 and 0.54, respectively (Figure 3). These two sequences significantly accumulated compared with HaCENH3CL124 ($P < 0.01$ via Student's t-test, $n = 4$). Similarly, HaCENH3CL124 showed the lowest RE (1.14) for H4Ac, and only the positive control (ubiquitin, RE = 13.36) increased significantly ($P = 0.003$ via Student's t-test, $n = 4$) compared with HaCENH3CL124 (Figure 3). Since leaf cells usually do not divide, these data suggest that all HaCENH3-positive sequences were involved with heterochromatin at interphase.

DISCUSSION

In this study, a CENH3-encoding cDNA was identified in sunflower. Based on the amino acid sequence deduced from the cDNA sequence, a peptide corresponding to the N-terminal 20 amino acids was synthesized. An antibody against the synthesized peptide recognized centromeres on all sunflower chromosomes. Using this antibody, ChIP-Seq analysis was applied and succeeded in isolating centromeric DNA sequences from sunflowers. Additionally, the epigenetic status at the centromeric DNA sequences was investigated by ChIP-qPCR with antibodies against modified histones.

Usually, centromeric DNA sequences in plant species consist of species-specific tandem repeats and retrotransposons, and these sequences are located on all centromeric regions (Zhong et al., 2002; Nagaki et al., 2003, 2009, 2011; Nagaki and Murata, 2005; Houben et al., 2007; Tek et al., 2010, 2011; Wang et al., 2011; Neumann et al., 2012; He et al., 2015). In this study of sunflower, we identified two types of centromeric DNA sequences (Figure 4). One was a LINE-like sequence, HaCEN-LINE, which showed centromeric signals on all of the chromosomes (Figures 4N–U). Although some tandem repeats evolved from retrotransposons were reported in potato (Gong et al., 2012), no centromere-specific LINE-like elements have been reported. The cluster graphs of HaCEN-LINEs in RepeatExplorer showed line shape, implying their retroelement form rather than modified tandem repeat form. Therefore, this is the first report describing a centromere-specific LINE. Although the centromere-targeting mechanisms of transposable elements remain unknown, the present study indicated that the LINE-like sequence can also target centromeric regions in the same manner as retrotransposons. Another centromeric DNA sequence in sunflowers was a 187-bp tandem repeat, which was located on a single pair of chromosomes (Figure 4). Such chromosome-specific centromeric DNA sequences have been reported in chickens (Shang et al., 2010), tobacco (Nagaki et al., 2012a), and potatoes (Gong et al., 2012); however, they are not very common. The enrichment of the tandem repeats in HaCENH3 ChIP was much higher than that of the LINE (Figure 3), suggesting that the tandem repeat may be more useful for building more stable centromeres than the LINE. As discussed previously for other species (Nagaki et al., 2004, 2012a; Shang et al., 2010; Gong et al., 2012), sunflower centromeres may also be undergoing centromeric DNA evolution to equalize centromeric DNA sequences among chromosomes; the tandem repeats may form stabilized centromeres on the all sunflower chromosomes after this evolutionary event.

With the exception of phosphorylation, other histone modifications on the centromeric regions are not well characterized in plants (Desvoyes et al., 2014; Sharma et al., 2015). According to Houben et al. (2003), plant species with a genome smaller than 510 Mb formed chromocenters at interphase, and H3K9me2 preferentially occurred on the chromocenters. In *Arabidopsis*, H3K9me2 signals appeared on all centromeric and pericentromeric regions of metaphase chromosomes. Euchromatin-specific histone modifications, such as H3K4me2 and H4Ac, were observed in an inverse pattern compared to that of the H3K9me2 (Houben et al., 2003; Jasencakova et al., 2003). In plant species with genomes larger than 530 Mb, such as barley, no chromocenters were observed, and dispersed H3K9me2 signals were observed at interphase; their low level of modification appears on centromeric and pericentromeric regions of metaphase chromosomes (Fuchs et al., 2006). Occasionally, H4Ac were detected at centromeric and pericentromeric regions of metaphase chromosomes in barley and field beans (Jasencakova et al., 2000, 2001). In this study, sunflowers showed dispersed modification patterns for H3K4me2, H3K9Ac and H3K9me2 at interphase and a lower level of H3K9me2 modification on metaphase centromeres (Figures 5, 6). However, the irregular modification patterns observed at metaphase in barley and field bean were not detected at metaphase in sunflower. The genome size of sunflower was estimated to be 2.43 Gb (Bennett et al., 1982), and the observed distribution patters of the modified histones in this study (Figures 5 and 6) coincided with those in plants with large genomes (Houben et al., 2003).

To determine histone modification events on each different centromeric DNA sequence, ChIP analysis has higher resolution and quantitative capacity than immunostaining. In rice, ChIP revealed that centromeric DNA sequences excepting the genic regions on chromosome 8 were heterochromatic (Nagaki et al., 2004; Yan et al., 2006). Subsequently, different histone modifications were observed in CentO sequences (Zhang et al., 2013). CentO sequences with regular size units (155 bp) were enriched in OsCENH3 ChIP, whereas 167-bp CentO variants were enriched in a euchromatic modification, H3K4me2, ChIP rather than OsCENH3. In the present study with sunflowers, two centromeric DNA sequences showed heterochromatic status (Figure 3), suggesting that the status of plant centromeres at interphase was heterochromatic. Additionally, immunohistochemical staining revealed the existence of H3K9me2 on the centromeres at metaphase (Figure 6). Furthermore, immunostaining showed that the level of H3K9Ac in the interphase nuclei was higher than that on metaphase chromosomes. Although we could not quantify the H3K9Ac level on centromeres, the increased level of H3K9Ac in the interphase nuclei implies that it increased on centromeres as well. In the case of human cells, the combination of rigid cell cycle control and ChIP made it possible to detect H3K9Ac within a short range of time during the cell cycle (Ohzeki et al., 2012). To investigate H3K9Ac and other modifications on plant centromeres, a more accurate quantification system, such as that utilized in human cells, is required in the future.

CONFLICT OF INTEREST STATEMENT

The authors declare that the research was conducted in the absence of any commercial or financial relationships that could be construed as a potential conflict of interest.

ACKNOWLEDGMENTS

This research was supported by the Cooperative Research Grant of the Genome Research for BioResource, NODAI Genome Research Center, Tokyo University of Agriculture.

REFERENCES

1. Amor, D. J., Kalitsis, P., Sumer, H., and Choo, K. H. (2004). Building the centromere: from foundation proteins to 3D organization. *Trends Cell Biol.* 14, 359–368. doi: 10.1016/j.tcb.2004.05.009

2. Bennett, M. D., Smith, J. B., and Heslopharrison, J. S. (1982). Nuclear-DNA amounts in angiosperms. *Proc. R. Soc. Lond. B Biol. Sci.* 216, 179–199. doi: 10.1098/rspb.1982.0069

3. Black, B. E., Foltz, D. R., Chakravarthy, S., Luger, K., Woods, V. L. Jr., and Cleveland, D. W. (2004). Structural determinants for generating centromeric chromatin. *Nature* 430, 578–582. doi: 10.1038/nature02766

4. Ceccarelli, M., Sarri, V., Natali, L., Giordani, T., Cavallini, A., Zuccolo, A., et al. (2007). Characterization of the chromosome complement of *Helianthus annuus* by *in situ* hybridization of a tandemly repeated DNA sequence. *Genome* 50, 429–434. doi: 10.1139/G07-019

5. Desvoyes, B., Fernández-Marcos, M., Sequeira-Mendes, J., Otero, S., Vergara, Z., and Gutierrez, C. (2014). Looking at plant cell cycle from the chromatin window. *Front. Plant Sci.* 5:369. doi: 10.3389/fpls.2014.00369

6. Dunemann, F., Schrader, O., Budahn, H., and Houben, A. (2014). Characterization of centromeric histone H3 (CENH3) variants in cultivated and wild carrots (*Daucus* sp.). *PLoS ONE* 9:e98504. doi: 10.1371/journal.pone.0098504

7. Earnshaw, W. C., and Rothfield, N. (1985). Identification of a family of human centromere proteins using autoimmune sera from patients with scleroderma. *Chromosoma* 91, 313–321. doi: 10.1007/BF00328227

8. Feng, J., Liu, Z., Cai, X., and Jan, C. C. (2013). Toward a molecular cytogenetic map for cultivated sunflower (*Helianthus annuus* L.) by landed BAC/BIBAC clones. *G3 (Bethesda)* 3, 31–40. doi: 10.1534/g3.112.004846

9. Fuchs, J., Demidov, D., Houben, A., and Schubert, I. (2006). Chromosomal histone modification patterns–from conservation to diversity. *Trends Plant Sci.* 11, 199–208. doi: 10.1016/j.tplants.2006.02.008

10. Gerlach, W. L., and Bedbrook, J. R. (1979). Cloning and characterization of ribosomal RNA genes from wheat and barley. *Nucleic Acids Res.* 7, 1869–1885. doi: 10.1093/nar/7.7.1869

11. Gong, Z., Wu, Y., Koblízková, A., Torres, G. A., Wang, K., Iovene, M., et al. (2012). Repeatless and repeat-based centromeres in potato: implications for centromere evolution. *Plant Cell* 24, 3559–3574. doi: 10.1105/tpc.112.100511

12. Han, F., Lamb, J. C., and Birchler, J. A. (2006). High frequency of centromere inactivation resulting in stable dicentric chromosomes of maize. *Proc. Natl. Acad. Sci. U.S.A.* 103, 3238–3243. doi: 10.1073/pnas.0509650103

13. He, Q., Cai, Z., Hu, T., Liu, H., Bao, C., Mao, W., et al. (2015). Repetitive sequence analysis and karyotyping reveals centromere-associated DNA sequences in radish (*Raphanus sativus* L.). *BMC Plant Biol.* 15:105. doi: 10.1186/s12870-015-0480-y

14. Houben, A., Demidov, D., Gernand, D., Meister, A., Leach, C. R., and Schubert, I. (2003). Methylation of histone H3 in euchromatin of plant chromosomes depends on basic nuclear DNA content. *Plant J.* 33, 967–973. doi: 10.1046/j.1365-313X.2003.01681.x

15. Houben, A., Schroeder-Reiter, E., Nagaki, K., Nasuda, S., Wanner, G., Murata, M., et al. (2007). CENH3 interacts with the centromeric retrotransposon cereba and GC-rich satellites and locates to centromeric substructures in barley. *Chromosoma* 116, 275–283. doi: 10.1007/s00412-007-0102-z

16. Jasencakova, Z., Meister, A., and Schubert, I. (2001). Chromatin organization and its relation to replication and histone acetylation during the cell cycle in barley. *Chromosoma* 110, 83–92. doi: 10.1007/s004120100132

17. Jasencakova, Z., Meister, A., Walter, J., Turner, B. M., and Schubert, I. (2000). Histone H4 acetylation of euchromatin and heterochromatin is cell cycle dependent and correlated with replication rather than with transcription. *Plant Cell* 12, 2087–2100. doi: 10.1105/tpc.12.11.2087

18. Jasencakova, Z., Soppe, W. J. J., Meister, A., Gernand, D., Turner, B. M., and Schubert, I. (2003). Histone modifications in Arabidopsis - high methylation of H3 lysine 9 is dispensable for constitutive heterochromatin. *Plant J.* 33, 471–480. doi: 10.1046/j.1365-313X.2003.01638.x

19. Lermontova, I., Schubert, V., Fuchs, J., Klatte, S., Macas, J., and Schubert, I. (2006). Loading of *Arabidopsis* centromeric histone CENH3 occurs mainly during G2 and requires the presence of the histone fold domain. *Plant Cell* 18, 2443–2451. doi: 10.1105/tpc.106.043174

20. Maheshawari, S. (2015). Naturally occurring differences in CENH3 affect chromosome segregation in Zygotic Mitosis of hybrid. *PLoS Genet.* 11:e1004970. doi: 10.1371/journal.pgen.1004970

21. Masonbrink, R. E., Gallagher, J. P., Jareczek, J. J., Renny-Byfield, S., Grover, C. E., Gong, L., et al. (2014). CenH3 evolution in diploids and polyploids of three angiosperm genera. *BMC Plant Biol.* 14:383. doi: 10.1186/s12870-014-0383-3

22. Nagaki, K., Cheng, Z., Ouyang, S., Talbert, P. B., Kim, M., Jones, K. M., et al. (2004). Sequencing of a rice centromere uncovers active genes. *Nat. Genet.* 36, 138–145. doi: 10.1038/ng1289

23. Nagaki, K., Kashihara, K., and Murata, M. (2005). Visualization of diffuse centromeres with centromere-specific histone H3 in the holocentric plant *Luzula nivea*. *Plant Cell* 17, 1886–1893. doi: 10.1105/tpc.105.032961

24. Nagaki, K., Kashihara, K., and Murata, M. (2009). A centromeric DNA sequence colocalized with a centromere-specific histone H3 in tobacco. *Chromosoma* 118, 249–257. doi: 10.1007/s00412-008-0193-1

25. Nagaki, K., and Murata, M. (2005). Characterization of CENH3 and centromere-associated DNA sequences in sugarcane. *Chromosome Res.* 13, 195–203. doi: 10.1007/s10577-005-0847-2

26. Nagaki, K., Shibata, F., Kanatani, A., Kashihara, K., and Murata, M. (2012a). Isolation of centromeric-tandem repetitive DNA sequences by chromatin affinity purification using a HaloTag7-fused centromere-

specific histone H3 in tobacco. *Plant Cell Rep.* 31, 771–779. doi: 10.1007/s00299-011-1198-4

27. Nagaki, K., Shibata, F., Suzuki, G., Kanatani, A., Ozaki, S., Hironaka, A., et al. (2011). Coexistence of NtCENH3 and two retrotransposons in tobacco centromeres. *Chromosome Res.* 19, 591–605. doi: 10.1007/s10577-011-9219-2

28. Nagaki, K., Talbert, P. B., Zhong, C. X., Dawe, R. K., Henikoff, S., and Jiang, J. M. (2003). Chromatin immunoprecipitation reveals that the 180-bp satellite repeat is the key functional DNA element of *Arabidopsis thaliana* centromeres. *Genetics* 163, 1221–1225.

29. Nagaki, K., Yamamoto, M., Yamaji, N., Mukai, Y., and Murata, M. (2012b). Chromosome dynamics visualized with an anti-centromeric histone H3 antibody in Allium. *PLoS ONE* 7:e51315. doi: 10.1371/journal.pone.0051315

30. Nasuda, S., Hudakova, S., Schubert, I., Houben, A., and Endo, T. R. (2005). Stable barley chromosomes without centromeric repeats. *Proc. Natl. Acad. Sci. U.S.A.* 102, 9842–9847. doi: 10.1073/pnas.0504235102

31. Neumann, P., Navrátilová, A., Schroeder-Reiter, E., Koblížková, A., Steinbauerová, V., Chocholová, E., et al. (2012). Stretching the rules: monocentric chromosomes with multiple centromere domains. *PLoS Genet.* 8:e1002777. doi: 10.1371/journal.pgen.1002777

32. Neumann, P., Pavlíková, Z., Koblížková, A., Fuková, I., Jedličková, V., Novák, P., et al. (2015). Centromeres off the hook: massive changes in centromere size and structure following duplication of CenH3 gene in Fabeae species. *Mol. Biol. Evol.* 32, 1862–1879. doi: 10.1093/molbev/msv070

33. Novák, P., Neumann, P., and Macas, J. (2010). Graph-based clustering and characterization of repetitive sequences in next-generation sequencing data.*BMC Bioinformatics* 11:378. doi: 10.1186/1471-2105-11-378

34. Ohzeki, J., Bergmann, J. H., Kouprina, N., Noskov, V. N., Nakano, M., Kimura, H., et al. (2012). Breaking the HAC Barrier: histone H3K9 acetyl/methyl balance regulates CENP-A assembly. *EMBO J.* 31, 2391–2402. doi: 10.1038/emboj.2012.82

35. Perpelescu, M., and Fukagawa, T. (2011). The ABCs of CENPs. *Chromosoma* 120, 425–446. doi: 10.1007/s00412-011-0330-0

36. Saitou, N., and Nei, M. (1987). The neighbor-joining method: a new method for reconstructing phylogenic trees. *Mol. Biol. Evol.* 4, 406–425.

37. Sanei, M., Pickering, R., Kumke, K., Nasuda, S., and Houben, A. (2011). Loss of centromeric histone H3 (CENH3) from centromeres precedes

uniparental chromosome elimination in interspecific barley hybrids. *Proc. Natl. Acad. Sci. U.S.A.* 108, E498–E505. doi: 10.1073/pnas.1103190108

38. Shang, W. H., Hori, T., Toyoda, A., Kato, J., Popendorf, K., Sakakibara, Y., et al. (2010). Chickens possess centromeres with both extended tandem repeats and short non-tandem-repetitive sequences. *Genome Res.* 20, 1219–1228. doi: 10.1101/gr.106245.110

39. Sharma, S. K., Yamamoto, M., and Mukai, Y. (2015). Immuno-cytogenetic manifestation of epigenetic chromatin modification marks in plants. *Planta* 241, 291–301. doi: 10.1007/s00425-014-2233-9

40. Talbert, P. B., Masuelli, R., Tyagi, A. P., Comai, L., and Henikoff, S. (2002). Centromeric localization and adaptive evolution of an *Arabidopsis* histone H3 variant. *Plant Cell* 14, 1053–1066. doi: 10.1105/tpc.010425

41. Talia, P., Greizerstein, E., Quijano, C. D., Peluffo, L., Fernandez, L., Fernández, P., et al. (2010). Cytological characterization of sunflower by *in situ* hybridization using homologous rDNA sequences and a BAC clone containing highly represented repetitive retrotransposon-like sequences. *Genome* 53, 172–179. doi: 10.1139/G09-097

42. Tek, A. L., Kashihara, K., Murata, M., and Nagaki, K. (2010). Functional centromeres in soybean include two distinct tandem repeats and a retrotransposon. *Chromosome Res.* 18, 337–347. doi: 10.1007/s10577-010-9119-x

43. Tek, A. L., Kashihara, K., Murata, M., and Nagaki, K. (2011). Functional centromeres in Astragalus sinicus include a compact centromere-specific histone H3 and a 20-bp tandem repeat. *Chromosome Res.* 19, 969–978. doi: 10.1007/s10577-011-9247-y

44. Thompson, J. D., Gibson, T. J., Plewniak, F., Jeanmougin, F., and Higgins, D. G. (1997). The CLUSTAL_X windows interface: flexible strategies for multiple sequence alignment aided by quality analysis tools. *Nucleic Acids Res.* 25, 4876–4882. doi: 10.1093/nar/25.24.4876

45. Vermaak, D., Hayden, H. S., and Henikoff, S. (2002). Centromere targeting element within the histone fold domain of Cid. *Mol. Cell. Biol.* 22, 7553–7561. doi: 10.1128/MCB.22.21.7553-7561.2002

46. Wang, G., He, Q., Cheng, Z., Talbert, P. B., and Jin, W. (2011). Characterization of CENH3 proteins and centromere-associated DNA sequences in diploid and allotetraploid Brassica species. *Chromosoma* 120, 353–365. doi: 10.1007/s00412-011-0315-z

47. Warburton, P. E., Cooke, C. A., Bourassa, S., Vafa, O., Sullivan, B. A., Stetten, G., et al. (1997). Immunolocalization of CENP-A suggests a distinct nucleosome structure at the inner kinetochore plate of active

centromeres. *Curr. Biol.* 7, 901–904. doi: 10.1016/S0960-9822(06)00382-4

48. Yamaji, N., and Ma, J. F. (2007). Spatial distribution and temporal variation of the rice silicon transporter Lsi1. *Plant Physiol.* 143, 1306–1313. doi: 10.1104/pp.106.093005

49. Yan, H., Ito, H., Nobuta, K., Ouyang, S., Jin, W., Tian, S., et al. (2006). Genomic and genetic characterization of rice Cen3 reveals extensive transcription and evolutionary implications of a complex centromere. *Plant Cell* 18, 2123–2133. doi: 10.1105/tpc.106.043794

50. Zhang, T., Talbert, P. B., Zhang, W., Wu, Y., Yang, Z., Henikoff, J. G., et al. (2013). The CentO satellite confers translational and rotational phasing on cenH3 nucleosomes in rice centromeres. *Proc. Natl. Acad. Sci. U.S.A.* 110, E4875–E4883. doi: 10.1073/pnas.1319548110

51. Zhong, C. X., Marshall, J. B., Topp, C., Mroczek, R., Kato, A., Nagaki, K., et al. (2002). Centromeric retroelements and satellites interact with maize kinetochore protein CENH3. *Plant Cell* 14, 2825–2836. doi: 10.1105/tpc.006106

Chapter 4

CHROMOSOME SEGREGATION IN PLANT MEIOSIS

Linda Zamariola[1], Choon Lin Tiang[2], Nico De Storme[1], Wojtek Pawlowski[2], and Danny Geelen[1]

[1]Department of Plant Production, Faculty of Bioscience Engineering, University of Ghent, Ghent, Belgium

[2]Department of Plant Breeding and Genetics, Cornell University, Ithaca, NY, USA

Faithful chromosome segregation in meiosis is essential for ploidy stability over sexual life cycles. In plants, defective chromosome segregation caused by gene mutations or other factors leads to the formation of unbalanced or unreduced gametes creating aneuploid or polyploid progeny, respectively. Accurate segregation requires the coordinated execution of conserved processes occurring throughout the two meiotic cell divisions. Synapsis and recombination ensure the establishment of chiasmata that hold homologous chromosomes together allowing their correct segregation in the first meiotic division, which is also tightly regulated by cell-cycle dependent release of cohesin and monopolar attachment of sister kinetochores to microtubules. In meiosis II, bi-orientation of sister kinetochores and proper spindle orientation correctly segregate chromosomes in four haploid cells. Checkpoint mechanisms acting at kinetochores control the accuracy of kinetochore-microtubule attachment, thus ensuring the completion of segregation. Here we review the current knowledge on the processes taking place during chromosome segregation in plant meiosis, focusing on the characterization of the molecular factors involved.

INTRODUCTION

Meiosis is a specialized cell division that generates four haploid daughter cells from a diploid parent cell after a single round of DNA replication and two consecutive rounds of nuclear division. In the first nuclear division, homologous chromosomes segregate (reductional cell division), and in the second one, sister chromatids segregate (equational cell division). As such, each daughter cell carries half the amount of the parental genetic material. The accurate segregation of chromosomes during meiosis is essential for the formation of haploid gametes. Failure in the proper execution of chromosome segregation inevitably leads to the formation of imbalanced gametes and aneuploid or polyploid progeny. In plants, aneuploidy is more tolerated than in animals and viable aneuploid plants have been observed, especially among the progeny of triploid individuals (Henry et al., 2005, 2010). Despite being affected in growth and reproduction (Birchler et al., 2001), aneuploids may have an evolutionary role, serving as a bridge to euploid polyploid plant formation through repeated generations of selfing (Ramsey and Schemske, 1998; Henry et al., 2005). Polyploid plants generated through aneuploids or by the polyploidization events of somatic doubling and unreduced gametes, are considered as a prominent driving force in plant genome evolution (Ramsey and Schemske, 2002; Adams and Wendel, 2005; Comai, 2005; Otto, 2007).

To ensure the correct completion of the meiotic cell division program, a sequence of coordinated steps must take place during the two phases of meiosis. In meiosis I, homologous chromosomes must pair and synapse and physically exchange genetic material through recombination. The resulting points of crossing-over, also termed chiasmata, form links between the two homologs in the bivalent configuration and ensure proper positioning of the bivalent relative to the division spindle and balanced segregation of homologs in anaphase I. Additionally, to achieve this, sister kinetochores from each homolog must attach to microtubules emanating from the same spindle pole, a process called monopolar kinetochore attachment, and cohesion must be lost in a stepwise manner. More specifically, at anaphase I, cohesion is released at chromosome arms but not at sister centromeres, allowing homologs to segregate without affecting the physical connection between both sister chromatids. In meiosis II, chromosome segregation in the two resulting haploid interphase nuclei occurs in an equational manner and hence strongly resembles the dynamics of a mitotic cell division. Cohesion at centromeres is retained until anaphase II to ensure bipolar attachment of sister kinetochores to microtubules and equational segregation of chromatids into four haploid daughter cells. Progression through the meiotic cell division is regulated at determined checkpoints by the activity of CDK (Cyclin-Dependent Kinase) - cyclin complexes and the Anaphase Promoting Complex/Cyclosome (APC/C) (Harper et al., 2002; Cooper and Strich, 2011). In particular, the Spindle Assembly Checkpoint (SAC) acts during the transition between metaphase and anaphase of the two meiotic cell divisions to ensure correct kinetochore-microtubule attachments and faithful chromosome segregation (Malmanche et al., 2006; Yamamoto et al., 2008). In plants, checkpoints appear to be less stringent compared to yeast and animals, since completion of meiosis is achieved in several meiotic mutants creating imbalanced gametes (Wijnker and Schnittger, 2013).

Most of the knowledge on the molecular biology of mitotic and meiotic chromosome segregation comes from studies in yeast (reviewed in Marston, 2014). However, the mechanisms of chromosome segregation are conserved in eukaryotes, including plants (Dawe, 1998; Bhatt et al., 2001). In the last decade, the increasing availability of genomic tools and the development of *Arabidopsis thaliana*, but also maize (*Zea mays*) and rice (*Oryza sativa*), as model systems, have led to the identification of a large number of conserved meiotic genes (Mercier and Grelon, 2008). Phenotypic and cytogenetic analyses of the corresponding mutants, have unraveled the function of several molecular factors required for proper chromosome segregation in plants (Bhatt et al., 2001; Ma, 2006). Therefore, the focus of this review will be on the major cellular processes that take place to ensure accurate chromosome segregation in plant meiosis and the related genes that have been yet identified in *Arabidopsis*, maize and rice. Other factors having an effect on chromosome segregation in plant meiosis, such as environmental stresses and changes in ploidy level have been described in recent reviews (Comai, 2005; Madlung and Wendel, 2013; De Storme and Geelen, 2014). After mentioning the importance of homologous chromosome pairing and recombination, two subjects extensively discussed in other reviews (Hamant et al., 2006; Edlinger and Schlögelhofer, 2011; Osman et al., 2011; Tiang et al., 2012; Da Ines et al., 2014), we describe the relevance of cohesion, focusing on the roles of the cohesin complex and on the cohesion dynamics (e.g., loading, release and protection) during meiotic cell division. Next, we discuss the role of centromeric and kinetochore proteins in establishing proper spindle attachment during meiosis I and II, and additionally describe what is currently known on the checkpoint control mechanisms acting at kinetochores. Finally, we report the molecular mechanisms underlying microtubule organization and we focus on the relevance of spindle orientation in plant meiosis.

HOMOLOGOUS PAIRING AND RECOMBINATION AS A BASIS FOR REDUCTIONAL CELL DIVISION IN MEIOSIS I

Homologous Chromosome Pairing and Synapsis

To ensure accurate segregation, chromosomes must first recognize their homologous partners and pair with them during early meiotic prophase I. This process leads to the formation of bivalents, which ensures correct bipolar attachment of homologous centromeres to the division spindle at metaphase I in a way that each of the chromosomes in the bivalent moves to a different pole at anaphase I. Bivalent formation is also required for proper positioning of chromosomes at the metaphase plate. Consequently, mutants with chromosome pairing problems exhibit chromosome segregation defects (Bozza and Pawlowski, 2008).

It is assumed that chromosome homology recognition is based on their DNA sequence. Although mechanisms that bring homologous chromosomes together have yet to be fully elucidated, studies in a variety of species, including plants, have shown that chromosome pairing is strongly dependent on their dynamics in early meiotic prophase as well as the initiation and progression through the early stages of the recombination pathway. Chromosome dynamics in prophase I is largely controlled by the behavior of telomeres, blocks of highly conserved repetitive DNA sequence at the ends of chromosomes (Siderakis and Tarsounas, 2007). Telomeres attach to the nuclear envelope before the onset of chromosome pairing, and gather on a small region, forming a unique structure that resembles a flower bouquet, the so called telomere bouquet (Bass et al., 2000; Golubovskaya et al., 2002; Harper et al., 2004; Richards et al., 2012). The bouquet arrangement has been observed in most eukaryotes (Klutstein and Cooper, 2014). The exact role of the bouquet is still being debated. However, mutants defective in bouquet formation are frequently also defective in chromosome pairing, which implies a role of the bouquet in this process (Harper et al., 2004; Klutstein and Cooper, 2014). One example of such mutant is *plural abnormality of meiosis 1* (*pam1*) in maize, which exhibits significant reduction in homologous pairing (Golubovskaya et al., 2002). In this mutant, telomeres attach to the nuclear envelope but fail to cluster. The bouquet formation has been, therefore, suggested to promote homologous paring by bringing chromosome ends together (Harper et al., 2004).

Alternative chromosome interaction mechanisms have been described in several species, including *Caenorhabditis elegans* and *Arabidopsis* (Armstrong et al., 2001; Phillips and Dernburg, 2006). In *C. elegans*, telomeres do not form the bouquet but pairing centers, short chromosome segments recognized by specific zinc-finger proteins, that attach to the nuclear envelope during early prophase I, also bringing homologous chromosomes together (Phillips and Dernburg, 2006). In *Arabidopsis*, telomeres cluster in meiotic interphase on the nucleolus rather than the nuclear envelope (Armstrong et al., 2001). Subtelomeric regions of *Arabidopsis* chromosomes start to pair before telomeres dissociate from the nucleolus, suggesting that the clustering on the nucleolus may play a role similar to that of the canonical bouquet. *Arabidopsis* telomeres establish their connections with the nuclear envelope during leptotene and zygotene, although without an obvious bouquet formation (Armstrong et al., 2001).

Interestingly, the connections used to attach chromosomes to the nuclear envelope in *C. elegans* and *Arabidopsis* are homologs of the same transmembrane proteins that are used in other species to tether telomeres to the nuclear envelope during bouquet formation. SUN domain proteins, identified in yeast, mammals, *C. elegans*, maize, as well as*Arabidopsis*, cross the inner nuclear membrane (Chikashige et al., 2007; Schmitt et al., 2007; Penkner et al., 2009; Sato et al., 2009; Graumann et al., 2010; Murphy et al., 2010). They interact at their N-termini with telomere binding proteins while their C-termini bind transmembrane proteins containing a conserved KASH domain that cross the outer membrane and interact with the cytoskeleton (Miki et al., 2004; Zhou et al., 2012). The commonality of the structures attaching telomeres to the nuclear envelope reinforces the notion that the telomere-nuclear membrane attachments in*C. elegans* and *Arabidopsis* may be functionally similar to the presence of the canonical bouquet.

It has been shown in several species that the cytoskeleton acts through the telomere-nuclear membrane attachments to induce dynamic motility of chromosomes (Bhalla and Dernburg, 2008; Koszul et al., 2009; Sheehan and Pawlowski, 2009; Woglar and Jantsch, 2013). The chromosome movements are thought to help the chromosomes to engage in finding their pairing partners as well as resolving their entanglements.

Another process, which is required for proper chromosome segregation, and closely follows chromosome pairing, is synapsis. Synapsis is installation of a proteinaceous structure, the synaptonemal complex (SC), between the paired homologous chromosomes, which stabilizes the pairing interactions. The SC consists of two lateral elements (LEs) which reside at the base of the chromosome loops and are held together in parallel by transverse filament proteins. In most eukaryotes, the LEs are derived from the axial elements (AEs) loaded on the chromosomal axis before synapsis. Installation of the synaptonemal complex is also closely linked with the formation of crossovers (see the following section), and so synapsis also affects chromosome segregation through its role in crossover formation. *Arabidopsis*mutants defective in synaptonemal complex formation exhibit univalents at metaphase I and improper chromosome segregation at anaphase I (Ross et al., 1997; Higgins et al., 2005).

Meiotic Recombination

Meiotic recombination affects segregation of chromosomes in at least two ways. First, studies in many species, including plants, mammals, and fungi, have indicated that homologous chromosome pairing is closely connected to meiotic recombination (Pawlowski and Cande, 2005). Second, crossovers, reciprocal chromosome segment exchanges formed as a result of meiotic recombination, form physical connections, known as chiasmata, between homologous chromosomes in each bivalents. Chiasmata keep bivalents together to ensure proper orientation and segregation of chromosomes during the first meiotic division.

Recombination in meiosis is initiated by the formation of double strand breaks (DSBs) in chromosomal DNA, triggered by Spo11, a conserved topoisomerase type-II-like protein (Keeney et al., 1997). The MRN complex (MRE11/RAD50/NBS1) then resects the breaks creating single-stranded DNA overhangs (Borde, 2007), which then invade appropriate regions on the homologous chromosomes. This process is promoted by two recombination proteins, Rad51 and Dmc1 (Masson and West, 2001). Rad51 is solely responsible for the repair of DNA breaks using sister chromatids as templates. However, this process is restrained and replaced by repair via the homologous chromosome when Dmc1 is localized to meiotic DNA break sites together with Rad51 (Bishop et al., 1992; Niu et al., 2009). In *Arabidopsis*, mutating *Rad51* results in chromosome fragmentation (Li et al., 2004). However, fragmentation is not observed in the *dmc1* mutant (Couteau et al., 1999). These observations suggest that the function of Dmc1 is distinct from Rad51, as Dmc1 promotes interhomolog recombination rather than intersister recombination (Kurzbauer et al., 2012; Pradillo et al., 2012).

Meiotic recombination results in formation of crossovers and non-crossovers (which include gene conversions). The number and location of crossovers are tightly regulated. In most plant species, only one to four crossovers are formed per bivalent (Crismani and Mercier, 2012). At least one crossover must be formed per bivalent to ensure correct chromosome segregation at anaphase I. However, the number of crossovers per chromosome is limited by crossover interference, a mechanism that prevents formation of crossovers next to each other (Jones, 1984). A group of proteins called ZMM, which contains Zip1, Zip2, Zip3, Zip4, Msh4, Msh5, and Mer3, have been identified as essential for the formation of interference-dependent crossovers in yeast (Börner et al., 2004). Homologs of several of these proteins have been studied in *Arabidopsis* and found to play similar roles in crossover formation (Higgins et al., 2004, 2005,2008; Chen et al., 2005; Mercier et al., 2005; Chelysheva et al., 2007). Loss of MSH4 in *Arabidopsis*, results in a reduction in crossover frequency to 15% of the wild-type level (Higgins et al., 2004). Similar effect was shown in the *Arabidopsis mer3* mutant (Chen et al., 2005; Mercier et al., 2005). Interestingly, the ZMM group includes proteins that are primary components of the synaptonemal complex, such as ZIP1. This interdependence indicates a link between crossover formation and synapsis. Overall, about 85% of *Arabidopsis* crossovers arise from the interference-dependent pathway (Higgins et al., 2004). The remaining crossovers are interference-independent, and are generated by a distinct group of proteins including MUS81 and EME1/MMS4 (Berchowitz et al., 2007).

Recombination events, including crossovers are not distributed randomly along chromosomes. Instead they tend to appear at certain chromosomal locations known as recombination hotspots (Drouaud et al., 2006). In plant species with large genomes, such as maize, barley, or wheat, crossovers are predominantly present in chromosome regions close to the telomeres (Akhunov et al., 2003; Gore et al., 2009). Crossover distribution affects the positions of chiasmata and may have implications for bivalent stability and chromosome segregation. However, neither mechanisms that control crossover distribution nor implications of crossover distribution for chromosome behavior in meiosis are well understood.

Early Defects in Chromatin Structure have an Impact on Homologous Chromosome Segregation: ASK1

ASK1 (*Arabidopsis* SKP1-like1) encodes one of the 21 predicted *Arabidopsis* homologs of the yeast and human Skp1 proteins (Yang et al., 1999; Zhao et al., 2003a,b). Skp proteins are an essential component of the Skp1-Cullin-F-box (SCF) complex, that belongs to a class of E3 ubiquitin ligases that target a variety of proteins for ubiquitin-mediated degradation via the 26S proteasome pathway (Petroski and Deshaies, 2005). ASK1 is the Skp homolog that has been best characterized in *Arabidopsis. ask1-1* mutants display defects in plant growth, flower development and male fertility (Yang et al., 1999; Zhao et al., 2001, 2003b). Male sterility arises from meiotic defects in prophase I that lead to erroneous homologous chromosome segregation in meiosis I and sister chromatid segregation in meiosis II, and to the subsequent formation of unbalanced spores. During prophase I, chromosomes maintain a leptotene-like structure with long and thin threads that do not synapse, as demonstrated by the absence of the typical SC structure (Wang et al., 2004). FISH experiments using a centromeric probe showed the presence of more than 5 signals in *ask1-1* meiocytes during pachytene, confirming lack of homologous pairing and bivalents formation (Zhao et al., 2006). The localization of the α-kleisin subunit of the cohesin complex SYN1 (described in the next paragraph) was also found to be altered in *ask1*meiocytes from zygotene to anaphase I. These observations together with a premature sister chromatid detachment detected by FISH in anaphase I, suggest that *ask1* mutation alters cohesin distribution and function, which is necessary for proper pairing and synapsis (Zhao et al., 2006). The abnormalities detected in *ask1* seem to derive from early defects in meiotic chromatin structure and chromosome reorganization in leptotene that cause a prolonged attachment of chromosomes to the nuclear membrane and the nucleolus, alterations in rDNA structure, prolonged attachment of the telomeres to the nucleolus, and defects in histone 3 acetylation, overall leading to the absence of homologous chromosome pairing (Yang et al., 2006). Hence, ASK1 is most likely required for chromosome conformation and remodeling of meiotic chromosomes by controlling the release of chromatin from the nucleolus and nuclear membrane starting from leptotene (Yang et al., 2006). Several hypotheses have been currently proposed to explain the potential role of ASK1 in meiosis, consistent with the meiotic defects observed in the mutant and the homology of ASK1 to Skp proteins (Yang et al., 2006; Zhao et al., 2006). ASK1 may control the degradation of a protein which inhibits the leptotene to zygotene transition, so that the alterations observed in chromatin structure and organization would be a consequence of the block of this transition. Alternatively, ASK1 might regulate the interaction of chromosomes to the nuclear membrane by degrading one or more proteins that link chromatin to the nuclear matrix, thus allowing a nuclear reorganization during leptotene and zygotene. ASK1 may also control chromatin structure by regulating chromatin remodeling proteins, as suggested by the alterations detected in histone 3 acetylation. However, the specific function of ASK1 in male meiosis is not yet defined.

SISTER CHROMATID COHESION IS ESSENTIAL FOR FAITHFUL CHROMOSOME SEGREGATION

The Cohesin Complex

Sister chromatids must be held together from the moment of their synthesis in S-phase until their separation in anaphase to ensure correct attachment of chromosomes to the spindle and accurate chromosome segregation in dividing cells. Cohesin is the multi-subunit protein complex that mediates sister chromatid cohesion in meiosis and mitosis by physically trapping them in a tripartite ring structure (Haering et al., 2008). The complex is highly conserved in eukaryotes and is composed of a core of four evolutionary conserved proteins, extensively studied in yeast and animals. In mitosis, the cohesin complex is composed of two members of the SMC family (structural maintenance of chromosomes), SMC1 and SMC3, and two auxiliary SCC subunits (sister chromatid cohesion), the α-kleisin RAD21/SCC1 and SCC3. In meiosis, the structure of the cohesin complex is highly similar, except for the RAD21/SSC1 component, which is replaced by its counterpart Rec8 (Stoop-Myer and Amon, 1999; Watanabe and Nurse, 1999). SMC1 and SMC3 consist, in their folded configuration, of a globular head and a hinge domain, connected by a long anti-parallel coiled coil. The proposed model of action of cohesin, the embrace model, requires the connection of the SMC hinge domains to form a SMC1/SMC3 heterodimer with a V-shaped structure, that can bind across sister chromatids and close, forming a ring, through a physical connection of the α-kleisin subunit to the C-terminal domain of SMC1 and the N-terminal domain of SMC3 (Gruber et al., 2003). The complex is stabilized by recruitment of SCC3 by the α-kleisin subunit (Figure 1) (for reviews on cohesin complex: Nasmyth and Haering, 2005; Onn et al., 2008; Peters et al., 2008).

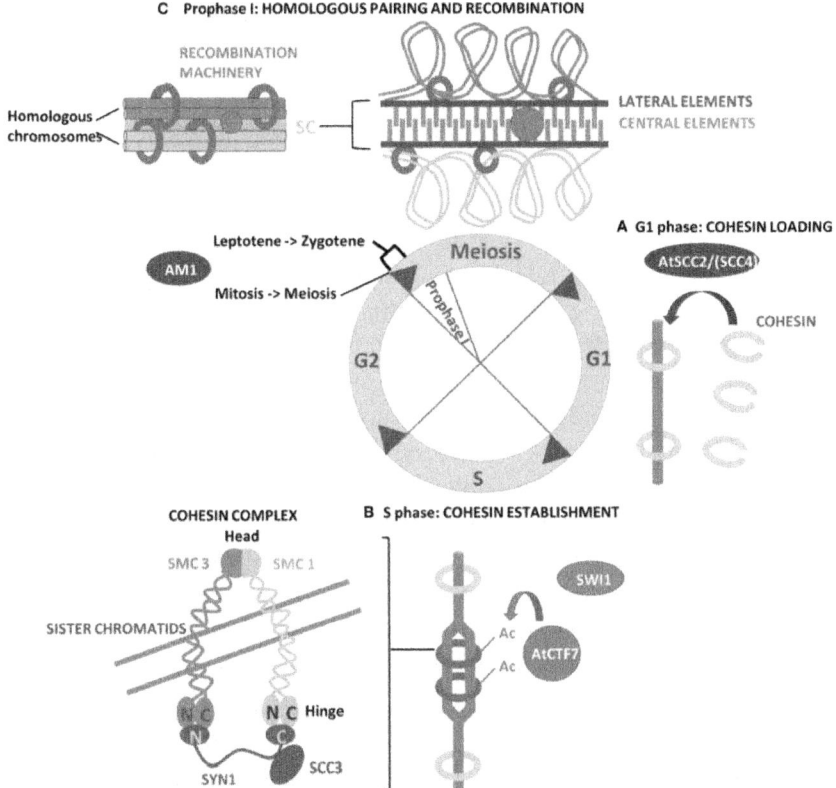

Figure 1. Overview of the events that allow the establishment of the cohesin complex on chromosomes, necessary for the successive steps of chromosome segregation, including homologous pairing and recombination in meiosis I. (A) Loading of cohesin on chromosomes requires the SCC2/SCC4 complex, only AtSCC2 has been characterized in *Arabidopsis*. (B) The establishment of chromosome cohesion takes place during DNA replication in S phase when Eco1/CTF7 acetylates SMC3 residues, ef-

fectively closing the cohesin ring. AtCTF7 has acetylatransferase activity *in vitro* and is required for the establishment of chromosome cohesion in *Arabidopsis*. SWI1 is an *Arabidopsis* protein with a role in cohesin establishment even if the mechanism of action is not yet known. Its maize homolog AM1 is required for the switch from mitosis to meiosis and for a putative checkpoint between leptotene and zygotene in prophase I. (C) Cohesion is required for SC elongation and polymerization and for meiotic recombination in prophase I.

SMC proteins

Similar as in yeast and animals, the sister chromatid connection in plants is also established through the cohesin complex. Homologs of the cohesin complex have been identified in some plant species and major progress on the understanding of their function has been achieved in the model plant Arabidopsis thaliana, in which all the components have been described. The Arabidopsis genome contains single copies of SMC1 and SMC3 cohesin subunits. Genetic studies revealed that loss of AtSMC1 or AtSMC3 functionality causes seedling lethality, hence impairing functional characterization (Liu et al., 2002). Localization studies using a specific antibody revealed that AtSMC3 is present in the cytoplasm and nucleus, on chromosomes and in the nuclear matrix of meiotic and mitotic cells, indicating a function in both types of cell divisions (Lam et al., 2005). At meiotic prophase, AtSMC3 localizes along sister chromatids to axial elements and lateral elements, similar to the Arabidopsis α-kleisin subunit SYN1. This observation confirms the conserved role of the cohesin complex in sister chromatid cohesion but also supports an additional function in SC formation, as proposed in yeast and mammals (Klein et al., 1999; Eijpe et al., 2000). By metaphase I, AtSMC3 localizes only to chromosome centromeres and, in addition, co-localizes to the spindle at metaphase I and anaphase I and II. The spindle localization is independent of SYN1 functionality and suggests that AtSMC3 might play an additional role as spindle associated protein, distinct from its conserved role in sister chromatid cohesion. This novel localization pattern is also conserved in mitosis and could be related to a role of AtSMC3 in spindle assembly and/or in the chromosome association with the spindle (Lam et al., 2005). A similar novel function has been suggested in human mitosis for the entire cohesin complex (Gregson et al., 2001).

Immunolocalization studies in tomato meiocytes (Solanum lycopersicum) revealed that SMC1 and SMC3 show a similar localization pattern as AtSMC3. In prophase I, SMC1 and SMC3 antibodies display a signal along AEs of the SC from leptotene to diplotene and a weak and diffuse signal on chromosomes at metaphase I and telophase II (Lhuissier et al., 2007). However, no localization to the spindle was documented, suggesting that the novel spindle function might be specific for Arabidopsis AtSMC3 and not conserved in other plant species.

Rec8 and SCC3

More intensive studies have been undertaken on the role of the meiotic α-kleisin subunit Rec8 in Arabidopsis (named SYN1 but also DIF1 and AtRec8), maize (AFD1), and rice (OsRad21-4/OsRec8) meiosis. In Arabidopsis, the homolog of Rec8, SYN1, is required for sister chromatid cohesion in meiosis (Cai et al., 2003; Chelysheva et al., 2005). SYN1 fully co-localizes with AtSCC3 at pachytene and is necessary for its proper loading on sister chromatids, confirming that they are indeed part of a complex (Chelysheva et al., 2005). FISH (fluorescence *in situ* hybridization) analysis on syn1 meiocytes using chromosome arm and centromeric probes show defective sister chromatid arm and centromere cohesion in meiosis I, confirming that SYN1 functions in cohesion (Cai et al., 2003).

In addition, SYN1 is required for synapsis of homologous chromosomes, being necessary for SC polymerization and elongation (Chelysheva et al., 2005). In syn1 meiocytes, synapsis is blocked and chromosome condensation and pairing are almost completely absent, leading to the presence of univalents at metaphase I (Bai et al., 1999; Bhatt et al., 1999). Localization of ASY1, a protein required for chromosome synapsis, recombination and SC assembly and widely used as a marker for chromosome axes in meiosis (Sanchez-Moran et al., 2008), is impaired in syn1 mutants, confirming the requirement of SYN1 for AE polymerization and elongation but not for their formation (Chelysheva et al., 2005). Synapsis is known to be closely related to meiotic recombination. Therefore, it is not surprising that SYN1 plays also a role in recombination, specifically in DSBs repair. Indeed, chromatin bridges and chromosome fragmentation are observed in syn1 meiosis I. They are suppressed by introducing into the syn1 mutant background the Atspo11 mutation, which abolishes DSBs formation and prevents recombination, confirming that SYN1 is required for DSBs repair (Chelysheva et al., 2005). Involvement of the cohesin complex in homologous chromosome pairing, assembly of the SC, and in meiotic recombination has been shown previously in other organisms. In yeast, Rec8 and SMC3 are required for SC formation and for repairing DSBs (Klein et al., 1999). In mouse, loss of Rec8 affects homologous recombination but does not affect SC formation and assembly. However, synapsis occurs between sister chromatids instead of homologous chromosomes, suggesting that Rec8 might define the chromosome unit and limit the SC binding sites to one single chromosome surface of a sister-chromatid pair in mammals (Xu et al., 2005).

Support for an additional role of Rec8 in homologous pairing and recombination in plants comes from studies on the maize α-kleisin subunit AFD1 and the rice OsRad21-4/OsRec8. A study on different afd1 alleles has revealed that AFD1 is required for AE installation, affecting the deposition of the recombination machinery on chromosomes (Golubovskaya et al., 2006). The rice OsRec8 regulates AE formation and may have a role in DNA DSBs repair, since localization of PAIR2 (homolog of Arabidopsis ASY1), ZEP1 (ZYP1 homolog), and MER3, involved in the formation of crossovers, is affected in Osrec8 mutants. As a consequence, no proper homologous pairing occurs (Zhang et al., 2006; Shao et al., 2011). Moreover, defective telomere bouquet formation is observed in Osrec8 and afd1 mutants, also preventing proper pairing of homologous chromosomes. Hence, OsRec8 regulates AE formation, homologous recombination and synapsis by affecting downstream proteins PAIR2, ZEP1, and MER3 (Shao et al., 2011).

Rec8 has a crucial role in the determination of kinetochore geometry for monopolar orientation in fission yeast, since rec8 mutants display loss of monopolar orientation at meiosis I and chromosome segregation defects (Yokobayashi et al., 2003; Sakuno et al., 2009). Similarly, Arabidopsis SYN1 is necessary for the monopolar attachment of sister kinetochores in meiosis I, as indicated by the observation of bipolar sister kinetochore attachment in meiosis I in the double syn1 Atspo11 mutant, in which syn1 chromosome fragmentation is suppressed allowing a clearer observation of chromosome segregation. However, the same defect in kinetochore orientation is observed for the other SCC cohesin subunit, AtSCC3, indicating that SYN1 is not sufficient for monopolar kinetochore orientation or, most likely, is inactive when the other members of the complex are not present. These data suggests that Rec8-containing cohesin complex is responsible for defining kinetochore geometry in meiosis I in plants, as proposed in yeast, Drosophila and mammals (Chelysheva et al., 2005; Watanabe, 2012).

AtSCC3 is the sole SCC3 homolog investigated in plants so far. It is required for normal plant growth and fertility and has a conserved role in proper sister chromatid cohesion, confirmed by the combination of univalents and bivalents observed in Atscc3 mutants (Chelysheva et al., 2005). However, in contrast to SYN1, AtSCC3 is not required for AE formation, since ASY1 localization in Atscc3 is normal and synapsis does not show major defects in the mutant. Moreover, only a low level of fragmentation is observed in Atscc3 and recombination is not defective, suggesting that the two SCC subunits, although being part of the same complex, may fulfill different additional functions (Chelysheva et al., 2005).

While AtSCC3 has no paralogs in the Arabidopsis genome, three α-kleisin homologs, SYN2, SYN3, and SYN4, are present that share about 38 % sequence similarity at their N-termini and 20 % at their C-termini with SYN1, and could partially compensate for each other (Schubert et al., 2009a). Two observations raise the hypothesis that the α-kleisin paralogs may be involved in cohesion in meiosis. First, in the syn1 Atspo11 double mutant, sister chromatid cohesion is only lost at anaphase I, suggesting that other homologs of the SYN1 family might be responsible for cohesion before that stage (Chelysheva et al., 2005). Second, SYN1 localization is only observed along chromosome axes but not at the core centromeres at metaphase I and metaphase II (Chelysheva et al., 2005; Cromer et al., 2013; Zamariola et al., 2013). It is known that SYN1, SYN2 and SYN4 may partially compensate for each other whereas SYN3 is required for plant viability, it localizes to the nucleolus and might have evolved a role in rDNA transcription and/or processing. A specific function in DNA repair in somatic cells has been suggested for SYN2, while SYN4 is required for centromere cohesion in mitosis (Schubert et al., 2009b). However, the role of the different paralogs is, at this time, not clear, and the creation of double or triple mutants might help unravelling their specific functions (Schubert et al., 2009b).

Loading and Establishment of Chromosome Cohesion

The loading of the cohesin complex onto chromosomes starts at telophase in humans and at the end of G1 in yeast and requires the evolutionary conserved SCC2/SCC4 complex (for reviews see Uhlmann, 2009; Ocampo-Hafalla and Uhlmann, 2011). Cohesin loading has been shown to be enriched at centromeric and pericentromeric regions promoting high fidelity chromosome segregation (Eckert et al., 2007). Recent studies in budding yeast have revealed that the observed enrichment is defined by the presence of the kinetochore subcomplex Ctf19, that promotes SCC2/SCC4 centromere association (Fernius et al., 2013). Also in Angiosperms, interphase nuclei show a preferred alignment of sister chromatids at centromeres, which might facilitate kinetochore bipolar orientation in mitosis, essential for correct chromosome segregation (Schubert et al., 2007). In plants, only the *Arabidopsis* homolog of the adherin SCC2 has been described. AtSCC2 is essential for plant viability and *Atscc2* plants show defects in embryogenesis and endosperm development (Schubert et al., 2009b; Sebastian et al., 2009). Using an inducible RNAi (RNA interference) system,Sebastian et al. (2009) demonstrated that AtSCC2 is required for sister chromatid cohesion and loading of the cohesin complex in meiosis, as indicated by defects in AtSCC3 localization. Furthermore, *Atscc2* mutants show an irregular localization of ASY1 and chromosome fragmentation, indicating that

AtSCC2 is required for axial development and most likely for repair of DNA DSBs, supporting the notion that sister chromatid cohesion is a prerequisite for axial development and DSBs resolution (Sebastian et al., 2009).

The loading of cohesin is the first step through the establishment of sister chromatid cohesion that takes place during DNA replication. After the loading, cohesin is unstable due to the activity of the Wapl-Pds5 complex that promotes cohesin dissociation (Rowland et al., 2009). In yeast, cohesion is established during S-phase by the Eco1/CTF7 protein, that acetylates the SMC3 residues, effectively closing the cohesin ring (Rowland et al., 2009). In *Arabidopsis*, AtCTF7 exhibits acetyltranferase activity *in vitro* like its yeast and human homologs (Jiang et al., 2010). *Atctf7* homozygous mutants display a dwarf phenotype and aberrant microsporogenesis due to defects in chromosome segregation in mitosis and PMCs (pollen mother cells). FISH performed with a centromeric and a chromosome 4 arm probes on male meiocytes of *Atctf7* and AtCTF7 RNAi plants, revealed that the protein is required for both centromere and arm cohesion in meiosis (Bolaños-Villegas et al., 2013; Singh et al., 2013). Furthermore, localization of the cohesin complex subunits AtSMC3, AtSYN1 and AtSCC3 is impaired in *Atctf7* male meiosis, indicating that AtCTF7 is necessary for association of cohesin on chromatin in meiosis (Bolaños-Villegas et al., 2013; Singh et al., 2013). In addition, the level of expression of genes required for DNA repair is significantly altered in *Atctf7* mitotic and meiotic tissues, and the mutant plants show a lower ability to repair DNA double strand breaks *in vivo* in mitotic cells (Bolaños-Villegas et al., 2013). Taken together, these observations suggest that AtCTF7 is also required for DNA repair in *Arabidopsis*, as shown for Eco1 in yeast mitosis (Lu et al., 2010).

SWITCH1/DYAD (SWI1/DYAD) is an *Arabidopsis* protein with an essential role in the establishment of sister chromatid cohesion during early meiosis (Mercier et al., 2001, 2003). Different allelic mutations have been investigated for the *SWI1/DYAD* gene, all of them showing an impact on fertility due to different mechanisms affecting megasporogenesis (*swi1-1* and *dyad*; Motamayor et al., 2000; Siddiqi et al., 2000; Mercier et al., 2001; Agashe et al., 2002) or both mega and microsporogenesis (*swi1-2* and *dsy10*; Mercier et al., 2003; Boateng et al., 2008). *Swi1-1* and *swi1-2* alleles have been shown to have an effect on the female mitosis-meiosis switch, so that meiosis is converted into a mitotic cell division (Motamayor et al., 2000; Mercier et al., 2001). However, analysis of the *dyad* allele by Agashe et al. (2002) and Siddiqi et al. (2000) with a meiotic marker, provided evidence that the female megaspore enters the meiotic programme but does not progress into further meiotic divisions. Detailed studies of male meiosis for *swi1-2* and *dsy10* alleles, have shown that the mutants loose cohesion in a stepwise manner already in meiosis I, leading to the presence of 20 chromatids at metaphase I which segregate randomly in meiosis II, forming polyads (Mercier et al., 2001). Furthermore, the mutant lacks AE formation, leading to incorrect pairing and synapsis, and does not initiate recombination. These defects probably all derive from defective establishment of cohesion before the initiation of meiosis, since the protein is expressed exclusively in meiotic G1 and S phase (Mercier et al., 2003). Specifically, the localization of SYN1 in *swi1-2* meiocytes, indicates that SWI1 performs its function after the loading of the cohesin complex (Mercier et al., 2003). However, its specific function in chromosome cohesion is not yet understood.

Maize AM1 and rice OsAM1 are proteins closely related to SWI1. Mutants in *AM1* and *OsAM1* genes show defective sister chromatid cohesion, absence of homologous pairing and synapsis, and lack of homologous recombination (Pawlowski et al., 2009; Che et al., 2011). However, while *Arabidopsis swi1* mutants affect meiotic processes downstream of meiotic initiation and do not affect entrance in meiosis, maize *am1* mutants show typical features of mitotic division in the early steps of meiosis, indicating that AM1 is required for the transition from the mitotic cell cycle into meiosis. Meiocytes of a specific *am1* allele arrest during early meiotic prophase at the transition between leptotene and zygotene, suggesting the presence of a novel checkpoint in maize required for progression through prezygotene (Pawlowski et al., 2009). Similarly, in rice, OsAM1 is also likely involved in a checkpoint mechanism that regulates the transition from leptotene to zygotene (Che et al., 2011).

A schematic overview of the processes of cohesin loading and establishment and homologous chromosome pairing and recombination, is shown in Figure 1.

Release of Chromosome Cohesion: Separase

Cleavage of the α-klesin subunit occurs in a stepwise manner during meiosis. In meiosis I, Rec8 is cleaved at chromosome arms, allowing the resolution of chiasmata and homologous chromosome segregation in meiosis I, whereas in meiosis II cohesin is released at centromeres, enabling sister chromatid separation (Nasmyth, 2001). Cleavage of Rec8 is performed by the cysteine protease Separase, which is conserved in various organisms, including yeast and vertebrates (Kitajima et al., 2003; Kudo et al., 2009). Separase function is inhibited by a protein called Securin, which is degraded at the onset of anaphase by ubiquitylation by the APC/C (Uhlmann, 2001). Homologs of separase are present in many plant species. However, the studies undertaken so far have only focused on the *Arabidopsis* separase AESP (Liu and Makaroff, 2006). AESP is an essential gene but RNA interference of AESP under the control of the meiotic DMC1 promoter, and the finding of the temperature permissive mutant *rsw4* (radially swollen 4), have allowed to investigate AESP function in meiosis (Liu and Makaroff, 2006; Wu et al., 2010; Yang et al., 2011). *Aesp* and *rsw4*mutants display defective chromosome segregation in meiosis I, in which entangled chromosomes and chromosome fragments are observed, and in meiosis II, where bivalents are still present, indicating persistence of cohesion (Liu and Makaroff, 2006; Yang et al., 2011). In support of this, SYN1 and SMC3 signals persist on *aesp* and *rsw4* chromosomes at later stages after metaphase I, demonstrating that AESP is responsible for removal of the cohesin complex from chromosomes. The creation of a double mutant between *aesp* and *ask1*, in which homologous chromosomes prematurely separate in meiosis I due to defects in homologous synapsis, showed that sister chromatids did not separate in meiosis II. This observation confirms that AESP is responsible for sister chromatid separation also in anaphase II (Yang et al., 2009). In *Arabidopsis*, a large amount of cohesin is released from chromosome arms in prophase I and the residual arm cohesin is released at anaphase I (Cai et al., 2003). While AESP is required for the release of cohesin at anaphase I and in meiosis II, it does not participate in the first step of release in prophase I, suggesting that a separase-independent mechanism might exist at early stages in *Arabidopsis*, similar to budding yeast, in which the condensin complex SMC2/SMC4 and a Polo kinase are responsible for cohesin removal at chromosome arms before metaphase I (Sumara et al., 2002; Yu and Koshland, 2005; Liu and Makaroff, 2006).

Separase is a multifunctional protein that in various organisms possesses additional roles to sister chromatid separation mechanistically less understood, such as proteolytic cleavage of other target proteins in yeast and spindle assembly in humans (Moschou and Bozhkov, 2012). Also in *Arabidopsis* additional functions of separase have been reported (Yang et al., 2009, 2011). *Aesp* mutants show alterations in non-homologous centromere associations at zygotene, suggesting that AESP might play a role in the control/release of the transient centromere associations that occur during zygotene in *Arabidopsis* (Armstrong et al., 2001). Furthermore, in *aesp* male meiocytes the radial microtubule array (RMA) is disturbed at telophase II and phragmoplast-like structures are observed, suggesting that AESP might have a function in microtubule organization or cell polarity (Yang et al., 2009). Absence of AESP also causes the formation of multinucleate microspores as a consequence of defective RMA (Yang et al., 2009). In contrast to yeast, where separase is required for normal meiotic spindle formation (Jensen et al., 2001; Baskerville et al., 2008), in*Arabidopsis* only RMA formation is defective while AESP might be required for the proper interaction of microtubules with the nuclear envelope at the tetrad stage (Yang et al., 2009).

Protection of Centromere Cohesion: Shugoshin and Patronus

In meiosis, sister chromatid cohesion is controlled in a time- and space-dependent manner, with chromosome arm cohesion release at the start of anaphase I, and maintenance of centromeric cohesion up till anaphase II. Meiosis-specific protection of Rec8 at pericentromeric regions from anaphase I to anaphase II is performed by Shugoshin (Sgo), a protein first described in *Drosophila* (MEI-S332; Kerrebrock et al., 1995), and successively identified in yeast, mammals and plants (Yao and Dai, 2012). Studies from yeast and vertebrates have elucidated the mechanism of action of Sgo, which is recruited at pericentromeric heterochromatin regions where it associates with the phosphatase PP2A to dephosphorylate Rec8 and prevent its cleavage in meiosis I (Lee et al., 2008a; Xu et al., 2009). In yeast, Sgo1 localizes at centromeres until the end of anaphase I (Kitajima et al., 2004), whereas in vertebrates SGOL2 persists on the chromosomes also in meiosis II (Lee et al., 2008a). Currently, two hypotheses are postulated to explain the dynamic association of Shugoshin with centromeres. On the one hand, Sgo function may be controlled by microtubule attachment and deactivated by a spatial change of its localization in the peri-centromeric domain in response to a change in microtubule tension (Lee et al., 2008a). Alternatively, a PP2A inhibitor may block dephosphorylation thereby conferring loss of protection of centromeric cohesion in meiosis II (Chambon et al., 2013). Flies and budding yeast possess a single copy of Sgo, while fission yeast, mammal and plant genomes have two Sgo paralogs, Sgo1 and Sgo2. In*Drosophila*, yeasts and plants, Sgo1 is responsible for the protection of centromere-specific sister chromatid cohesion in meiosis I, while in mammals SGOL2 performs the function of protector (Gutiérrez-Caballero et al., 2012). Though they are homologs, Sgo genes share limited sequence similarity and display in the different organisms somewhat different functions which have been acquired during evolution (for a recent review on the Shugoshin protein family and the additional roles of Shugoshin see Clift and Marston, 2011; Gutiérrez-Caballero et al., 2012). The *Sgo1* paralog *Sgo2*possesses different properties depending on the species examined. In fission yeast, Sgo2 plays a role in chromosome segregation in mitosis (Kitajima et al., 2004), in particular it has been shown to control the localization of the CPC, a protein complex that senses lack of tension between kinetochores and microtubules (Kawashima et al., 2007;Vanoosthuyse et al., 2007; Tsukahara et al., 2010). In addition, fission yeast Sgo2 also plays a role in meiosis, as *Sgo2*deletion leads to a modest increase in non-disjunction of homologs at meiosis I (Kitajima et al., 2004). In humans, hSGOL1 protects centromeric cohesion in mitosis (Salic et al., 2004; McGuinness et al., 2005), whereas hSGOL2 is dispensable for sister chromatid cohesion in mitotic cell division but is essential for correcting erroneous kinetochore attachments by recruiting the microtubule depolymerase MCAK to the centromeres (Huang et al., 2007), a role that is consistent with the one shown for fission yeast Sgo2 (Kawashima et al., 2007).

In plants, the role of Sgo as protector of centromere cohesion in meiosis has been described for the maize ZmSGO1, the rice OsSGO1 as well AtSGO1 and AtSGO2 of *Arabidopsis* (Hamant et al., 2005; Wang et al., 2011; Cromer et al., 2013;Zamariola et al., 2013, 2014). FISH analysis performed on *sgo1* meiocytes with a centromeric probe revealed a premature detachment of sister chromatid centromeres in anaphase I, resulting in random chromosome segregation in meiosis II. However, monopolar orientation of sister kinetochores in meiosis I is not affected in the mutants and chromosomes normally segregate in the reductional division, indicating that SGO proteins are required for protection of cohesion at anaphase I but not for monopolar orientation of sister kinetochores. In fission yeast and mammals, which possess two Sgo homologs, one copy is generally required for protection of sister chromatid cohesion in meiosis, while the other has evolved additional roles, as previously mentioned. So far, no function in somatic cells has been described for any of the plant Sgo proteins. *Arabidopsis* is the only species in which the role of both Sgo paralogs has been investigated. Single mutants show no vegetative phenotype and a meiotic phenotype is detected exclusively for *Atsgo1*. However, *Atsgo1 Atsgo2* double mutants reveal a partially redundant role for the two SGOs, opposite to yeast and vertebrate (Cromer et al., 2013; Zamariola et al., 2014). Immunolocalization of ZmSGO1 and OsSGO1 has revealed that SGO1 is loaded on chromosomes at leptotene, earlier than in other organisms such as yeast or mammals in which loading occurs during late prophase I or at diplotene, respectively (Kitajima et al., 2004; Gómez et al., 2007). Thus, plant SGO proteins might have a function in prophase I. In support of this hypothesis, ZEP1 localization is defective in*Ossgo1* mutants in about 21% of meiocytes, indicating that OsSGO1 may be required for the timely assembly of the SC, even if not for its initial assembly (Wang et al., 2011). In contrast, *Arabidopsis* ZYP1 localizes normally in *Atsgo1*mutants (Zamariola et al., 2013).

Recently, a novel protein involved in the protection of sister chromatid cohesion during meiosis II has been identified in*Arabidopsis*, named PATRONUS (PANS1) (Cromer et al., 2013; Zamariola et al., 2014). PANS1 is a plant specific protein that shares homology with genes belonging to the Eudicots family. *Pans1* meiocytes show a premature release of sister chromatid cohesion at metaphase II but not at meiosis I, indicating that the protein is required for protection of cohesion during interkinesis, at a later stage than SGOs. Moreover, similar to SGOs, PANS1 is not required for monopolar attachment of sister kinetochores in meiosis I. TAP-TAG and Y2H experiments have revealed that PANS1 may be a regulator of the APC/C complex because of the interaction with some of the APC/C subunits. In addition, the presence of two destruction boxes in the PANS1 sequence may indicate that PANS1 is at the same time also targeted by the APC/C complex. Currently, three hypotheses have been suggested to explain how PANS1 maintains sister chromatid cohesion at interkinesis: (1) by protecting SGOs from destruction by the APC/C; (2) by protecting sister chromatid cohesion from Separase independently of SGOs, in the case SGOs are no longer present after anaphase I, and (3) by inhibiting via APC/C regulation the Wapl-dependent process of cohesin release, which is usually activated at the end of mitosis/G1 phase to allow dynamic cohesin renewal and that could be present also at the end of meiotic telophase I (Cromer et al., 2013). At the moment, AtSGOs and PANS1 localization, that could help unraveling the function of PANS1 in meiosis and the relation among the protectors, is lacking. Besides its role as protector of cohesion, PANS1 has also been shown to be required for spindle organization in meiosis since *pans1* meiocytes display defective spindles starting from telophase I. Defective spindles is probably the cause of the formation of an aberrant internuclear organelle band at interkinesis, detected in 7% of *pans1* meiocytes. Taken together, these phenotypes and the premature separation of sister chromatids observed in meiosis II, suggest a function of PANS1 in ensuring the coordinate organization of the cell organelles in accordance with the meiotic cell cycle phase and chromosome cohesion (Zamariola et al., 2014), which is in agreement with the interaction of PANS1 with the APC/C.

CENTROMERES AND KINETOCHORES

Role of Centromeres and Kinetochores in Chromosome Segregation

Centromeres are DNA-protein structures necessary to direct chromosome movement in cell division. Centromere DNA sequences are fast evolving and highly variable among species. However, centromeric regions in most plant species encompass mainly two domains. One is the core centromere, which contains satellite tandem repeats, usually 150–180 bp long, and specialized nucleosomes in which histone H3 is replaced by a centromere-specific H3 histone variant, CENH3. This region is required for the assembly of the kinetochore, a protein structure that binds to spindle microtubules allowing faithful chromosome segregation. The core centromere is flanked by pericentromeric heterochromatin domains containing retroelements and other transposons. In yeast the pericentromeric domains have been shown to have mainly a role in the recruitment of Shugoshin (Pidoux and Allshire, 2005; Yamagishi et al., 2008). In addition, epigenetic mechanisms may be involved in the specification of centromeric chromatin and propagation of centromeres (Houben and Schubert, 2003; Ekwall, 2007; Torras-Llort et al., 2009; Wang et al., 2009).

The specific centromeric variant Histone 3, CENH3, was first identified in human as CENP-A and subsequently in all eukaryotic model systems (De Rop et al., 2012), including *Arabidopsis* (also called HTR12; Talbert, 2002). Despite its essential and conserved role in ensuring proper chromosome segregation, CENH3 proteins are highly variable in their sequences and fast evolving, especially their N-terminal tail domain and a loop 1 region at the C-terminal domain, which are necessary for CENH3 localization to centromeres in *Arabidopsis* (Ravi et al., 2010; Moraes et al., 2011). The C-terminal part of the protein is sufficient for the centromeric localization of CENH3 in mitotic cells even when the N-terminal part is absent (Lermontova et al., 2006). In meiosis, a different loading mechanism for CENH3 is present, in which the N-terminal tail plays a critical role. *Arabidopsis* plants transformed with a N-terminally truncated YFP-CENH3(C) protein show meiotic defects and partial sterility and the YFP signal cannot be detected in meiotic nuclei (Lermontova et al., 2011). Similarly, the replacement of the N-terminal tail with a GFP tagged variant, GFP-tailswap, causes sterility due to defects during sporogenesis (Ravi et al., 2011). In GFP-tailswap plants, meiosis is disturbed starting from metaphase I, in which bivalents align on the division plate but are not subjected to tension from the spindle, which is confirmed by decreased interkinetochore distance and by defective spindles (Ravi et al., 2011). CENH3 protein signal is reduced or not detected in GFP-tailswap meiocytes and is again detected after meiosis on mitotic chromosomes at the microspore stage, indicating the existence of distinct mechanisms for CENH3 loading in meiosis and mitosis (Ravi et al., 2011). The work of Lermontova et al. (2011) also suggests a different loading mechanism in meiosis and mitosis, since the YFP-CENH3(C) variant is deposited to the centromeres in mitosis but not in meiotic nuclei.

Recently, the *Arabidopsis* homolog of KNL2 has been identified. It represents one of the components of the Mis18 complex, responsible for the initiation of CENH3 deposition at the centromeres in humans (Hayashi et al., 2004), *C. elegans* (De Rop et al., 2012) and fission yeast (Hayashi et al., 2004). In *Arabidopsis*, KNL2 is associated with centromeres at all stages of the cell cycle except from metaphase to mid-anaphase. *Arabidopsis* KNL2 knockout mutants show defects in mitosis and meiosis and reduced CENH3 loading at the centromeres (Lermontova et al., 2013). Furthermore, CENH3 gene expression is decreased in *knl2* mutants but KNL2 expression is stable in CENH3 RNAi transformants, indicating that KNL2 acts upstream of CENH3 and has a function in the assembly of CENH3 at the centromeres (Lermontova et al., 2013). Moreover, KNL2 is co-expressed with H3K9 histone methyltransferases genes, whose expression is reduced in *knl2* mutants. Also DNA methylation levels are lower in *knl2* mutant plants. The requirement of KNL2 for CENH3 expression and for DNA methylation, suggests that KNL2 may interact with methyltransferases to allow the maintenance of DNA methylation, in order to control the epigenetic status of centromeric chromatin and to control CENH3 loading (Lermontova et al., 2013).

Sister kinetochores must behave differently in meiosis I and II: in meiosis I are oriented toward the same pole (mono-orientation) to allow homologous chromosomes segregation, while in meiosis II they face opposite poles (bi-orientation) (Brar and Amon, 2009). The tension exerted at the kinetochores by microtubules during division, and the kinetochore geometry, defined in meiosis and mitosis by cohesion, are fundamental for stabilizing the monopolar attachment in MI and the bipolar in MII (for review see Watanabe, 2012). In contrast to the high variability of centromeric sequences, more than 20 active kinetochore proteins are conserved between humans and yeasts (Lampert and Westermann, 2011), and are specific either for the inner kinetochore, where they directly recognize and bind DNA, or for the outer kinetochore, being responsible for the interaction with microtubules (Santaguida and Musacchio, 2009; Wang et al., 2009). However, to date, only 7 kinetochore proteins have been reported to be conserved in *A. thaliana*, the majority of which has not been yet functionally characterized (Murata, 2013). The inability to identify homologs of many human and yeast kinetochore proteins in plants, may suggest the existence of different kinetochore structure in plants (Murata, 2013).

Kinetochore functionality depends on the presence of a functional centromere in meiosis. Indeed MIS-12, a kinetochore protein which co-localizes with CENH3 at the centromere regions (Sato et al., 2005), does not do it in *Arabidopsis* GFP-tailswap meiocytes. In contrast, CENP-C, another kinetochore protein which localizes at the centromeres in mitotic cells (Ogura et al., 2004), is not affected in CENH3 RNAi transformants, suggesting that its localization does not depend on the presence of a functional CENH3 (Lermontova et al., 2011).

In maize, kinetochore proteins have been more thoroughly investigated. CENPC is part of the inner kinetochore and interacts at one side with the DNA repeats located at the centromeric regions, and, on the other side, with the members of the outer kinetochore (Dawe et al., 1999; Zhong et al., 2002). At the outer kinetochore NCD80 and MIS12 are present. Homologs of these two proteins are known to be parts of the KMN (KNL-1/Mis12/Ndc80) complex that constitutes the core microtubule-binding site of the kinetochore in *C.*

elegans (Cheeseman et al., 2006). NDC80 is a constitutive kinetochore protein which localizes at kinetochores in all meiotic and mitotic stages (Du and Dawe, 2007). It does not bind DNA directly and interacts with MIS12, which is also present at kinetochores during all stages of the cell cycle. NCD80 and MIS12 form at metaphase I a bridge structure that links sister kinetochores, while CENH3 and CENPC appear at the inner side of sister kinetochores as two distinct signals (Li and Dawe, 2009). MIS12 has an important role in sister chromatid connection at meiosis I and is required for the initiation of reductional division (Li and Dawe, 2009). Knock-down of MIS12 by RNAi leads to a weakening of the MIS12-NCD80 bridge and aberrant chromosome segregation in meiosis I, where in 30% of the cells sister kinetochores separate and segregate in an equational division instead of reductional (Li and Dawe, 2009). In MIS12 RNAi cells, the signal of the centromere protector ZmSGO1 does not weaken (Li and Dawe, 2009). The protein lies between sister kinetochores but cannot restore kinetochore co-orientation, confirming that Shugoshin is not required for the monopolar orientation of kinetochores (Hamant et al., 2005; Li and Dawe, 2009). A model, in which axial elements and cohesin hold sister chromatids together during prophase I and create the base for fused sister kinetochore formation promoted by the MIS12-NCD80 bridge has been proposed (Li and Dawe, 2009). This structure would cooperate with Shugoshin to induce reductional segregation by co-orienting sister kinetochores (Figure 2). MIS12 and NCD80 are thought to be similar to the monopolin complex, which promotes sister kinetochore co-orientation in budding yeast (Corbett and Harrison, 2012).

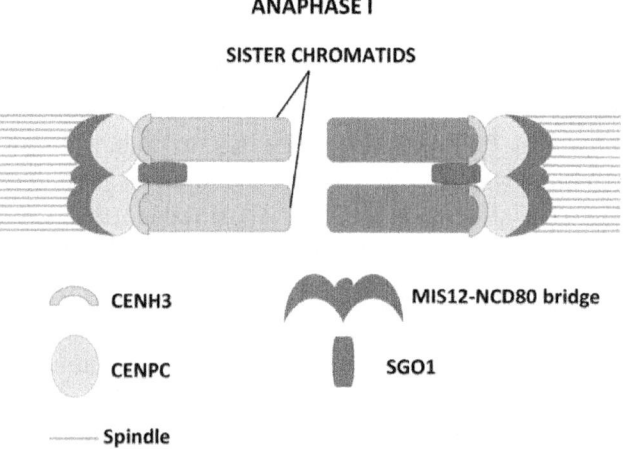

Figure 2. Model proposed in maize by Li and Dawe (2009) for reductional segregation in meiosis I. Sister kinetochores are fused in meiosis I by formation of the MIS12-

NCD80 bridge that, together with SGO1, allows monopolar attachment of sister chromatids to the spindle pole. The inner kinetochore proteins CENPC and CENH3 are visualized as two distinct signals.

Checkpoint Mechanisms Acting at Kinetochores

In eukaryotes, checkpoint mechanisms are present in meiosis and mitosis to prevent chromosome mis-segregation that would result in aneuploidy or apoptosis (Murray, 1994). The SAC is a conserved protein complex that controls proper attachment of microtubules to kinetochores in the metaphase to anaphase transition. In case of lacking or improper kinetochore-microtubule attachment, SAC creates a "wait anaphase" signal that stops anaphase progression. This response is promoted by APC/C together with its co-activator Cdc20 protein (for reviews see Peters, 2006; Vader et al., 2008; Musacchio, 2011). When all kinetochores are properly attached to microtubules, APC/C targets the destruction of Securin, the inhibitor of Separase as well as other cyclins, promoting chromosome segregation and exit from meiosis or mitosis. Evolutionary conserved proteins of SAC are MAD1, MAD2 (mitotic arrest deficient), Bub1, Bub3 (budding unhibited by benomyl), BubR1 kinase (bub-related1, Mad3 in yeast), and Mps1 (Monopolar kinase1) (May and Hardwick, 2006). The SAC proteins BubR1, Bub3, and MAD2 are also members of the Mitotic Checkpoint Complex (MCC), which is the effector of SAC that physically inhibits APC/C by binding to its co-activator Cdc20 until the moment all chromosomes are properly attached to kinetochores (Sudakin et al., 2001). SAC function has been investigated in depth in mitosis. However, a similar control mechanism is active also during meiosis (Malmanche et al., 2006; Sun and Kim, 2012).

Homologs of SAC proteins have been described in plants. MAD2 was first identified in maize where it localizes to the outer kinetochore in prometaphase I and II of meiosis, next to the inner kinetochore protein CENPC (Yu et al., 1999). During meiosis I and II, microtubule attachment is not sufficient for MAD2 dissociation from kinetochores, and the dissociation might occur in response to tension applied to the kinetochores (Yu et al., 1999). This hypothesis is supported by the concomitant staining of the MAD2 and 3F3/2 antibodies in maize meiosis. 3F3/2 recognizes a kinetochore phosphoepitope that is known to disappear in animal cells when tension is applied to the kinetochore (Nicklas et al., 1995). MAD2 homologs have been identified also in wheat and *Arabidopsis*, and their roles have been mainly investigated in mitotic checkpoint control. In wheat, intense MAD2 signal was observed at all centromeres in colchicine treated cells but not in untreated cells, confirming the function of MAD2 in the spindle checkpoint (Kimbara et al., 2004). In

Arabidopsis, MAD2 localization was studied together with BubR1 and Bub3.1, the others SAC proteins identified in the model plant (Caillaud et al., 2009). Interactions between the three proteins were observed in the nuclei of tobacco cells using bimolecular fluorescence complementation (Caillaud et al., 2009). During normal mitosis, localization of the SAC proteins to the kinetochores was not detected. However, by application of microtubule destabilizing drugs or of the proteasome inhibitor MG132, MAD2, BubR1, and Bub3.1 localized at the kinetochores, suggesting that SAC proteins are only recruited at kinetochores in case of defective spindle assembly in *Arabidopsis*(Caillaud et al., 2009). In contrast, a study by Ding et al. (2012) on the *Arabidopsis* MAD2 protein, showed that MAD2-GFP localizes at kinetochores also during normal mitotic progression from prophase to metaphase, as shown in maize. Moreover, AtMAD2 binds to AtMAD1, which interacts with the nucleoporin NUA, showing that SAC components interact with the nuclear pore. This interaction has been found in several other organisms, and it seems that the presence of SAC proteins at the nuclear pore mediates mitotic spindle checkpoint (Lee et al., 2008b).

Mps1 (Monopolar kinase 1) is also required for SAC function in the mitotic checkpoint in several eukaryotes and has been shown to be responsible for the recruitment of Mad1 and Mad2 at kinetochores in humans (Hewitt et al., 2010). Mps1 *Arabidopsis* homolog has conserved motifs which could mediate its interaction with MAD2 but also with cyclins, the APC/C and MAPK (mitogen-activated protein kinases), however, proof of its biological role in the checkpoint mechanism is still required (De Oliveira et al., 2012).

In most organisms, SAC is controlled by the chromosome passenger complex (CPC). In general, the CPC consists of the core enzyme Aurora B kinase, and three non-enzymatic subunits that control the targeting, enzymatic activity and stability of Aurora B: inner centromeric protein (INCENP), borealin and survivin (for review see Ruchaud et al., 2007). The major role of CPC is sensing incorrect kinetochore-microtubule attachments and generating, in response, unattached kinetochores, which allows new rounds of attachment until the correct configuration is obtained. The presence of unattached kinetochores activates the SAC that blocks the progression of cell divisions until all chromosomes are under tension. In plants, little is known about the role of the CPC in meiosis and few components of the complex have been identified. Like animals, *Arabidopsis* possesses three Aurora kinase homologs, which share a similar structure to the ones of other species (Kawabe et al., 2005). AtAurora1 and AtAurora2 display similar localization dynamics to Aurora B kinase in *Arabidopsis* mitosis, suggesting that they could function as chromosomal passenger proteins (Demidov et al., 2005). AtAurora1 interacts with SAC proteins BubR1 and MAD2 *in vivo* and phosphorylate them *in vitro*, which suggests that it functions in checkpoint mechanisms (Demidov D., personal communication). Furthermore, deregulation of AtAurora kinases activity, either by mutagenesis or by chemical treatment, results in defects in microsporogenesis and generation of polyploid and aneuploid progeny, suggesting that AtAurora may regulate correct chromosome segregation in *Arabidopsis* meiosis (Demidov D., personal communication).

A putative ortholog of the CPC subunit INCENP, WYR, has been identified in *Arabidopsis* (Kirioukhova et al., 2011). WYR shares with the INCENP homolog proteins a characteristic C-terminal domain, a coiled coil domain and a IN-box at the C-terminus, required for the binding of Aurora kinase. WYR is an essential gene with a role in cell cycle control and, independently, in cell fate and differentiation in *Arabidopsis*, since is required for both female and male gametogenesis. Similar functions have been reported also for the orthologs of INCENP in animals (Ruchaud et al., 2007). However, further genetic and biochemical analyses on WYR and Aurora kinases are required to establish the role of CPC proteins in plants.

MICROTUBULE ORGANIZATION AND SPINDLE DYNAMICS

In all eukaryotic cells, faithful chromosome segregation is accomplished by microtubule-based movement and requires a bipolar structure, the spindle, which consists of an antiparallel array of microtubules. The microtubules have their minus-end anchored at the spindle pole and their plus-end toward the chromosomes (Wittmann et al., 2001). They are highly dynamic polar polymers of noncovalently bound α and β tubulin heterodimers and represent the major components of the cytoskeleton in eukaryotic cells (Nogales, 2000). They rapidly polymerize and depolymerize while being continually translocated toward the poles. In animal and yeast cells, microtubules nucleate from microtubule-organizing centers (MTOC), such as the centrosome and the spindle pole body, which are responsible for the organization of the cortical astral arrays in interphase and mitotic spindles during cell division (Pereira and Schiebel, 1997; Jaspersen and Winey, 2004). γ-tubulin is enriched at the nucleation centers where it is recruited as a ring-shaped complex together with associated proteins, enhancing the nucleation of microtubules (O'Toole et al., 2012). In contrast to animals and yeast, plant microtubules lack conspicuous organizing centers. However, they are organized into ordered arrays that are associated with a growth pattern of the plant cell and relocate in a cell-cycle specific manner (Azimzadeh et al., 2001). During cell division, a succession of microtubule arrays is identified: radial arrays from the nuclear surface and cortical arrays of interphase, preprophase bands, spindles, and phragmoplasts (Wasteneys, 2002; De Storme and Geelen, 2013a). Like in animal and yeast, γ-tubulin is also required for microtubule nucleation in plants, being essential for the organization of the microtubule structures in interphase and cell division (Canaday et al., 2000; Shimamura et al., 2004; Pastuglia et al., 2006).

Microtubule motor proteins have an essential role in spindle assembly in both centrosomal and acentrosomal systems (Walczak et al., 1998). The best studied class of microtubule motor proteins are kinesins, proteins that participate in a variety of biological processes, including transport of vesicles, chromosomes or organelles, and organization of spindle microtubules, and chromosome segregation (Woehlke and Schliwa, 2000). They move unidirectionally along microtubules toward their plus or minus-ends. They use energy derived from ATP hydrolysis, in a processive or non-processive way, depending on their capacity of moving cargo long or only short distances before detaching from the microtubules. Several kinesins are known to be required for the structure, assembly and positioning of the mitotic and meiotic spindles in animals and fungi (Endow, 1999; Sharp et al., 2000). The *Arabidopsis* genome contains 61 predicted kinesins, one-third of them belonging to the kinesin-14 family that includes minus end-directed motor proteins (Reddy and Day, 2001). ATK1 is a member of this family and has been shown to support microtubule movement in an ATP-dependent manner and to be a non-processive, minus-end motor protein (Marcus et al., 2002). ATK1 has a specific role in male meiosis, in which *atk1-1* meiocytes display defective chromosome alignment and segregation in meiosis I and II due to aberrant formation of metaphase and anaphase spindles, leading to spore and pollen abortion and decreased plant fertility (Chen et al., 2002). ATK1 is involved in the assembly of the meiotic spindle and is needed for organizing microtubules at the two poles at metaphase and anaphase I and II, but not for the organization of microtubules for other structures, such as the interzonal microtubule array formed at telophase I (Chen et al., 2002). Studies in yeast and *Drosophila* have suggested that minus and plus-ended motor proteins could produce counteracting forces within the spindle to maintain its structure (Sharp et al., 1999, 2000). Thus, ATK1 might have a similar function in plant male meiosis, by producing inward-acting forces necessary for the assembly and maintenance of a bipolar spindle (Chen et al., 2002). The creation of a double heterozygote mutant between ATK1 and its homolog ATK5 (also named AtKIN14a and AtKIN14b, respectively), has shown that both proteins are required for proper chromosome segregation in female and male meiosis and for normal spindle morphogenesis in male meiosis (Quan et al., 2008). In addition to its male meiotic function, ATK1 localizes to the midzone of the mitotic spindle from metaphase through anaphase, suggesting a function also in the mitotic spindle apparatus (Liu et al., 1996).

AtPRD2/MPS1 (Multi-polar spindle1) is a putative *Arabidopsis* coiled-coil protein with homologs only among Embryophytes. Although having been identified as AtPRD2, an essential protein for DSBs formation, due to the presence of univalents in *Atprd2* mutant meiosis (De Muyt et al., 2009), the protein has also been found to be required for spindle organization and determination of spindle polarity in male meiosis (MPS1; Jiang et al., 2009). *Mps1*meiocytes display multiple focused spindles at metaphase I, indicating that spindle assembly is not defective, in contrast to *atk1* and *atk1/atk5* mutants, but spindle bipolarity is compromised in meiosis I and II, and chromosome segregation results more affected than in the kinesin mutants. This observations suggest that MPS1, ATK1, and ATK5 play a role in different mechanisms in plant meiosis. It has been proposed that MPS1 might guide microtubule minus-end migration in meiosis, maybe through binding to an unknown MAP (microtubule associated proteins) or, alternatively, could be a component of the spindle pole transmitting the signal to attract the minus-end of the spindle microtubules before spindle assembly (Jiang et al., 2009). However, whether the spindle defects observed in *mps1*meiocytes correspond to a primary function of the protein in spindle organization and polarity, or to a secondary effect caused by univalents formation in meiosis I, is not clear since conflicting observations on the relationship between unpaired chromosomes and spindle aberrations have been reported (Chan and Cande, 1998; Dawe, 1998).

In rice, a Kinesin-1-like protein, Pollen Semisterility 1 (PSS1), has been shown to have microtubule-stimulated ATPase activity and to be required for proper chromosome alignment and segregation in meiosis. However, spindle morphology is only slightly affected in *pss1* mutants, indicating that PSS1 might have a minor and not essential role in the formation of the meiotic spindle or alternatively might be involved in the regulation of chromosome movements along the spindles, as suggested by the delayed chromosomes observed in meiosis in *pss1* (Zhou et al., 2011).

Recently, the identification of a MATH-BTB domain protein, MAB1 (MATH-BTB1) in maize has been reported. This protein is required for organizing microtubule spindles and nuclei positioning in meiosis II and in the first mitotic division in both male and female germlines. Since no direct interaction between MAB1 and the spindles has been observed, it has been proposed that MAB1 may act through the control of a spindle apparatus regulator(s) (Juranič et al., 2012). Six MATH-BTB proteins have been currently identified in the *Arabidopsis* genome, however, no similar function has been reported (Weber and Hellmann, 2009).

The correct orientation of spindles in the second meiotic division is an essential requirement for faithful chromosome segregation. Alterations in the orthogonal configuration of the division planes in meiosis II lead to co-orientation of the spindles producing unreduced gametes, that represent the major route to polyploidization in plants (Brownfield and Köhler, 2011). Co-orientation can lead to the formation of three types of MII spindle defects which usually occur together in cells: parallel, tripolar or fused (Conicella et al., 2003; De Storme and Geelen, 2013b). This phenomenon only takes place in PMCs (pollen mother cells) of plants with simultaneous cytokinesis. In this type of cytokinesis, as opposed to the successive type, no cell plate is formed at the end of meiosis I and the two sets of chromosomes stay in the same cytoplasm and need to be perpendicularly oriented to create the tetrahedral configuration observed at the end of meiosis II (De Storme and Geelen, 2013a). They have been documented in many plant species, however, the molecular mechanisms behind their occurrence are still largely unknown. Two proteins involved in spindle orientation specifically in male meiosis II have been identified in *Arabidopsis*: AtPS1 and JASON (D'Erfurth et al., 2008; Erilova et al., 2009; De Storme and Geelen, 2011). Mutations in these genes produce at the end of meiosis II a high number of unreduced gametes (i.e., dyads and triads) instead of normal haploid gametes, leading to diploid pollen formation and triploid offspring. The biological mechanism causing 2n gamete formation in the mutants has been elucidated by tubulin immunostainings, which have shown the formation of parallel, tripolar and fused spindles in meiosis II. The defective spindles lead to 2n spores that retain parental heterozygosity at the centromeres, indicative of a FDR-type (first division restitution) of meiotic restitution (D'Erfurth et al., 2008; De Storme and Geelen, 2011). The introduction of *Atps1* or *jason* mutations into the *Atspo11* mutant background has confirmed the model of 2n gametes formation through co-oriented spindles, since the unbalanced segregation caused by *Atspo11* at meiosis I is nullified by parallel spindles in meiosis II, leading to the formation of mainly balanced dyads as result of meiosis in the double mutants.*Atps1* and *jason* meiocytes lack the characteristic interzonal microtubule array (IMA) observed in simultaneous PMCs at telophase I, which physically separates the two new formed nuclei. They mostly show fused nuclei at metaphase II. In potato, the absence of IMA has also been proposed to cause alterations in cell polarity and the formation of fused spindles (Conicella et al., 2003), suggesting that also in the *Arabidopsis* mutants depending on the total, partial, or unipolar loss of IMA fused, parallel or tripolar spindles are formed (De Storme and Geelen, 2013b).

AtPS1 is a protein conserved in the plant kingdom (Cigliano et al., 2011), which contains two conserved domains in its structure: an N-terminal

Forkhead-associated (FHA) domain required for phosphoprotein interaction in many signaling pathways (Li et al., 2000) and a PINc domain that has RNA-binding properties associated with RNAse activity, and which is generally found in proteins involved in RNAi and in nonsense-mediated mRNA decay (NMRD) (Clissold and Ponting, 2000). JASON encodes a protein of unknown function and no known domains that is conserved in plants (Erilova et al., 2009). Expression analysis have demonstrated that *JASON* controls the *AtPS1* transcript level specifically in meiotic flower buds, suggesting the existence of a regulatory mini-network for the control of spindle orientation in meiosis II (De Storme and Geelen, 2011).

Defects in spindle orientation in the second meiotic division have been also reported in mutants in one of the *Arabidopsis* formins, AFH14 (Li et al., 2010). Formins are a class of proteins known to regulate the microfilament cytoskeleton (Blanchoin and Staiger, 2010), but have been recently shown to have also a prominent role in microtubule regulation and in the crosstalk between actin filaments and microtubules in higher eukaryotes (Bartolini and Gundersen, 2010). Indeed, microtubules and microfilaments have been shown to co-distribute and interact in the meiotic spindle and in the phragmoplast in maize (Staiger and Cande, 1991). AFH14 co-localizes with MTs and MFs arrays during cell division in *Arabidopsis* suspension cells and with MTs in meiotic cells, affecting their arrangement during microsporogenesis. *Afh14* mutants display abnormal MTs structures including defective RMS at telophase I, parallel spindles at metaphase II and the absence of phragmoplast structures at late cytokinesis.

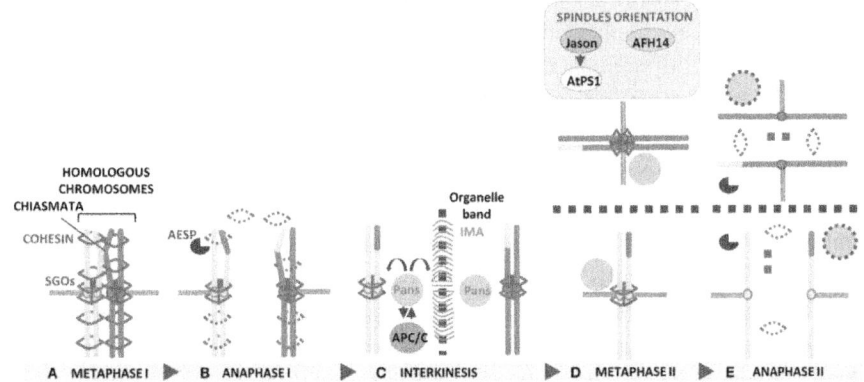

Figure 3. Overview of chromosome segregation in *Arabidopsis* meiosis. (A) At metaphase I, homologous chromosomes are connected by chiasmata and SHUGOSHINs (AtSGOs) are present at the centromeres. (B) At anaphase I, Separase AESP is activated and cleaves the cohesin at chromosome arms but not at centromeres, allowing

resolution of chiasmata and homologous chromosomes segregation by monopolar at-tachment to the spindles. (C) At interkinesis, an internuclear microtubule array (IMA) is formed at the site of the organelle band, to physically separate homologous chromo-somes. PANS1 is active and protects centromere cohesin, probably in conjunction with SGOs. PANS1 also interacts with the APC/C, and it is probably also an APC/C target. In addition, PANS1 plays a role in spindle organization from telophase I to telophase II. (D) At metaphase II, the chromosomes orient perpendicularly to the metaphase plate through the perpendicular orientation of spindles regulated by Jason and AtPS1. Also the formin AFH14 influences spindle orientation by linking MTs and MFs. (E) Releasing or degradation of SGOs and PANS allows cleavage of centromeric cohesin by separase and sister chromatids segregation.

AFH14 has been shown to preferentially bind MTs and to link MTs and MFs *in vitro*, thus playing a key role in cytoskeletal dynamics and organization required for cell division, including MII spindle orientation (Li et al., 2010; De Storme and Geelen, 2013b).

An overview of the process of chromosome segregation between metaphase I and anaphase II, and of the molecular factors playing an essential role in *Arabidopsis* chromosome segregation, is displayed in Figure 3.

CONCLUSIONS AND PERSPECTIVES

In the past 15 years the identification and characterization of plant meiotic genes has seen a remarkable acceleration due to the forward and reverse genetics strategies used in the model plants *Arabidopsis*, maize and rice. In addition, investigation of the molecular mechanisms regulating meiosis in other kingdoms has enormously contributed to the development of plant research in this field. The coordinate events leading to accurate chromosome segregation have been elucidated in budding yeast and studies in plants have confirmed the conserved role of many proteins in the steps of meiotic chromosome segregation, such as cohesin and the dynamics of cohesin removal and protection, the machinery of homologous pairing and recombination, and the function of kinetochores and microtubules. However, even if the main genes have been identified and their function in chromosome segregation confirmed, not much is known about their regulation in accordance with the cell cycle. Further research should focus on investigating the molecular mechanisms regulating protein functions and the interaction between the proteins to define their role in the broader context of chromosome segregation.

CONFLICT OF INTEREST STATEMENT

The authors declare that the research was conducted in the absence of any commercial or financial relationships that could be construed as a potential conflict of interest.

REFERENCES

1.   Adams, K. L., and Wendel, J. F. (2005). Polyploidy and genome evolution in plants. *Curr. Opin. Plant Biol.* 8, 135–141. doi: 10.1016/j. pbi.2005.01.001

2.   Agashe, B., Prasad, C. K., and Siddiqi, I. (2002). Identification and analysis of DYAD: a gene required for meiotic chromosome organisation and female meiotic progression in Arabidopsis. *Development* 129, 3935–3943.

3.   Akhunov, E. D., Goodyear, A. W., Geng, S., Qi, L.-L., Echalier, B., Gill, B. S., et al. (2003). The organization and rate of evolution of wheat genomes are correlated with recombination rates along chromosome arms. *Genome Res.* 13, 753–763. doi: 10.1101/gr.808603

4.   Armstrong, S. J., Franklin, F. C. H., and Jones, G. H. (2001). Nucleolus-associated telomere clustering and pairing precede meiotic chromosome synapsis in Arabidopsis thaliana. *J. Cell Sci.* 114, 4207–4217.

5.   Azimzadeh, J., Traas, J., and Pastuglia, M. (2001). Molecular aspects of microtubule dynamics in plants. *Curr. Opin. Plant Biol.* 4, 513–519. doi: 10.1016/S1369-5266(00)00209-0

6.   Bai, X., Peirson, B. N., Dong, F., Xue, C., and Makaroff, C. A (1999). Isolation and characterization of SYN1, a RAD21-like gene essential for meiosis in Arabidopsis. *Plant Cell* 11, 417–430. doi: 10.1105/tpc.11.3.417

7.   Bartolini, F., and Gundersen, G. G. (2010). Formins and microtubules. *Biochim. Biophys. Acta* 1803, 164–173. doi: 10.1016/j. bbamcr.2009.07.006

8.   Baskerville, C., Segal, M., and Reed, S. I. (2008). The protease activity of yeast separase (esp1) is required for anaphase spindle elongation independently of its role in cleavage of cohesin. *Genetics* 178, 2361–2372. doi: 10.1534/genetics.107.085308

9.   Bass, H. W., Riera-Lizarazu, O., Ananiev, E. V, Bordoli, S. J., Rines, H. W., Phillips, R. L., et al. (2000). Evidence for the coincident initiation of homolog pairing and synapsis during the telomere-clustering (bouquet) stage of meiotic prophase. *J. Cell Sci.* 113, 1033–1042.

10. Berchowitz, L. E., Francis, K. E., Bey, A. L., and Copenhaver, G. P. (2007). The role of AtMUS81 in interference-insensitive crossovers in *A. thaliana*. *PLoS Genet*. 3:e132. doi: 10.1371/journal.pgen.0030132

11. Bhalla, N., and Dernburg, A. F. (2008). Prelude to a division. *Annu. Rev. Cell Dev. Biol*. 24, 397–424. doi: 10.1146/annurev.cellbio.23.090506.123245

12. Bhatt, A. M., Canales, C., and Dickinson, H. G. (2001). Plant meiosis: the means to 1N. *Trends Plant Sci*. 6, 114–121. doi: 10.1016/S1360-1385(00)01861-6

13. Bhatt, A. M., Lister, C., Page, T., Fransz, P., Findlay, K., Jones, G. H., et al. (1999). The DIF1 gene of Arabidopsis is required for meiotic chromosome segregation and belongs to the REC8/RAD21 cohesin gene family. *Plant J*. 19, 463–472. doi: 10.1046/j.1365-313X.1999.00548.x

14. Birchler, J. A., Bhadra, U., Bhadra, M. P., and Auger, D. L. (2001). Dosage-dependent gene regulation in multicellular eukaryotes: implications for dosage compensation, aneuploid syndromes, and quantitative traits. *Dev. Biol*. 234, 275–288. doi: 10.1006/dbio.2001.0262

15. Bishop, D. K., Park, D., Xu, L., and Kleckner, N. (1992). DMC1: a meiosis-specific yeast homolog of E. coli recA required for recombination, synaptonemal complex formation, and cell cycle progression. *Cell* 69, 439–456. doi: 10.1016/0092-8674(92)90446-J

16. Blanchoin, L., and Staiger, C. J. (2010). Plant formins: diverse isoforms and unique molecular mechanism. *Biochim. Biophys. Acta* 1803, 201–206. doi: 10.1016/j.bbamcr.2008.09.015

17. Boateng, K. A., Yang, X., Dong, F., Owen, H. A., and Makaroff, C. A. (2008). SWI1 is required for meiotic chromosome remodeling events. *Mol. Plant* 1, 620–633. doi: 10.1093/mp/ssn030

18. Bolaños-Villegas, P., Yang, X., Wang, H.-J., Juan, C.-T., Chuang, M.-H., Makaroff, C. A., et al. (2013). Arabidopsis CHROMOSOME TRANSMISSION FIDELITY 7 (AtCTF7/ECO1) is required for DNA repair, mitosis and meiosis. *Plant J*. 75, 927–940. doi: 10.1111/tpj.12261

19. Borde, V. (2007). The multiple roles of the Mre11 complex for meiotic recombination. *Chromosome Res*. 15, 551–563. doi: 10.1007/s10577-007-1147-9

20. Börner, G. V., Kleckner, N., and Hunter, N. (2004). Crossover/noncrossover differentiation, synaptonemal complex formation, and regulatory surveillance at the leptotene/zygotene transition of meiosis. *Cell* 117, 29–45. doi: 10.1016/S0092-8674(04)00292-2

21. Bozza, C. G., and Pawlowski, W. P. (2008). The cytogenetics of homologous chromosome pairing in meiosis in plants. *Cytogenet. Genome Res.* 120, 313–319. doi: 10.1159/000121080

22. Brar, G. A., and Amon, A. (2009). Emerging roles for centromeres in meiosis I chromosome segregation. *Nat. Rev. Genet.* 9, 899–910. doi: 10.1038/nrg2454

23. Brownfield, L., and Köhler, C. (2011). Unreduced gamete formation in plants: mechanisms and prospects. *J. Exp. Bot.* 62, 1659–1668. doi: 10.1093/jxb/erq371

24. Cai, X., Dong, F., Edelmann, R. E., and Makaroff, C. A. (2003). The Arabidopsis SYN1 cohesin protein is required for sister chromatid arm cohesion and homologous chromosome pairing. *J. Cell Sci.* 116, 2999–3007. doi: 10.1242/jcs.00601

25. Caillaud, M.-C., Paganelli, L., Lecomte, P., Deslandes, L., Quentin, M., Pecrix, Y., et al. (2009). Spindle assembly checkpoint protein dynamics reveal conserved and unsuspected roles in plant cell division. *PLoS ONE* 4:e6757. doi: 10.1371/journal.pone.0006757

26. Canaday, J., Stoppin-Mellet, V., Mutterer, J., Lambert, A. M., and Schmit, A. C. (2000). Higher plant cells: gamma-tubulin and microtubule nucleation in the absence of centrosomes. *Microsc. Res. Tech.* 49, 487–495. doi: 10.1002/(SICI)1097-0029(20000601)49:5<487::AID-JEMT11>3.0.CO;2-I

27. Chambon, J.-P., Touati, S. A., Berneau, S., Cladière, D., Hebras, C., Groeme, R., et al. (2013). The PP2A inhibitor I2PP2A is essential for sister chromatid segregation in oocyte meiosis II. *Curr. Biol.* 23, 485–490. doi: 10.1016/j.cub.2013.02.004

28. Chan, A., and Cande, W. Z. (1998). Maize meiotic spindles assemble around chromatin and do not require paired chromosomes. *J. Cell Sci.* 111, 3507–3515.

29. Che, L., Tang, D., Wang, K., Wang, M., Zhu, K., Yu, H., et al. (2011). OsAM1 is required for leptotene-zygotene transition in rice. *Cell Res.* 21, 654–665. doi: 10.1038/cr.2011.7

30. Cheeseman, I. M., Chappie, J. S., Wilson-Kubalek, E. M., and Desai, A. (2006). The conserved KMN network constitutes the core microtubule-binding site of the kinetochore. *Cell* 127, 983–997. doi: 10.1016/j.cell.2006.09.039

31. Chelysheva, L., Diallo, S., Vezon, D., Gendrot, G., Vrielynck, N., Belcram, K., et al. (2005). AtREC8 and AtSCC3 are essential to the

monopolar orientation of the kinetochores during meiosis. *J. Cell Sci.* 118, 4621–4632. doi: 10.1242/jcs.02583

32. Chelysheva, L., Gendrot, G., Vezon, D., Doutriaux, M.-P., Mercier, R., and Grelon, M. (2007). Zip4/Spo22 is required for class I CO formation but not for synapsis completion in Arabidopsis thaliana. *PLoS Genet.* 3:e83. doi: 10.1371/journal.pgen.0030083

33. Chen, C., Marcus, A., Li, W., Hu, Y., Calzada, J.-P. V., Grossniklaus, U., et al. (2002). The Arabidopsis ATK1 gene is required for spindle morphogenesis in male meiosis. *Development* 129, 2401–2409.

34. Chen, C., Zhang, W., Timofejeva, L., Gerardin, Y., and Ma, H. (2005). The Arabidopsis ROCK-N-ROLLERS gene encodes a homolog of the yeast ATP-dependent DNA helicase MER3 and is required for normal meiotic crossover formation. *Plant J.* 43, 321–334. doi: 10.1111/j.1365-313X.2005.02461.x

35. Chikashige, Y., Haraguchi, T., and Hiraoka, Y. (2007). Another way to move chromosomes. *Chromosoma* 116, 497–505. doi: 10.1007/s00412-007-0114-8

36. Cigliano, R. A., Sanseverino, W., Cremona, G., Consiglio, F. M., and Conicella, C. (2011). Evolution of parallel spindles like genes in plants and highlight of unique domain architecture. *BMC Evol. Biol.* 11:78. doi: 10.1186/1471-2148-11-78

37. Clift, D., and Marston, A. L. (2011). The role of shugoshin in meiotic chromosome segregation. *Cytogenet. Genome Res.* 133, 234–242. doi: 10.1159/000323793

38. Clissold, P. M., and Ponting, C. P. (2000). PIN domains in nonsense-mediated mRNA decay and RNAi. *Curr. Biol.* 10, 888–890. doi: 10.1016/S0960-9822(00)00858-7

39. Comai, L. (2005). The advantages and disadvantages of being polyploid. *Nat. Rev. Genet.* 6, 836–846. doi: 10.1038/nrg1711

40. Conicella, C., Capo, A., Cammareri, M., Errico, A., Shamina, N., and Monti, L. M. (2003). Elucidation of meiotic nuclear restitution mechanisms in potato through analysis of microtubular cytoskeleton. *Euphytica*, 107–115. doi: 10.1023/A:1025636321757

41. Cooper, K. F., and Strich, R. (2011). Meiotic control of the APC/C: similarities & differences from mitosis. *Cell Div.* 6:16. doi: 10.1186/1747-1028-6-16

42. Corbett, K. D., and Harrison, S. C. (2012). Molecular Architecture of the Yeast Monopolin Complex. *Cell. Rep.* 1, 583–589. doi: 10.1016/j.celrep.2012.05.012

43. Couteau, F., Belzile, F., Horlow, C., Grandjean, O., Vezon, D., and Doutriaux, M. P. (1999). Random chromosome segregation without meiotic arrest in both male and female meiocytes of a dmc1 mutant of Arabidopsis. *Plant Cell* 11, 1623–1634. doi: 10.1105/tpc.11.9.1623

44. Crismani, W., and Mercier, R. (2012). What limits meiotic crossovers? *Cell Cycle* 11, 3527–3528. doi: 10.4161/cc.21963

45. Cromer, L., Jolivet, S., Horlow, C., Chelysheva, L., Heyman, J., De Jaeger, G., et al. (2013). Centromeric cohesion is protected twice at meiosis, by SHUGOSHINs at anaphase I and by PATRONUS at interkinesis. *Curr. Biol.* 23, 2090–2099. doi: 10.1016/j.cub.2013.08.036

46. Da Ines, O., Gallego, M. E., and White, C. I. (2014). Recombination-independent mechanisms and pairing of homologous chromosomes during meiosis in plants. *Mol. Plant* 7, 492–501. doi: 10.1093/mp/sst172

47. Dawe, R. K. (1998). Meiotic chromosome organization and segregation in plants. *Annu. Rev. Plant Physiol. Plant Mol. Biol.* 49, 371–395. doi: 10.1146/annurev.arplant.49.1.371

48. Dawe, R. K., Reed, L. M., Yu, H.-G., Muszynski, G., and Hiatt, E. N. (1999). A maize homolog of mammalian CENPC is a constitutive component of the inner kinetochore. *Plant Cell* 11, 1227–1238. doi: 10.1105/tpc.11.7.1227

49. Demidov, D., Van Damme, D., Geelen, D., Blattner, F. R., and Houben, A. (2005). Identification and dynamics of two classes of aurora-like kinases in arabidopsis and other plants. *Plant* 17, 836–848. doi: 10.1105/tpc.104.029710.1

50. De Muyt, A., Pereira, L., Vezon, D., Chelysheva, L., Gendrot, G., Chambon, A., et al. (2009). A high throughput genetic screen identifies new early meiotic recombination functions in Arabidopsis thaliana. *PLoS Genet.* 5:e1000654. doi: 10.1371/journal.pgen.1000654

51. De Oliveira, E. A. G., Romeiro, N. C., Ribeiro, E. D. S., Santa-Catarina, C., Oliveira, A. E. A., Silveira, V., et al. (2012). Structural and functional characterization of the protein kinase Mps1 in Arabidopsis thaliana. *PLoS ONE* 7:e45707. doi: 10.1371/journal.pone.0045707

52. D'Erfurth, I., Jolivet, S., Froger, N., Catrice, O., Novatchkova, M., Simon, M., et al. (2008). Mutations in AtPS1 (Arabidopsis thaliana parallel spindle 1) lead to the production of diploid pollen grains. *PLoS Genet.* 4:e1000274. doi: 10.1371/journal.pgen.1000274

53. De Rop, V., Padeganeh, A., and Maddox, P. S. (2012). CENP-A: the key player behind centromere identity, propagation, and kinetochore assembly.*Chromosoma* 121, 527–538. doi: 10.1007/s00412-012-0386-5

54. De Storme, N., and Geelen, D. (2011). The Arabidopsis mutant jason produces unreduced first division restitution male gametes through a parallel/fused spindle mechanism in meiosis II. *Plant Physiol.* 155, 1403–1415. doi: 10.1104/pp.110.170415

55. De Storme, N., and Geelen, D. (2013a). Cytokinesis in plant male meiosis. *Plant Signal. Behav.* 8:e23394. doi: 10.4161/psb.23394

56. De Storme, N., and Geelen, D. (2013b). Sexual polyploidization in plants—cytological mechanisms and molecular regulation. *New Phytol.* 198, 670–684. doi: 10.1111/nph.12184

57. De Storme, N., and Geelen, D. (2014). The impact of environmental stress on male reproductive development in plants: biological processes and molecular mechanisms. *Plant. Cell Environ.* 37, 1–18. doi: 10.1111/pce.12142

58. Ding, D., Muthuswamy, S., and Meier, I. (2012). Functional interaction between the Arabidopsis orthologs of spindle assembly checkpoint proteins MAD1 and MAD2 and the nucleoporin NUA. *Plant Mol. Biol.* 79, 203–216. doi: 10.1007/s11103-012-9903-4

59. Drouaud, J., Camilleri, C., Bourguignon, P.-Y., Canaguier, A., Bérard, A., Vezon, D., et al. (2006). Variation in crossing-over rates across chromosome 4 of Arabidopsis thaliana reveals the presence of meiotic recombination "hot spots." *Genome Res.* 16, 106–114. doi: 10.1101/gr.4319006

60. Du, Y., and Dawe, R. K. (2007). Maize NDC80 is a constitutive feature of the central kinetochore. *Chromosome Res.* 15, 767–775. doi: 10.1007/s10577-007-1160-z

61. Eckert, C. A., Gravdahl, D. J., and Megee, P. C. (2007). The enhancement of pericentromeric cohesin association by conserved kinetochore components promotes high-fidelity chromosome segregation and is sensitive to microtubule-based tension. *Genes Dev.* 21, 278–291. doi: 10.1101/gad.1498707.somes

62. Edlinger, B., and Schlögelhofer, P. (2011). Have a break: determinants of meiotic DNA double strand break (DSB) formation and processing in plants. *J. Exp. Bot.* 62, 1545–1563. doi: 10.1093/jxb/erq421

63. Eijpe, M., Heyting, C., Gross, B., and Jessberger, R. (2000). Association of mammalian SMC1 and SMC3 proteins with meiotic chromosomes and synaptonemal complexes. *J. Cell Sci.* 113(Pt 4), 673–682.

64. Ekwall, K. (2007). Epigenetic control of centromere behavior. *Annu. Rev. Genet.* 41, 63–81. doi: 10.1146/annurev.genet.41.110306.130127

65. Endow, S. A. (1999). Microtubule motors in spindle and chromosome motility. *Eur. J. Biochem.* 262, 12–18. doi: 10.1046/j.1432-1327.1999.00339.x

66. Erilova, A., Brownfield, L., Exner, V., Rosa, M., Twell, D., Mittelsten Scheid, O., et al. (2009). Imprinting of the polycomb group gene MEDEA serves as a ploidy sensor in Arabidopsis. *PLoS Genet.* 5:e1000663. doi: 10.1371/journal.pgen.1000663

67. Fernius, J., Nerusheva, O. O., Galander, S., Alves, F. D. L., Rappsilber, J., and Marston, A. L. (2013). Cohesin-dependent association of scc2/4 with the centromere initiates pericentromeric cohesion establishment. *Curr. Biol.* 23, 599–606. doi: 10.1016/j.cub.2013.02.022

68. Golubovskaya, I. N., Hamant, O., Timofejeva, L., Wang, C.-J. R., Braun, D., Meeley, R., et al. (2006). Alleles of afd1 dissect REC8 functions during meiotic prophase I. *J. Cell Sci.* 119, 3306–3315. doi: 10.1242/jcs.03054

69. Golubovskaya, I. N., Harper, L. C., Pawlowski, W. P., Schichnes, D., and Cande, W. Z. (2002). The pam1 gene is required for meiotic bouquet formation and efficient. *Genetics* 1993, 1979–1993.

70. Gómez, R., Valdeolmillos, A., Parra, M. T., Viera, A., Carreiro, C., Roncal, F., et al. (2007). Mammalian SGO2 appears at the inner centromere domain and redistributes depending on tension across centromeres during meiosis II and mitosis. *EMBO Rep.* 8, 173–180. doi: 10.1038/sj.embor.7400877

71. Gore, M. A., Chia, J. M., Elshire, R. J., Sun, Q., Ersoz, E. S., Hurwitz, B. L., et al. (2009). A first-generation haplotype map of maize. *Science* 326, 1115–1117. doi: 10.1126/science.1177837

72. Graumann, K., Runions, J., and Evans, D. E. (2010). Characterization of SUN-domain proteins at the higher plant nuclear envelope. *Plant J.* 61, 134–144. doi: 10.1111/j.1365-313X.2009.04038.x

73. Gregson, H. C., Schmiesing, J. A., Kim, J. S., Kobayashi, T., Zhou, S., and Yokomori, K. (2001). A potential role for human cohesin in mitotic spindle aster assembly. *J. Biol. Chem.* 276, 47575–47582. doi: 10.1074/jbc.M103364200

74. Gruber, S., Haering, C. H., and Nasmyth, K. (2003). Chromosomal cohesin forms a ring. *Cell* 112, 765–777. doi: 10.1016/S0092-8674(03)00162-4

75. Gutiérrez-Caballero, C., Cebollero, L. R., and Pendás, A. M. (2012). Shugoshins: from protectors of cohesion to versatile adaptors at the centromere.*Trends Genet.* 28, 351–360. doi: 10.1016/j.tig.2012.03.003

76. Haering, C. H., Farcas, A.-M., Arumugam, P., Metson, J., and Nasmyth, K. (2008). The cohesin ring concatenates sister DNA molecules. *Nature* 454, 297–301. doi: 10.1038/nature07098

77. Hamant, O., Golubovskaya, I., Meeley, R., Fiume, E., Timofejeva, L., Schleiffer, A., et al. (2005). A REC8-dependent plant Shugoshin is required for maintenance of centromeric cohesion during meiosis and has no mitotic functions. *Curr. Biol.* 15, 948–954. doi: 10.1016/j.cub.2005.04.049

78. Hamant, O., Ma, H., and Cande, W. Z. (2006). Genetics of meiotic prophase I in plants. *Annu. Rev. Plant Biol.* 57, 267–302. doi: 10.1146/annurev.arplant.57.032905.105255

79. Harper, J. W., Burton, J. L., and Solomon, M. J. (2002). The anaphase-promoting complex: it's not just for mitosis any more. *Genes Dev.* 16, 2179–2206. doi: 10.1101/gad.1013102

80. Harper, L., Golubovskaya, I., and Cande, W. Z. (2004). A bouquet of chromosomes. *J. Cell Sci.* 117, 4025–4032. doi: 10.1242/jcs.01363

81. Hayashi, T., Fujita, Y., Iwasaki, O., Adachi, Y., Takahashi, K., and Yanagida, M. (2004). Mis16 and Mis18 are required for CENP-A loading and histone deacetylation at centromeres. *Cell* 118, 715–729. doi: 10.1016/j.cell.2004.09.002

82. Henry, I. M., Dilkes, B. P., Miller, E. S., Burkart-Waco, D., and Comai, L. (2010). Phenotypic consequences of aneuploidy in Arabidopsis thaliana. *Genetics* 186, 1231–1245. doi: 10.1534/genetics.110.121079

83. Henry, I. M., Dilkes, B. P., Young, K., Watson, B., Wu, H., and Comai, L. (2005). Aneuploidy and genetic variation in the Arabidopsis thaliana triploid response. *Genetics* 170, 1979–1988. doi: 10.1534/genetics.104.037788

84. Hewitt, L., Tighe, A., Santaguida, S., White, A. M., Jones, C. D., Musacchio, A., et al. (2010). Sustained Mps1 activity is required in mitosis to recruit O-Mad2 to the Mad1-C-Mad2 core complex. *J. Cell Biol.* 190, 25–34. doi: 10.1083/jcb.201002133

85. Higgins, J. D., Armstrong, S. J., Franklin, F. C. H., and Jones, G. H. (2004). The Arabidopsis MutS homolog AtMSH4 functions at an early step in recombination: evidence for two classes of recombination in Arabidopsis. *Genes Dev.* 18, 2557–2570. doi: 10.1101/gad.317504

86. Higgins, J. D., Sanchez-Moran, E., Armstrong, S. J., Jones, G. H., and Franklin, F. C. H. (2005). The Arabidopsis synaptonemal complex protein ZYP1 is required for chromosome synapsis and normal fidelity of crossing over. *Genes Dev.* 19, 2488–2500. doi: 10.1101/gad.354705

87. Higgins, J. D., Vignard, J., Mercier, R., Pugh, A. G., Franklin, F. C. H., and Jones, G. H. (2008). AtMSH5 partners AtMSH4 in the class I meiotic crossover pathway in Arabidopsis thaliana, but is not required for synapsis. *Plant J.* 55, 28–39. doi: 10.1111/j.1365-313X.2008.03470.x

88. Houben, A., and Schubert, I. (2003). DNA and proteins of plant centromeres. *Curr. Opin. Plant Biol.* 6, 554–560. doi: 10.1016/j.pbi.2003.09.007

89. Huang, H., Feng, J., Famulski, J., Rattner, J. B., Liu, S. T., Kao, G. D., et al. (2007). Tripin/hSgo2 recruits MCAK to the inner centromere to correct defective kinetochore attachments. *J. Cell Biol.* 177, 413–424. doi: 10.1083/jcb.200701122

90. Jaspersen, S. L., and Winey, M. (2004). The budding yeast spindle pole body: structure, duplication, and function. *Annu. Rev. Cell Dev. Biol.* 20, 1–28. doi: 10.1146/annurev.cellbio.20.022003.114106

91. Jensen, S., Segal, M., Clarke, D. J., and Reed, S. I. (2001). A novel role of the budding yeast separin Esp1 in anaphase spindle elongation: evidence that proper spindle association of Esp1 is regulated by Pds1. *J. Cell Biol.* 152, 27–40. doi: 10.1083/jcb.152.1.27

92. Jiang, H., Wang, F.-F., Wu, Y.-T., Zhou, X., Huang, X.-Y., Zhu, J., et al. (2009). MULTIPOLAR SPINDLE 1 (MPS1), a novel coiled-coil protein of Arabidopsis thaliana, is required for meiotic spindle organization. *Plant J.* 59, 1001–1010. doi: 10.1111/j.1365-313X.2009.03929.x

93. Jiang, L., Yuan, L., Xia, M., and Makaroff, C. A. (2010). Proper levels of the Arabidopsis cohesion establishment factor CTF7 are essential for embryo and megagametophyte, but not endosperm, development. *Plant Physiol.* 154, 820–832. doi: 10.1104/pp.110.157560

94. Jones, G. H. (1984). The control of chiasma distribution. *Symp. Soc. Exp. Biol.* 38, 293–320.

95. Juranič, M., Srilunchang, K., Krohn, N. G., Leljak-Levanic, D., Sprunck, S., and Dresselhaus, T. (2012). Germline-specific MATH-BTB substrate adaptor MAB1 regulates spindle length and nuclei identity in maize. *Plant Cell* 24, 4974–4991. doi: 10.1105/tpc.112.107169

96. Kawabe, A., Matsunaga, S., Nakagawa, K., Kurihara, D., Yoneda, A., Hasezawa, S., et al. (2005). Characterization of plant Aurora kinases during mitosis. *Plant Mol. Biol.* 58, 1–13. doi: 10.1007/s11103-005-3454-x

97. Kawashima, S. A., Tsukahara, T., Langegger, M., Hauf, S., Kitajima, T. S., and Watanabe, Y. (2007). Shugoshin enables tension-generating

attachment of kinetochores by loading Aurora to centromeres. *Genes Dev.* 21, 420–435. doi: 10.1101/gad.1497307

98. Keeney, S., Giroux, C. N., and Kleckner, N. (1997). Meiosis-specific DNA double-strand breaks are catalyzed by Spo11, a member of a widely conserved protein family. *Cell* 88, 375–384. doi: 10.1016/S0092-8674(00)81876-0

99. Kerrebrock, A. W., Moore, D. P., Wu, J. S., and Orr-Weaver, T. L. (1995). Mei-S332, a Drosophila protein required for sister-chromatid cohesion, can localize to meiotic centromere regions. *Cell* 83, 247–256. doi: 10.1016/0092-8674(95)90166-3

100. Kimbara, J., Endo, T. R., and Nasuda, S. (2004). Characterization of the genes encoding for MAD2 homologues in wheat. *Chromosome Res.* 12, 703–714. doi: 10.1023/B:CHRO.0000045760.42880.8c

101. Kirioukhova, O., Johnston, A. J., Kleen, D., Kägi, C., Baskar, R., Moore, J. M., et al. (2011). Female gametophytic cell specification and seed development require the function of the putative Arabidopsis INCENP ortholog WYRD. *Development* 138, 3409–3420. doi: 10.1242/dev.060384

102. Kitajima, T. S., Kawashima, S. A., and Watanabe, Y. (2004). The conserved kinetochore protein shugoshin protects centromeric cohesion during meiosis.*Nature* 427, 510–517. doi: 10.1038/nature02312

103. Kitajima, T. S., Miyazaki, Y., Yamamoto, M., and Watanabe, Y. (2003). Rec8 cleavage by separase is required for meiotic nuclear divisions in fission yeast.*EMBO J.* 22, 5643–5653. doi: 10.1093/emboj/cdg527

104. Klein, F., Mahr, P., Galova, M., Buonomo, S. B., Michaelis, C., Nairz, K., et al. (1999). A central role for cohesins in sister chromatid cohesion, formation of axial elements, and recombination during yeast meiosis. *Cell* 98, 91–103. doi: 10.1016/S0092-8674(00)80609-1

105. Klutstein, M., and Cooper, J. P. (2014). The chromosomal courtship dance-homolog pairing in early meiosis. *Curr. Opin. Cell Biol.* 26, 123–131. doi: 10.1016/j.ceb.2013.12.004

106. Koszul, R., Kim, K. P., Prentiss, M., Kleckner, N., and Kameoka, S. (2009). Actin-mediated motion of meiotic chromosomes. *NIH* 133, 1188–1201. doi: 10.1016/j.cell.2008.04.050

107. Kudo, N. R., Anger, M., Peters, A. H., Stemmann, O., Theussl, H.-C., Helmhart, W., et al. (2009). Role of cleavage by separase of the Rec8 kleisin subunit of cohesin during mammalian meiosis I. *J. Cell Sci.* 122, 2686–2698. doi: 10.1242/jcs.035287

108. Kurzbauer, M.-T., Uanschou, C., Chen, D., and Schlögelhofer, P. (2012). The recombinases DMC1 and RAD51 are functionally and spatially separated during meiosis in Arabidopsis. *Plant Cell* 24, 2058–2070. doi: 10.1105/tpc.112.098459

109. Lam, W. S., Yang, X., and Makaroff, C. A. (2005). Characterization of Arabidopsis thaliana SMC1 and SMC3: evidence that AtSMC3 may function beyond chromosome cohesion. *J. Cell Sci.* 118, 3037–3048. doi: 10.1242/jcs.02443

110. Lampert, F., and Westermann, S. (2011). A blueprint for kinetochores— new insights into the molecular mechanics of cell division. *Nat. Rev. Mol. Cell. Biol.* 12, 407–412. doi: 10.1038/nrm3133

111. Lee, J., Kitajima, T. S., Tanno, Y., Yoshida, K., Morita, T., Miyano, T., et al. (2008a). Unified mode of centromeric protection by shugoshin in mammalian oocytes and somatic cells. *Nat. Cell Biol.* 10, 42–52. doi: 10.1038/ncb1667

112. Lee, S. H., Sterling, H., Burlingame, A., and McCormick, F. (2008b). Tpr directly binds to Mad1 and Mad2 and is important for the Mad1-Mad2-mediated mitotic spindle checkpoint. *Genes Dev.* 22, 2926–2931. doi: 10.1101/gad.1677208

113. Lermontova, I., Koroleva, O., Rutten, T., Fuchs, J., Schubert, V., Moraes, I., et al. (2011). Knockdown of CENH3 in Arabidopsis reduces mitotic divisions and causes sterility by disturbed meiotic chromosome segregation. *Plant J.* 68, 40–50. doi: 10.1111/j.1365-313X.2011.04664.x

114. Lermontova, I., Kuhlmann, M., Friedel, S., Rutten, T., Heckmann, S., Sandmann, M., et al. (2013). Arabidopsis KINETOCHORE NULL2 is an upstream component for centromeric histone H3 variant cenH3 deposition at centromeres. *Plant Cell* 25, 3389–3404. doi: 10.1105/tpc.113.114736

115. Lermontova, I., Schubert, V., Fuchs, J., Klatte, S., Macas, J., and Schubert, I. (2006). Loading of Arabidopsis centromeric histone CENH3 occurs mainly during G2 and requires the presence of the histone fold domain. *Plant Cell* 18, 2443–2451. doi: 10.1105/tpc.106.043174

116. Lhuissier, F. G. P., Offenberg, H. H., Wittich, P. E., Vischer, N. O. E., and Heyting, C. (2007). The mismatch repair protein MLH1 marks a subset of strongly interfering crossovers in tomato. *Plant Cell* 19, 862–876. doi: 10.1105/tpc.106.049106

117. Li, J., Lee, G. I., Van Doren, S. R., and Walker, J. C. (2000). The FHA domain mediates phosphoprotein interactions. *J. Cell Sci.* 113(Pt 23), 4143–4149.

118. Li, W., Chen, C., Markmann-Mulisch, U., Timofejeva, L., Schmelzer, E., Ma, H., et al. (2004). The Arabidopsis AtRAD51 gene is dispensable for vegetative development but required for meiosis. *Proc. Natl. Acad. Sci. U.S.A.* 101, 10596–10601. doi: 10.1073/pnas.0404110101

119. Li, X., and Dawe, R. K. (2009). Fused sister kinetochores initiate the reductional division in meiosis I. *Nat. Cell Biol.* 11, 1103–1108. doi: 10.1038/ncb1923

120. Li, Y., Shen, Y., Cai, C., Zhong, C., Zhu, L., Yuan, M., et al. (2010). The type II Arabidopsis formin14 interacts with microtubules and microfilaments to regulate cell division. *Plant Cell* 22, 2710–2726. doi: 10.1105/tpc.110.075507

121. Liu, B., Cyr, R. J., and Palevitz, B. A. (1996). A kinesin-like protein, KatAp, in the cells of arabidopsis and other plants. *Plant Cell* 8, 119–132. doi: 10.1105/tpc.8.1.119

122. Liu, C. M., McElver, J., Tzafrir, I., Joosen, R., Wittich, P., Patton, D., et al. (2002). Condensin and cohesin knockouts in Arabidopsis exhibit a titan seed phenotype. *Plant J.* 29, 405–415. doi: 10.1046/j.1365-313x.2002.01224.x

123. Liu, Z., and Makaroff, C. A. (2006). Arabidopsis separase AESP is essential for embryo development and the release of cohesin during meiosis. *Plant Cell*18, 1213–1225. doi: 10.1105/tpc.105.036913.2001

124. Ma, H. (2006). A molecular portrait of Arabidopsis meiosis. *Arabidopsis Book* 4:e0095. doi: 10.1199/tab.0095

125. Madlung, A., and Wendel, J. F. (2013). Genetic and epigenetic aspects of polyploid evolution in plants. *Cytogenet. Genome Res.* 140, 270–285. doi: 10.1159/000351430

126. Malmanche, N., Maia, A., and Sunkel, C. E. (2006). The spindle assembly checkpoint: preventing chromosome mis-segregation during mitosis and meiosis. *FEBS Lett.* 580, 2888–2895. doi: 10.1016/j.febslet.2006.03.081

127. Marcus, A. I., Ambrose, J. C., Blickley, L., Hancock, W. O., and Cyr, R. J. (2002). Arabidopsis thaliana protein, ATK1, is a minus-end directed kinesin that exhibits non-processive movement. *Cell Motil. Cytoskeleton* 52, 144–150. doi: 10.1002/cm.10045

128. Marston, A. L. (2014). Chromosome segregation in budding yeast: sister chromatid cohesion and related Mechanisms. *Genetics* 196, 31–63. doi: 10.1534/genetics.112.145144

129. Masson, J. Y., and West, S. C. (2001). The Rad51 and Dmc1 recombinases: a non-identical twin relationship. *Trends Biochem. Sci.* 26, 131–136. doi: 10.1016/S0968-0004(00)01742-4

130. May, K. M., and Hardwick, K. G. (2006). The spindle checkpoint. *J. Cell Sci.* 119, 4139–4142. doi: 10.1242/jcs.03165

131. McGuinness, B. E., Hirota, T., Kudo, N. R., Peters, J.-M., and Nasmyth, K. (2005). Shugoshin prevents dissociation of cohesin from centromeres during mitosis in vertebrate cells. *PLoS Biol.* 3:e86. doi: 10.1371/journal.pbio.0030086

132. Mercier, R., Armstrong, S. J., Horlow, C., Kackson, N. P., Makaroff, C. A., Vezon, D., et al. (2003). The meiotic protein SWI1 is required for axial element formation and recombination initiation in Arabidopsis. *Development* 130, 3309–3318. doi: 10.1242/dev.00550

133. Mercier, R., and Grelon, M. (2008). Meiosis in plants: ten years of gene discovery. *Cytogenet. Genome Res.* 120, 281–290. doi: 10.1159/000121077

134. Mercier, R., Jolivet, S., Vezon, D., Huppe, E., Chelysheva, L., Giovanni, M., et al. (2005). Two meiotic crossover classes cohabit in Arabidopsis: one is dependent on MER3,whereas the other one is not. *Curr. Biol.* 15, 692–701. doi: 10.1016/j.cub.2005.02.056

135. Mercier, R., Vezon, D., Bullier, E., Motamayor, J. C., Sellier, A., Lefèvre, F., et al. (2001). SWITCH1 (SWI1): a novel protein required for the establishment of sister chromatid cohesion and for bivalent formation at meiosis. *Genes Dev.* 15, 1859–1871. doi: 10.1101/gad.203201

136. Miki, F., Kurabayashi, A., Tange, Y., Okazaki, K., Shimanuki, M., and Niwa, O. (2004). Two-hybrid search for proteins that interact with Sad1 and Kms1, two membrane-bound components of the spindle pole body in fission yeast. *Mol. Genet. Genomics* 270, 449–471. doi: 10.1007/s00438-003-0938-8

137. Moraes, I. C. R., Lermontova, I., and Schubert, I. (2011). Recognition of *A.* thaliana centromeres by heterologous CENH3 requires high similarity to the endogenous protein. *Plant Mol. Biol.* 75, 253–261. doi: 10.1007/s11103-010-9723-3

138. Moschou, P. N., and Bozhkov, P. V (2012). Separases: biochemistry and function. *Physiol. Plant.* 145, 67–76. doi: 10.1111/j.1399-3054.2011.01550.x

139. Motamayor, J. C., Vezon, D., Bajon, C., Sauvanet, A., Grandjean, O., Marchand, M., et al. (2000). Switch (swi1), an Arabidopsis thaliana mutant affected in the female meiotic switch. *Sex. Plant Reprod.* 12, 209–218. doi: 10.1007/s004970050002

140. Murata, M. (2013). "Arabidopsis centromeres," in *Plant Centromere Biology*, eds J. Jiang and J. A. Birchler (Oxford, UK: Wiley-Blackwell). doi: 10.1002/9781118525715.ch1

141. Murphy, S. P., Simmons, C. R., and Bass, H. W. (2010). Structure and expression of the maize (Zea mays L.) SUN-domain protein gene family: evidence for the existence of two divergent classes of SUN proteins in plants. *BMC Plant Biol.* 10:269. doi: 10.1186/1471-2229-10-269

142. Murray, A. (1994). Cell cycle checkpoints. *Curr. Opin. Cell Biol.* 6, 872–876. doi: 10.1016/0955-0674(94)90059-0

143. Musacchio, A. (2011). Spindle assembly checkpoint: the third decade. *Philos. Trans. R. Soc. Lond. B. Biol. Sci.* 366, 3595–604. doi: 10.1098/rstb.2011.0072

144. Nasmyth, K. (2001). Disseminating the genome: joining, resolving, and separating sister chromatids during mitosis and meiosis. *Annu. Rev. Genet.* 35, 673–745. doi: 10.1146/annurev.genet.35.102401.091334

145. Nasmyth, K., and Haering, C. H. (2005). The structure and function of SMC and kleisin complexes. *Annu. Rev. Biochem.* 74, 595–648. doi: 10.1146/annurev.biochem.74.082803.133219

146. Nicklas, R. B., Ward, S. C., and Gorbsky, G. J. (1995). Kinetochore chemistry is sensitive to tension and may link mitotic forces to a cell cycle checkpoint.*J. Cell Biol.* 130, 929–939. doi: 10.1083/jcb.130.4.929

147. Niu, H., Wan, L., Busygina, V., Kwon, Y., Allen, J. A., Li, X., et al. (2009). Regulation of meiotic recombination via Mek1-mediated Rad54 phosphorylation. *Mol. Cell* 36, 393–404. doi: 10.1016/j.molcel.2009.09.029

148. Nogales, E. (2000). Structural insights into microtubule function. *Annu. Rev. Biochem.* 69, 277–302. doi: 10.1146/annurev.biochem.69.1.277

149. Ocampo-Hafalla, M. T., and Uhlmann, F. (2011). Cohesin loading and sliding. *J. Cell Sci.* 124, 685–691. doi: 10.1242/jcs.073866

150. Ogura, Y., Shibata, F., Sato, H., and Murata, M. (2004). Characterization of a CENP-C homolog in Arabidopsis thaliana. *Genes Genet. Syst.* 79, 139–144. doi: 10.1266/ggs.79.139

151. Onn, I., Heidinger-Pauli, J. M., Guacci, V., Unal, E., and Koshland, D. E. (2008). Sister chromatid cohesion: a simple concept with a complex reality.*Annu. Rev. Cell Dev. Biol.* 24, 105–129. doi: 10.1146/annurev.cellbio.24.110707.175350

152. Osman, K., Higgins, J. D., Sanchez-Moran, E., Armstrong, S. J., and Franklin, F. C. H. (2011). Pathways to meiotic recombination in

Arabidopsis thaliana.*New Phytol.* 190, 523–544. doi: 10.1111/j.1469-8137.2011.03665.x

153. O'Toole, E., Greenan, G., Lange, K. I., Srayko, M., and Müller-Reichert, T. (2012). The role of γ-tubulin in centrosomal microtubule organization. *PLoS ONE* 7:e29795. doi: 10.1371/journal.pone.0029795

154. Otto, S. P. (2007). The evolutionary consequences of polyploidy. *Cell* 131, 452–462. doi: 10.1016/j.cell.2007.10.022

155. Pastuglia, M., Azimzadeh, J., Magali, G., Christine, C., Belcram, K., Evrard, J., et al. (2006). g-Tubulin is essential for microtubule organization and development in Arabidopsis. *Plant Cell* 18, 1412–1425. doi: 10.1105/tpc.105.039644.2

156. Pawlowski, W. P., and Cande, W. Z. (2005). Coordinating the events of the meiotic prophase. *Trends Cell Biol.* 15, 674–681. doi: 10.1016/j.tcb.2005.10.005

157. Pawlowski, W. P., Wang, C.-J. R., Golubovskaya, I. N., Szymaniak, J. M., Shi, L., Hamant, O., et al. (2009). Maize AMEIOTIC1 is essential for multiple early meiotic processes and likely required for the initiation of meiosis. *Proc. Natl. Acad. Sci. U.S.A.* 106, 3603–3608. doi: 10.1073/pnas.0810115106

158. Penkner, A. M., Fridkin, A., Gloggnitzer, J., Baudrimont, A., Machacek, T., Woglar, A., et al. (2009). Meiotic chromosome homology search involves modifications of the nuclear envelope protein Matefin/SUN-1. *Cell* 139, 920–933. doi: 10.1016/j.cell.2009.10.045

159. Pereira, G., and Schiebel, E. (1997). Centrosome-microtubule nucleation. *J. Cell Sci.* 110(Pt 3), 295–300.

160. Peters, J.-M. (2006). The anaphase promoting complex/cyclosome: a machine designed to destroy. *Nat. Rev. Mol. Cell Biol.* 7, 644–656. doi: 10.1038/nrm1988

161. Peters, J.-M., Tedeschi, A., and Schmitz, J. (2008). The cohesin complex and its roles in chromosome biology. *Genes Dev.* 22, 3089–3114. doi: 10.1101/gad.1724308

162. Petroski, M. D., and Deshaies, R. J. (2005). Function and regulation of cullin-RING ubiquitin ligases. *Nat. Rev. Mol. Cell Biol.* 6, 9–20. doi: 10.1038/nrm1547

163. Phillips, C. M., and Dernburg, A. F. (2006). A family of zinc-finger proteins is required for chromosome-specific pairing and synapsis during meiosis in *C. elegans*. *Dev. Cell* 11, 817–829. doi: 10.1016/j.devcel.2006.09.020

164. Pidoux, A. L., and Allshire, R. C. (2005). The role of heterochromatin in centromere function. *Philos. Trans. R. Soc. Lond. B. Biol. Sci.* 360, 569–579. doi: 10.1098/rstb.2004.1611

165. Pradillo, M., López, E., Linacero, R., Romero, C., Cuñado, N., Sánchez-Morán, E., et al. (2012). Together yes, but not coupled: new insights into the roles of RAD51 and DMC1 in plant meiotic recombination. *Plant J.* 69, 921–933. doi: 10.1111/j.1365-313X.2011.04845.x

166. Quan, L., Xiao, R., Li, W., Oh, S.-A., Kong, H., Ambrose, J. C., et al. (2008). Functional divergence of the duplicated AtKIN14a and AtKIN14b genes: critical roles in Arabidopsis meiosis and gametophyte development. *Plant J.* 53, 1013–1026. doi: 10.1111/j.1365-313X.2007.03391.x

167. Ramsey, J., and Schemske, D. W. (1998). Pathways, mechanisms, and rates of polyploid formation in flowering plants. *Annu. Rev. Ecol. Syst.* 29, 467–501.

168. Ramsey, J., and Schemske, D. W. (2002). Neopolyploidy in flowering plants. *Annu. Rev. Ecol. Syst.* 33, 589–639. doi: 10.1146/annurev. ecolsys.33.010802.150437

169. Ravi, M., Kwong, P. N., Menorca, R. M. G., Valencia, J. T., Ramahi, J. S., Stewart, J. L., et al. (2010). The rapidly evolving centromere-specific histone has stringent functional requirements in Arabidopsis thaliana. *Genetics* 186, 461–471. doi: 10.1534/genetics.110.120337

170. Ravi, M., Shibata, F., Ramahi, J. S., Nagaki, K., Chen, C., Murata, M., et al. (2011). Meiosis-specific loading of the centromere-specific histone CENH3 in Arabidopsis thaliana. *PLoS Genet.* 7:e1002121. doi: 10.1371/ journal.pgen.1002121

171. Reddy, A. S. N., and Day, I. S. (2001). Kinesins in the Arabidopsis genome?: a comparative analysis among eukaryotes. *BMC Genomics* 2:2. doi: 10.1186/1471-2164-2-2

172. Richards, D. M., Greer, E., Martin, A. C., Moore, G., Shaw, P. J., and Howard, M. (2012). Quantitative dynamics of telomere bouquet formation. *PLoS Comput. Biol.* 8:e1002812. doi: 10.1371/journal. pcbi.1002812

173. Ross, K. J., Fransz, P., Armstrong, S. J., Vizir, I., Mulligan, B., Franklin, F. C., et al. (1997). Cytological characterization of four meiotic mutants of Arabidopsis isolated from T-DNA-transformed lines. *Chromosome Res.* 5, 551–559. doi: 10.1023/A:1018497804129

174. Rowland, B. D., Roig, M. B., Nishino, T., Kurze, A., Uluocak, P., Mishra, A., et al. (2009). Building sister chromatid cohesion: smc3 acetylation

counteracts an antiestablishment activity. *Mol. Cell* 33, 763–774. doi: 10.1016/j.molcel.2009.02.028

175. division. *Nat. Rev. Mol. Cell Biol.* 8, 798–812. doi: 10.1038/nrm2257

176. Sakuno, T., Tada, K., and Watanabe, Y. (2009). Kinetochore geometry defined by cohesion within the centromere. *Nature* 458, 852–858. doi: 10.1038/nature07876

177. Salic, A., Waters, J. C., and Mitchison, T. J. (2004). Vertebrate shugoshin links sister centromere cohesion and kinetochore microtubule stability in mitosis. *Cell* 118, 567–578. doi: 10.1016/j.cell.2004.08.016

178. Sanchez-Moran, E., Osman, K., Higgins, J. D., Pradillo, M., Cuñado, N., Jones, G. H., et al. (2008). ASY1 coordinates early events in the plant meiotic recombination pathway. *Cytogenet. Genome Res.* 120, 302–312. doi: 10.1159/000121079

179. Santaguida, S., and Musacchio, A. (2009). The life and miracles of kinetochores. *EMBO J.* 28, 2511–2531. doi: 10.1038/emboj.2009.173

180. Sato, A., Isaac, B., Phillips, C. M., Rillo, R., Carlton, P. M., Wynne, D. J., et al. (2009). Cytoskeletal forces span the nuclear envelope to coordinate meiotic chromosome pairing and synapsis. *Cell* 139, 907–919. doi: 10.1016/j.cell.2009.10.039

181. Sato, H., Shibata, F., and Murata, M. (2005). Characterization of a Mis12 homologue in Arabidopsis thaliana. *Chromosome Res.* 13, 827–834. doi: 10.1007/s10577-005-1016-3

182. Schmitt, J., Benavente, R., Hodzic, D., Hoog, C., Stewart, C. L., and Alsheimer, M. (2007). Transmembrane protein Sun2 is involved in tethering mammalian meiotic telomeres to the nuclear envelope. *Proc. Natl. Acad. Sci. U.S.A.* 104, 7426–7431. doi: 10.1073/pnas.0609198104

183. Schubert, V., Kim, Y.-M., Berr, A., Fuchs, J., Meister, A., Marschner, S., et al. (2007). Random homologous pairing and incomplete sister chromatid alignment are common in angiosperm interphase nuclei. *Mol. Genet. Genomics* 278, 167–176. doi: 10.1007/s00438-007-0242-0

184. Schubert, V., Weissleder, A., Ali, H., Fuchs, J., Lermontova, I., Meister, A., et al. (2009a). Cohesin gene defects may impair sister chromatid alignment and genome stability in Arabidopsis thaliana. *Chromosoma* 118, 591–605. doi: 10.1007/s00412-009-0220-x

185. Schubert, V., Weissleder, A., Ali, H., Fuchs, J., Lermontova, I., Meister, A., et al. (2009b). Cohesin gene defects may impair sister chromatid alignment and genome stability in Arabidopsis thaliana. *Chromosoma* 118, 591–605. doi: 10.1007/s00412-009-0220-x

186. Sebastian, J., Ravi, M., Andreuzza, S., Panoli, A. P., Marimuthu, M. P. A., and Siddiqi, I. (2009). The plant adherin AtSCC2 is required for embryogenesis and sister-chromatid cohesion during meiosis in Arabidopsis. *Plant J.* 59, 1–13. doi: 10.1111/j.1365-313X.2009.03845.x

187. Shao, T., Tang, D., Wang, K., Wang, M., Che, L., Qin, B., et al. (2011). OsREC8 is essential for chromatid cohesion and metaphase I monopolar orientation in rice meiosis. *Plant Physiol.* 156, 1386–1396. doi: 10.1104/pp.111.177428

188. Sharp, D. J., Rogers, G. C., and Scholey, J. M. (2000). Microtubule motors in mitosis. *Nature* 407, 41–47. doi: 10.1038/35024000

189. Sharp, D. J., Yu, K. R., Sisson, J. C., Sullivan, W., and Scholey, J. M. (1999). Antagonistic microtubule-sliding motors position mitotic centrosomes in*Drosophila* early embryos. *Nat. Cell Biol.* 1, 51–54. doi: 10.1038/9025

190. Sheehan, M. J., and Pawlowski, W. P. (2009). Live imaging of rapid chromosome movements in meiotic prophase I in maize. *Proc. Natl. Acad. Sci. U.S.A.* 106, 20989–20994. doi: 10.1073/pnas.0906498106

191. Shimamura, M., Brown, R. C., Lemmon, B. E., Akashi, T., Mizuno, K., Nishihara, N., et al. (2004). γ-Tubulin in basal land plants?: characterization, localization, and implication in the evolution of acentriolar microtubule. *Plant Cell* 16, 45–59. doi: 10.1105/tpc.016501. bules

192. Siddiqi, I., Ganesh, G., Grossniklaus, U., and Subbiah, V. (2000). The dyad gene is required for progression through female meiosis in Arabidopsis.*Development* 127, 197–207.

193. Siderakis, M., and Tarsounas, M. (2007). Telomere regulation and function during meiosis. *Chromosome Res.* 15, 667–679. doi: 10.1007/s10577-007-1149-7

194. Singh, D. K., Andreuzza, S., Panoli, A. P., and Siddiqi, I. (2013). AtCTF7 is required for establishment of sister chromatid cohesion and association of cohesin with chromatin during meiosis in Arabidopsis. *BMC Plant Biol.* 13:117. doi: 10.1186/1471-2229-13-117

195. Staiger, C. J., and Cande, W. Z. (1991). Microfilament distribution in maize meiotic mutants correlates with microtubule organization. *Plant Cell* 3, 637–644. doi: 10.1105/tpc.3.6.637

196. Stoop-Myer, C., and Amon, A. (1999). Meiosis: rec8 is the reason for cohesion. *Nat. Cell Biol.* 1, E125–E127. doi: 10.1038/12956

197. Sudakin, V., Chan, G. K., and Yen, T. J. (2001). Checkpoint inhibition of the APC/C in HeLa cells is mediated by a complex of BUBR1,

BUB3, CDC20, and MAD2. *J. Cell Biol.* 154, 925–936. doi: 10.1083/jcb.200102093

198. Sumara, I., Vorlaufer, E., Stukenberg, P. T., Kelm, O., Redemann, N., Nigg, E. A., et al. (2002). The dissociation of cohesin from chromosomes in prophase is regulated by Polo-like kinase. *Mol. Cell* 9, 515–525. doi: 10.1016/S1097-2765(02)00473-2

199. Sun, S.-C., and Kim, N.-H. (2012). Spindle assembly checkpoint and its regulators in meiosis. *Hum. Reprod. Update* 18, 60–72. doi: 10.1093/humupd/dmr044

200. Talbert, P. B. (2002). Centromeric localization and adaptive evolution of an arabidopsis histone H3 variant. *Plant Cell Online* 14, 1053–1066. doi: 10.1105/tpc.010425

201. Tiang, C.-L., He, Y., and Pawlowski, W. P. (2012). Chromosome organization and dynamics during interphase, mitosis, and meiosis in plants. *Plant Physiol.* 158, 26–34. doi: 10.1104/pp.111.187161

202. Torras-Llort, M., Moreno-Moreno, O., and Azorín, F. (2009). Focus on the centre: the role of chromatin on the regulation of centromere identity and function. *EMBO J.* 28, 2337–2348. doi: 10.1038/emboj.2009.174

203. Tsukahara, T., Tanno, Y., and Watanabe, Y. (2010). Phosphorylation of the CPC by Cdk1 promotes chromosome bi-orientation. *Nature* 467, 719–723. doi: 10.1038/nature09390

204. Uhlmann, F. (2001). Chromosome cohesion and segregation in mitosis and meiosis. *Curr. Opin. Cell Biol.* 13, 754–761. doi: 10.1016/S0955-0674(00)00279-9

205. Uhlmann, F. (2009). A matter of choice: the establishment of sister chromatid cohesion. *EMBO Rep.* 10, 1095–1102. doi: 10.1038/embor.2009.207

206. Vader, G., Maia, A. F., and Lens, S. M. (2008). The chromosomal passenger complex and the spindle assembly checkpoint: kinetochore-microtubule error correction and beyond. *Cell Div.* 3:10. doi: 10.1186/1747-1028-3-10

207. Vanoosthuyse, V., Prykhozhij, S., and Hardwick, K. G. (2007). Shugoshin 2 regulates localization of the chromosomal passenger proteins in fission yeast mitosis. *Mol. Biol. Cell* 18, 1657–1669. doi: 10.1091/mbc.E06

208. Walczak, C. E., Vernos, I., Mitchison, T. J., Karsenti, E., and Heald, R. (1998). A model for the proposed roles of different microtubule-based motor proteins in establishing spindle bipolarity. *Curr. Biol.* 8, 903–913. doi: 10.1016/S0960-9822(07)00370-3

209. Wang, G., Zhang, X., and Jin, W. (2009). An overview of plant centromeres. *J. Genet. Genomics* 36, 529–537. doi: 10.1016/S1673-8527(08)60144-7

210. Wang, M., Tang, D., Wang, K., Shen, Y., Qin, B., Miao, C., et al. (2011). OsSGO1 maintains synaptonemal complex stabilization in addition to protecting centromeric cohesion during rice meiosis. *Plant J.* 67, 583–594. doi: 10.1111/j.1365-313X.2011.04615.x

211. Wang, Y., Wu, H., Liang, G., and Yang, M. (2004). Defects in nucleolar migration and synapsis in male prophase I in the ask1-1 mutant of Arabidopsis.*Sex. Plant Reprod.* 16, 273–282. doi: 10.1007/s00497-004-0206-z

212. Wasteneys, G. O. (2002). Microtubule organization in the green kingdom: chaos or self-order? *J. Cell Sci.* 115, 1345–1354.

213. Watanabe, Y. (2012). Geometry and force behind kinetochore orientation: lessons from meiosis. *Nat. Rev. Mol. Cell Biol.* 13, 370–382. doi: 10.1038/nrm3349

214. Watanabe, Y., and Nurse, P. (1999). Cohesin Rec8 is required for reductional chromosome segregation at meiosis. *Nature* 400, 461–464. doi: 10.1038/22774

215. Weber, H., and Hellmann, H. (2009). Arabidopsis thaliana BTB/ POZ-MATH proteins interact with members of the ERF/AP2 transcription factor family.*FEBS J.* 276, 6624–6635. doi: 10.1111/j.1742-4658.2009.07373.x

216. Wijnker, E., and Schnittger, A. (2013). Control of the meiotic cell division program in plants. *Plant Reprod.* 26, 143–158. doi: 10.1007/s00497-013-0223-x

217. Wittmann, T., Hyman, A., and Desai, A. (2001). The spindle: a dynamic assembly of microtubules and motors. *Nat. Cell Biol.* 3, E28–E34. doi: 10.1038/35050669

218. Woehlke, G., and Schliwa, M. (2000). Walking on two heads: the many talents of kinesin. *Nat. Rev. Mol. Cell Biol.* 1, 50–58. doi: 10.1038/35036069

219. Woglar, A., and Jantsch, V. (2013). Chromosome movement in meiosis I prophase of Caenorhabditis elegans. *Chromosoma* 123, 15–24. doi: 10.1007/s00412-013-0436-7

220. Wu, S., Scheible, W.-R., Schindelasch, D., Van Den Daele, H., De Veylder, L., and Baskin, T. I. (2010). A conditional mutation in Arabidopsis thaliana separase induces chromosome non-disjunction, aberrant morphogenesis and cyclin B1;1 stability. *Development* 137, 953–961. doi: 10.1242/dev.041939

221. Xu, H., Beasley, M. D., Warren, W. D., van der Horst, G. T. J., and McKay, M. J. (2005). Absence of mouse REC8 cohesin promotes synapsis of sister chromatids in meiosis. *Dev. Cell* 8, 949–961. doi: 10.1016/j. devcel.2005.03.018

222. Xu, Z., Cetin, B., Anger, M., Cho, U. S., Helmhart, W., Nasmyth, K., et al. (2009). Structure and function of the PP2A-shugoshin interaction. *Mol. Cell* 35, 426–441. doi: 10.1016/j.molcel.2009.06.031

223. Yamagishi, Y., Sakuno, T., Shimura, M., and Watanabe, Y. (2008). Heterochromatin links to centromeric protection by recruiting shugoshin. *Nature* 455, 251–255. doi: 10.1038/nature07217

224. Yamamoto, A., Kitamura, K., Hihara, D., Hirose, Y., Katsuyama, S., and Hiraoka, Y. (2008). Spindle checkpoint activation at meiosis I advances anaphase II onset via meiosis-specific APC/C regulation. *J. Cell Biol.* 182, 277–288. doi: 10.1083/jcb.200802053

225. Yang, M., Hu, Y., Lodhi, M., McCombie, W. R., and Ma, H. (1999). The Arabidopsis SKP1-LIKE1 gene is essential for male meiosis and may control homologue separation. *Proc. Natl. Acad. Sci. U.S.A.* 96, 11416–11421. doi: 10.1073/pnas.96.20.11416

226. Yang, X., Boateng, K. A., Strittmatter, L., Burgess, R., and Makaroff, C. A. (2009). Arabidopsis separase functions beyond the removal of sister chromatid cohesion during meiosis. *Plant Physiol.* 151, 323–333. doi: 10.1104/pp.109.140699

227. Yang, X., Boateng, K. A., Yuan, L., Wu, S., Baskin, T. I., and Makaroff, C. A. (2011). The radially swollen 4 separase mutation of Arabidopsis thaliana blocks chromosome disjunction and disrupts the radial microtubule system in meiocytes. *PLoS ONE* 6:e19459. doi: 10.1371/journal.pone.0019459

228. Yang, X., Timofejeva, L., Ma, H., and Makaroff, C. A. (2006). The Arabidopsis SKP1 homolog ASK1 controls meiotic chromosome remodeling and release of chromatin from the nuclear membrane and nucleolus. *J. Cell Sci.* 119, 3754–3763. doi: 10.1242/jcs.03155

229. Yao, Y., and Dai, W. (2012). Shugoshins function as a guardian for chromosomal stability in nuclear division. *Cell Cycle* 11, 2631–2642. doi: 10.4161/cc.20633

230. Yokobayashi, S., Yamamoto, M., and Watanabe, Y. (2003). Cohesins determine the attachment manner of kinetochores to spindle microtubules at meiosis I in fission yeast. *Mol. Cell Biol.* 23, 3965–3973. doi: 10.1128/MCB.23.11.3965-3973.2003

231. Yu, H.-G., and Koshland, D. (2005). Chromosome morphogenesis: condensin-dependent cohesin removal during meiosis. *Cell* 123, 397–407. doi: 10.1016/j.cell.2005.09.014

232. Yu, H., Muszynski, M. G., and Dawe, R. K. (1999). The maize homologue of the cell cycle checkpoint protein MAD2 localization Patterns. *J. Cell Biol.* 145, 425–435.

233. Zamariola, L., De Storme, N., Tiang, C. L., Armstrong, S. J., Franklin, F. C. H., and Geelen, D. (2013). SGO1 but not SGO2 is required for maintenance of centromere cohesion in Arabidopsis thaliana meiosis. *Plant Reprod.* 26, 197–208. doi: 10.1007/s00497-013-0231-x

234. Zamariola, L., De Storme, N., Vannerum, K., Vandepoele, K., Armstrong, S. J., Franklin, F. C., et al. (2014). SHUGOSHINs and PATRONUS protect meiotic centromere cohesion in Arabidopsis thaliana. *Plant J.* 77, 782–794. doi: 10.1111/tpj.12432

235. Zhang, L., Tao, J., Wang, S., Chong, K., and Wang, T. (2006). The rice OsRad21-4, an orthologue of yeast Rec8 protein, is required for efficient meiosis.*Plant Mol. Biol.* 60, 533–554. doi: 10.1007/s11103-005-4922-z

236. Zhao, D., Han, T., Risseeuw, E., Crosby, W. L., and Ma, H. (2003a). Conservation and divergence of ASK1 and ASK2 gene functions during male meiosis in Arabidopsis thaliana. *Plant Mol. Biol.* 53, 163–173. doi: 10.1023/B:PLAN.0000009273.81702.b5

237. Zhao, D., Ni, W., Feng, B., Han, T., Petrasek, M. G., and Ma, H. (2003b). Members of the arabidopsis-SKP1-like gene family exhibit a variety of expression patterns and may play diverse roles in arabidopsis 1. 133, 203–217. doi: 10.1104/pp.103.024703.to

238. Zhao, D., Yang, X., Quan, L., Timofejeva, L., Rigel, N. W., Ma, H., et al. (2006). ASK1, a SKP1 homolog, is required for nuclear reorganization, presynaptic homolog juxtaposition and the proper distribution of cohesin during meiosis in Arabidopsis. *Plant Mol. Biol.* 62, 99–110. doi: 10.1007/s11103-006-9006-1

239. Zhao, D., Yu, Q., Chen, M., and Ma, H. (2001). The ASK1 gene regulates B function gene expression in cooperation with UFO and LEAFY in Arabidopsis.*Development* 128, 2735–2746.

240. Zhong, C. X., Marshall, J. B., Topp, C., Mroczek, R., Kato, A., Nagaki, K., et al. (2002). Centromeric retroelements and satellites interact with maize kinetochore Protein CENH3. *Plant Cell* 14, 2825–2836. doi: 10.1105/tpc.006106

241. Zhou, S., Wang, Y., Li, W., Zhao, Z., Ren, Y., Wang, Y., et al. (2011). Pollen semi-sterility1 encodes a kinesin-1-like protein important for male

meiosis, anther dehiscence, and fertility in rice. *Plant Cell* 23, 111–129. doi: 10.1105/tpc.109.073692

242. Zhou, X., Graumann, K., Evans, D. E., and Meier, I. (2012). Novel plant SUN-KASH bridges are involved in RanGAP anchoring and nuclear shape determination. *J. Cell Biol.* 196, 203–211. doi: 10.1083/jcb.201108098

Chapter 5

IMMUNOLOCALIZATION OF CHROMOSOME-ASSOCIATED PROTEINS IN PLANTS – PRINCIPLES AND APPLICATIONS

Cristina Maria Pinto de Paula and Vânia Helena Techio
Department of Biology, Federal University of Lavras, Lavras, Minas Gerais, Brazil

ABSTRACT

The use of the immunolocalization technique combined with cytogenetic and epigenetic studies is an indispensable tool and has contributed significantly to the analysis of the structure and function of chromosomes, since it can provide information about the spatial or temporal distribution of a given protein in the nucleus and chromosomes. Several chromosome-associated proteins in plant cells have already been identified by immunolocalization, such as histone and non-histone proteins and cell division-related protein (mitosis and meiosis). The principle of the immunolocalization technique in plants basically involves fixation and permeabilization of cells, the use of monoclonal or polyclonal antibodies attached to a signaling molecule, usually a fluorochrome and detection of the target molecule by using an epifluorescence microscope.

INTRODUCTION

Considerable advances in plant biology have been accompanied by the use of microscopy of higher resolution and improvement of the detection limit of analyses through use of specific antibodies and fluorochromes, that provide an interaction between knowledge of cell biology, classical cytogenetics and molecular cytogenetics. From the development of molecular cytogenetic techniques, understanding the organization of genomes had a large impact on taxonomic and evolutionary studies (Fedak and Kim 2008; Figueroa and Bass 2010), in breeding and genetic engineering (Seijo et al. 2010).

Recent research on the architecture of the interphase nucleus and chromatin remodeling, mediated primarily by changes in histones, has prospected the era of epigenetics and epigenomics. In this new scientific background, the use of the technique of immunolocalization in chromosome analysis has become an indispensable tool for understanding the organization and expression of the genome of species. As a result of this interaction, significant improvements were obtained for the analysis of chromosome structure and function, as it can provide information about the spatial and temporal distribution of a given protein in the nucleus and chromosomes.

These analyses have been widely applied in recent decades to study and locate proteins in cells or cell populations and comprise the use of specific antibodies to the protein of interest (Hoshi et al. 2008). In the context of chromosomal studies, advances are more recent, but have already made possible to identify several chromosome-associated proteins in plant cells, including histones (Fuchs et al. 1998), non-histones (Stepinski 2009), proteins associated with mitosis (Caperta et al. 2006) and with meiosis (Franklin et al. 1999; Armstrong et al. 2002; Qiao et al. 2011).

Given the importance of the topic and potential use, this review aims to present the principles of the immunolocalization technique and its application in the detection of proteins associated with chromosomes in plants.

REVIEW

Principles of Immunolocalization

The application of the immunolocalization technique was facilitated through the development of fluorochrome-labeled antibodies, in the pioneering work of Albert Coons (Coons et al. 1941), thus allowing visualization of specific molecules in a cell. Therefore, it has been possible to observe the spatial or temporal distribution of a given protein in the whole cell, in the nucleus, in specific chromosome or organelles.

The procedures started to be used in plants from the experiences observed in animal cells. One of the first descriptions of the use of immunolocalization in plant cytology occurred in 1970, with the location of antigens of the cell wall in pollen (Knox 1970) and proteins of cotyledons (Graham and Gunning 1970). Currently, the immunolocalization has allowed, among other studies, the identification of chromosomal proteins, that is, the study of how the activity of these proteins is regulated during cell divisions (Seijo et al. 2010; Houben et al.2013), and also the assessment of post-translational changes in histones, which are reflected in the chromatin structure, known as epigenetic regulation (Chen et al. 2010).

The immunolocalization technique is based on the principles of antigen-antibody reactions to find highly specific molecules in a cell or tissue. The common reagent for immunolocalization are polyclonal or monoclonal antibodies, both produced in animals such as rabbits, goats, rats, mice, among others, by means of specific procedures (Boenisch 2009).

Polyclonal antibodies are a heterogeneous mixture of antibodies that bind to different epitopes of the same protein. Monoclonal antibodies are homogeneous populations, more specific for binding to only one epitope of the protein (Boenisch 2009).

There are many commercially available antibodies, some of universal use for chromosomal proteins, such as anti-tubulin (α, β and γ) or for modified DNA regions, such as the anti-5-methylcytosine (Guerra 2012; Birchler and Han 2013). Other antibodies are specific for a species or group of species, e.g., phosphorylated H2AThr120, which is used for cytological detection of centromeric chromatin of plants (Dong and Han2012; Demidov et al. 2014).

In addition to factors related to the type of antibody, storage conditions and dilution of antibodies, which are usually indicated by the manufacturers, the immunolocalization reactions must take into account factors like affinity, avidity and cross-reactivity.

Affinity refers to the interaction strength between the antigen epitope and the paratope of an antibody. Similarly, avidity is the term used to describe the overall strength of the interactions between the antigen and the various antibodies that recognize it (Steward and Steensgaard1983; Boenisch 2009). High affinity antigen-antibody means greater tendency to hold them together (Boenisch 2009). However, factors such as high salt concentration, high temperature and low pH during the washing cycles applied in the technique, may result in the dissociation of the antigen-antibody complex.

Cross-reactivity occurs when an antibody reacts with two or more antigens, or when an antigen reacts with several different antibodies. In this case there is a sharing of at least one common epitope among the various antigens (Boenisch 2009).

For detection of proteins by immunolocalization, the cells must be fixed and permeabilized to facilitate the access of antibodies to the cytoplasm and nucleus. Fixation of the nucleus or chromosomes must preserve the arrangement of molecules as they occur in vivo, and treatment should not affect the cytological integrity of the material to be analyzed. In general, aldehyde fixatives are used, which make a link between proteins, nucleic acids and phospholipids, thus forming a network which retains molecules in their original position. Formaldehyde, paraformaldehyde and glutaraldehyde are the most frequently used in immunolocalization (Marttila and Santén 2007).

There are some cases in which acid fixatives can be effective to the meiosis, for example, Carnoy (ethyl alcohol: acetic acid 3:1) (Chelysheva et al. 2010, 2013). In this last case, the fixation is associated with microwave procedures. This method combines a strong fixation to preserve the chromosomal structures, acetic acid spreading to remove the cytoplasm and treatment in microwave (850 W) to increase the access of the antibody to the protein.

Another possibility for meiosis is use of the methanol/acetone fixative, which extracts the lipids and rapidly dehydrates the cell, thus disrupting hydrophobic interactions and resulting in aggregation and precipitation of proteins. This fixative preserves cell architecture, but can result in reduced antigenicity of some proteins. Nevertheless, it can also allow better access of the antibody to certain antigens without requiring cell permeabilization (Yang et al. 2013).

Permeabilization is usually carried out by incubation with detergents or organic solvents, which solubilize or remove lipids from the plasma membrane and thus allow these antibodies to have access to the structure of interest. The most commonly used permeabilizing agents include Triton X-100, Tween 20 and also Lipsol. These are hydrophilic nonionic detergents, considered mild surfactants, in other words they break lipid-protein and lipid-lipid interactions, but not protein-protein bonds, maintaining their active form (Johnson 2013).

In the case of plant cells, it is necessary to overcome the barrier of the cell wall, so that permeabilization usually requires a prior cell wall digestion using enzyme cocktails containing pectinase, cellulase, pectoliase and others, whose concentration and exposure time are adjusted according to the species analyzed.

The usual procedure for plants is indirect immunolocalization. This technique includes the use of at least two antibodies: the primary which recognizes and binds to epitopes of the antigen (protein) to be located, and the secondary, often conjugated to a fluorochrome, which recognizes the corresponding primary antibody, the reason why the procedure is also referred to immunofluorescence. Alternatively, gold conjugated secondary antibodies can be used, in this case, the visualization is performed in a transmission electron microscope and the procedure is called immunogold (Holgate et al. 1983; Marttila and Santén 2007). The direct method employs only one antibody linked to a fluorochrome. The advantage of indirect labeling is the amplification of the signal obtained, since several secondary antibodies may bind to a single primary antibody (Brown and Lemmon 1995; Marttila and Santén 2007).

Fluorochromes are chemicals that, upon excitation at a specific wave length, fluoresce at another wavelength, and the signal is observed with an epifluorescence microscope (Brown and Lemmon 1995). Fluorescein (FITC) is the fluorochrome most used in immunolocalization, it can be excited by UV light (wavelength below 400 nm) or blue light (490 nm) and emits green light (500–550 nm), as well as Rhodamine (TRITC) and Texas red, fluorochromes which are excited between 520 and 590 nm and emit red light (550–620 nm) (McNamara 2006). There are other types of fluorochromes with higher fluorescence intensity and photostability, compared with those aforementioned and may also be used in immunolocalization such as Dylight, Alexa Fluor (Jensen 2012) and quantum dots, the latter a class of inorganic fluorochromes made up of nanoscale crystals of a semiconductor material (Ioannou et al. 2009).

An alternative methodology proposed by (Wang 2013) analyzes meiotic chromosomes by three-dimensional structured illumination microscopy (3D-SIM) using high resolution images and can be applied for the localization of proteins or fine structures by immunofluorescence. The protocol (Acrylamide sandwich) was developed aimed at better preservation of 3D chromosome structure and spatial organization of the nucleus; in turn the purpose of 3D-SIM is to illuminate a sample with a light distribution pattern produced by a diffraction grating. With this technique, several optical sections are captured and a 3D image is reconstructed by computer analysis.

Major Chromosome-Associated Proteins Localized in Plants

Study of Non-Histone Protein

A large number of studies have already identified several proteins associated with chromosomes in plant cells so far. The study of the behavior of these proteins provide information about the function of chromosomes, since part of the control of the genetic information encoded by DNA depends on interactions with proteins. The immunolocalization of chromosomal proteins in plants has been widely used for different purposes in cytogenetic and epigenetic studies such as identification of proteins that are directly or indirectly involved with mitosis and meiosis and associated with chromatin modeling.

The proper chromosome condensation during mitosis and meiosis is essential for the correct segregation of the genomic information contained in the DNA. Besides histones, other non-histone proteins are also involved in the process of condensation and cohesion, such as a topoisomerase II, condensins and cohesin (Moser and Swedlow 2011).

Cohesin complex proteins are essential for sister chromatid cohesion and proper chromosome segregation during mitosis and meiosis. Cohesin proteins are also components of axial elements (AEs) and lateral elements (LEs) of the synaptonemal complex during meiosis. The cytological behavior of four cohesin proteins (SMC1, SMC3, SCC and REC8/SYN1), assessed by immunofluorescence during prophase I in tomato microsporocyte revealed that the four cohesins are distributed unevenly and are not co-located along the AE/LEs in diplotene (Qiao et al. 2011). Nevertheless, based on current models of the cohesin complex, these proteins must be present at the same time and place in equivalent amounts. These results indicate that cohesin proteins studied can form different complexes and/or perform additional functions during meiosis in plants.

The success of meiosis depends on the regulation of several cellular processes that ensure proper chromosome segregation. By employing the immunolocalization technique it is possible to visualize a large number of proteins related to different aspects of meiosis, including sister chromatid cohesion, synapsis, recombination, and chromosome segregation (Yang et al. 2013).

A number of proteins involved in the pairing and recombination are expressed during meiosis, including RAD51 and MLH1. Studies with maize have shown that RAD51 is involved in the homology search during chromosomal pairing and recombination. Through immunolocalization, it was observed, in zygotene, 500 RAD51 signals per nucleus, clearly associated with the partially paired and paired chromosomes, which is consistent with its role in the homology search. In pachytene, the number of RAD51 signals decreased, ranging from 7 to 22 per nucleus, with the largest number corresponding approximately to the number of chiasmata found in maize, thus coinciding with the expected number of recombination during meiosis (Franklin et al. 1999).

Likewise, RAD51 protein in *Arabidopsis* is related to homology search and meiotic recombination (Kurzbauer et al. 2012). *Atrad51-2* mutants express low amounts of the RAD51 protein, meiosis with partial synapsis and 51% of bivalents between non-homologous chromosomes, thus demonstrating that RAD51 is required for effective chromosome pairing (Pradillo et al. 2012).

Similar studies were developed with *Lily*, being detected the presence of RAD51 and LIM15 proteins as discrete signals in leptotene and zygotene. This localization suggests that meiotic recombination is initiated in leptotene with the cooperation of these two proteins and remains in zygotene. The location of the signals on chromosomes or in their vicinity suggests that these proteins must also act in the pairing of homologous chromosomes in this species (Terasawa et al. 1995; Anderson et al. 1997).

The ASY1 gene is essential for synapsis in homologous chromosomes (Caryl et al. 2000). Studies using specific antibodies against the ASY1 protein were applied to investigate the temporal expression and localization of this protein in *Arabidopsis thaliana* (L.) Heynh. Detection of the immuno signal associated with electron microscopy data showed that ASY1 is not a component of the synaptonemal complex, but is associated with the axial/lateral elements of meiotic chromosomes. The authors suggest that this protein possibly defines the regions of chromatin associated with the development of the synaptonemal complex structure (Armstrong et al. 2002).

Cytoskeleton microtubules play a crucial role during the cell cycle and meiosis. The use of antimitotic substances, such as colchicine, promotes depolymerization of microtubules and prevents the formation of spindle fibers, this methodology is used to block cell cycle progression and thus induce polyploidy (Pereira et al. 2012). Through immunolocalization of alpha-tubulin, (Caperta et al. 2006) analyzed the effect of different colchicine concentrations on spindle fibers in *Secale cereale* L. At a low concentration of colchicine (0.5 mM) most of the cells showed arrays of discontinuous microtubule or no microtubule were detected at C-metaphase. Cells exposed to high concentrations (5.0 mM) showed a different effect, the C- metaphase presented fibrous and branched cortical filaments that allowed reconstitution of 4C nuclei and cell cycle progression. These results demonstrate the contrasting and opposite effects at different concentrations of colchicine.

Study of Histone Modifications

Histones form another group of proteins associated with eukaryotic DNA, comprising the nucleosome, which consists of an octamer of the four major histones (H3, H4, H2A, H2B) (Luger et al. 1997; Kouzarides 2007).

Each histone that comprises the nucleosome has a terminal amino acid chain, called N-terminal tail, which is subjected to various post-translational modifications (Kouzarides 2007; Jin et al. 2008). These epigenetic marks affect the structure and function of chromatin, thereby regulating gene expression and can be identified by immunolocalization (Bannister and Kouzarides 2011).

There are at least eight different types of alterations in histones, standing out acetylation, methylation and phosphorylation, which are the most studied (Kouzarides 2007; Chen et al. 2010).

Plant chromosomes usually exhibit a lower amount of acetylated histones in heterochromatin and often form significant markings in the nucleolar organizing region (NOR), which allows a correlation with active gene transcription. In *Vicia faba* L., antibodies against acetylated H4 at lysine 5, 8 and 12 labeled the entire chromosome complement, with the exception of large blocks of heterochromatin (Fuchs et al. 1998), since the NOR was the most strongly acetylated of all lysine residues investigated (Houben et al. 1996).

One of the most studied post-translational modification is the phosphorylation of histone H3, which seems to be decisive for cell cycle in relation to the chromosome condensation and segregation, activation of apoptosis, transcription and repair of DNA damage (Houben et al.2007). In plants, phosphorylation of histone H3 at serine 10 (H3S10ph) and 28 (H3S28ph) is restricted to the pericentromeric region in mitosis (Germand et al. 2003). In meiosis, the chromosomes are phosphorylated along the entire arm during the first division, however, the phosphorylation is restricted to the pericentromeric region during the second division (Manzanero et al. 2000).

In studies with forage grasses Paula et al. (2013) evaluated the phosphorylation pattern of H3S10 in diploid and tetraploid genotypes of*Brachiaria* (Trin.) Griseb. species in order to investigate the dynamics of this post-translational modification and correlate it with its regulatory function during mitosis and meiosis. In meiosis, the chromosomes are phosphorylated at H3S10 along its entire length during the first division; in the second division, phosphorylation is restricted to the pericentromeric region, as well as in mitosis. Alltogether, these observations indicated that H3S10ph is necessary for the maintenance of sister chromatid cohesion in mitosis and meiosis.

Phosphorylation of threonine in histone H3 in *Secale cereale* L. and threonine 11 in H3 in *Vicia faba* L. was observed throughout the chromosome during mitosis and meiosis. Phosphorylation began in prophase and ended in telophase, thus correlating with chromosome condensation (Houben et al. 2005, 2007).

By means of immunolocalization, it was possible to detect that methylation of some isoforms of histones in plants may be associated with heterochromatin (H3K9, H3K27 and H4K20) and also euchromatin (H3K4, H3K36 and H3K79) (Fuchs et al. 2006).

Different types of chromatin associated with histone modifications were analyzed by immunolocalization in *Costus spiralis*. Antibodies against components characteristic of histones of euchromatin (acetylated H4 histone at lysine 5 - H4K5ac and dimethylated H3 histone at lysine 4 - H3K4me2) and heterochromatin (dimethylated H3 histone at lysine 9 - H3K9me2) were used to characterize the centromeric chromatin during meiosis. The centromeric region was highly enriched only with H4k5ac and only in pachytene. Still in pachytene, the terminal decondensed euchromatin of chromosome arms were labeled with H4K5ac and H3K4me2, whereas the condensed proximal region was labeled with H3K9me2 (Feitoza and Guerra 2011a).

In plant species *Costus spiralis* and *Eleutherine bulbosa* during mitosis, labeling with antibodies against components characteristic of histones of euchromatin were observed in the regions of late condensation while the components of heterochromatin was more intense in the early condensation chromatin. These data indicate that there are different chromatin domains in these species, the late condensing prophase chromatin localized on the distal region and strongly enriched in H4K5ac and H3K4me2 and early condensing chromatin, localized on the proximal chromosome region in *C. spiralis* and mainly in the larger pair I of *E. bulbosa* (Feitoza and Guerra 2011b).

Intraspecific hybrids and parental acessions (Col-0 and C24) of *Arabidopsis thaliana* were used as a model to investigate the relationship between changes in DNA methylation, chromatin structure, endopolyploidy and gene expression in heterotic genotypes. Through immunolocalization, the distribution of methylation of histone H3 (H3K27me3, H3K4me2 and H3K9me2) was compared between parental and hybrid plants. For all modifications of the chromatin analyzed, no clear difference was found in the distribution and intensity of signals between Col-0, C24 and their reciprocal hybrids, revealing that the parental pattern profile was kept in the hybrid progeny (Moghaddam et al. 2010).

Besides that, the correlation between the distribution of histone H3 methylated at lysine 4 (typical of transcriptionally active euchromatin) and lysine 9 (inactive heterochromatin) was analyzed in relation to genome size for 24 species. In species with small genome, strong H3K9me was restricted to constitutive heterochromatin, whereas species with large genomes exhibited uniform distribution of this modification. Unlike and independent of the genome size, H3k4me was enriched in euchromatin of all species. The authors concluded that large genomes with large amounts of dispersed repetitive sequences (mainly retroelements) silence these sequences through epigenetic modifications such as H3K9me (Houben et al. 2003).

Centromeres are responsible for several important events in meiosis and mitosis. In contrast to the telomeres, centromeres do not have highly conserved DNA sequence and are determined by epigenetic modifications (Houben and Schubert 2003). In eukaryotes, the centromeres have a variant of histone H3, the CENH3 (CENPA in humans) which determines the chromosomal position of kinetochore assembly, forming a bond between the centromeric DNA and the kinetochore (Blower et al. 2002). Thus, the anti-CENH3 antibody has been used in studies related to the centromeric regions in plants like corn (Jin et al. 2005; Zhang et al. 2005), wheat (Zhang et al. 2010) and rye (Houben et al.2011).

Recently, Demidov et al. (2014) evidenced that the antibody against phosphorylated histone H2A at threonine 120 can be used as a universal maker for the cytological detection of centromeres in species of monocentric and holocentric plants. These results were obtained from studies on 20 different species of mono- and eudicots.

Although histones and their modifications are highly conserved, their distribution in chromosomes, observed by immunolocalization, showed a variation over the differentiation and cell cycle processes, as well as within and between groups of eukaryotes (Hans and Dimitrov 2001).

CONCLUSION

Immunolocalization is an important and efficient tool in studies involving the understanding of genome organization and analysis of the structure and function of chromosomes. Furthermore, immunolocalization studies allowed the understanding of epigenetic mechanisms involved in chromatin modeling and modifications related to different states of condensation, such as transcriptional activity and silencing.

AUTHORS' CONTRIBUTIONS

All authors were involved in drafting the manuscript, providing guidelines for the review, modification and preparation the final version of the manuscript. Both authors read and approved the final manuscript.

REFERENCES

1. Anderson LK, Offenberg HH, Verkuijlen WMHC, Heyting C (1997) RecA-like proteins are components of early meiotic nodules in Lily. Proc Natl Acad Sci U S A 94:6868–6873

2. Armstrong SJ, Caryl AP, Jones GH, Franklin FCH (2002) Asy1, a protein required for meiotic chromosome synapsis, localizes to axis-associated chromatin in *Arabidopsis* and *Brassica*. J Cell Sci 18:3645–3655

3. Bannister AJ, Kouzarides T (2011) Regulation of chromatin by histone modifications. Cell Res 21(3):381–395

4. Birchler JA, Han F (2013) Centromere epigenetics in plants. J Genet Genomics 40:201–204

5. Blower MD, Sullivan BA, Karpen GH (2002) Conserved organization of centromeric chromatin in flies and humans. Dev Cell 2:319–330

6. Boenisch T (2009) Antibodies. In: Kummar LG, Rudbeck L (eds) Immunohistochemical (IHC) Staining Methods. Dako North America, California, pp 1–9

7. Brown RC, Lemmon BE (1995) Methods in plant immunolight microscopy. Methods Cell Biol 49:85–107

8. Caperta AD, Delgado M, Ressurreição F, Meister A, Jones RN, Viegas W et al (2006) Colchicine-induced polyploidization depends on tubulin polymerization in c-metaphase cells. Protoplasma 227:147–153

9. Caryl AP, Armstrong SJ, Jones GH, Franklin FCH (2000) A homologue of the yeast *HOP1* gene is inactivated in the *Arabidopsis* meiotic mutant *asy1*. Chromosoma 159:62–71

10. Chelysheva L, Grandont L, Vrielynck N, Le Guin S, Mercier R, Grelon M (2010) An Easy Protocol for studying chromatin and recombination protein dynamics during *Arabidopsis thaliana* meiosis: Immunodetection of cohesins, histones and MLH1. Cytogenet Genome Res 129:143–153

11. Chelysheva LA, Grandont L, Grelon M (2013) Immunolocalization of meiotic proteins in Brassicaceae: Method 1. In: Pawlowski WP, Grelon M, Armstrong S (eds) Plant Meiosis: Methods and Protocols. Springer Science+Business Media, New York, pp 103–108

12. Chen M, Shaolei LV, Meng Y (2010) Epigenetic performers in plants. Dev Growt Differ 52:555–566

13. Coons AH, Creech HJ, Jones RN (1941) Immunological properties of an antibody containing a fluorescent group. Proc Soc Exp Biol Med 47:200–202

14. Demidov D, Schubert V, Kumbe K, Weiss O, Karimi-Ashtiyani R, Butlar J et al (2014) Anti-phosphorylated histone H2AThr120 - a universal marker for centromeric chromatin of mono- and holocentric plant species. Cytogenet Genome Res. doi:10.1159/000360018

15. Dong Q, Han F (2012) Phosphorylation of histone H2A is associated with centromere function and maintenance in meiosis. Plant J 71:800–809

16. Fedak G, Kim NS (2008) Tools and methodologies for cytogenetic studies of plant chromosomes. Cytology Genet 42:189–203

17. Feitoza L, Guerra M (2011a) The centromeric heterochromatin of *Costus spiralis*: poorly methylated and transiently acetylated during meiosis. Cytogenet Genome Res 135:160–166

18. Feitoza L, Guerra M (2011b) Different types of plant chromatin associated with modified histones H3 and H4 and methylated DNA. Genetica 139(3):305–314

19. Figueroa DM, Bass HW (2010) A historical and modern perspective on plant cytogenetics. Brief Funct Genomics 9:95–102

20. Franklin AE, Mcelver J, Sunjevaric I, Rothstein R, Bowen B, Cande WZ (1999) Three-dimensional microscopy of the Rad51 recombination protein during meiotic prophase. Plant Cell 11:809–824

21. Fuchs J, Strehl S, Brandes A, Schweizer D, Schubert I (1998) Molecular-cytogenetic characterization of *Vicia faba* genome-heterochromatin differentiation, replication patterns and sequence localization. Chromosome Res 6:219–230

22. Fuchs J, Demidov D, Houben A, Schubert I (2006) Chromosomal histone modification patterns: from conservation to diversity. Trends Plant Sci 11:199–208

23. Germand D, Demidov D, Houben A (2003) The temporal and spatial pattern of histone H3 phosphorylation at serine 28 and serine 10 is similar in plants but differs between mono- and polycentric chromosomes. Cytogenet Genome Res 101:172–176

24. Graham TA, Gunning BES (1970) Localization of legumin and vicilin in bean cotyledon cells using fluorescent antibodies. Nature 228:81–82

25. Guerra M (2012) Citogenética molecular: Protocolos comentados. Sociedade Brasileira de Genética, Ribeirão Preto

26. Hans F, Dimitrov S (2001) Histone H3 phosphorylation and cell division. Oncogene 20:3021–3027

27. Holgate CS, Jackson P, Cowen PN, Bird CC (1983) Immunogold-silver staining: New method of immunostaining with enhanced sensitivity. J Histochem Cytochem 31:938–944

28. Hoshi O, Hirota T, Kimura E, Komatsubara N, Ushiki T (2008) Immunocytochemistry for analyzing chromosomes. In: Fukui K, Ushiki T (eds) Chromosome nanoscience and technology. CRC press, New York, pp 81–89

29. Houben A, Schubert I (2003) DNA and proteins of plant centromeres. Curr Opin Plant Biol 6:554–560

30. Houben A, Belyaev ND, Turner BM, Schubert I (1996) Differential immunostaining of plant chromosomes by antibodies recognizing acetylated histone H4 variants. Chromosome Res 4:191–194

31. Houben A, Demidov D, Gernand D, Meister A, Leach CR, Schubert I (2003) Methylation of histone H3 in euchromatin of plant chromosomes depends on basic nuclear DNA content. Plant J 33:967–973

32. Houben A, Demidov D, Rutten T, Scheidtmann KH (2005) Novel phosphorylation of histone H3 at threonine 11 that temporally correlates with condensation of mitotic and meiotic chromosomes in plant cells. Cytogenet Genome Res 109:148–155

33. Houben A, Demidov D, Caperta AD, Karimi R, Aqueci F, Vlasenko L (2007) Phosphorylation of histone H3 in plants: a dynamic affair. Biochim Biophys Acta 1769:308–315

34. Houben A, Kumke K, Nagaki K, Hause G (2011) CENH3 distribution and differential chromatin modifications during pollen development in rye (*Secale cereale* L.). Chromosome Res 19:471–480

35. Houben A, Demidov D, Karimi-Ashtiyani R (2013) Epigenetic control of cell division. In: Grafi G, Ohad N (eds) Epigenetic Memory and Control in Plants. Springer-Verlag, Berlin, pp 155–175

36. Ioannou D, Tempest HG, Skinner BM, Thornhill AR, Ellis M, Griffin DK (2009) Quantum dots as new-generation fluorochromes for FISH: an appraisal. Chromosome Res 17:519–530

37. Jensen EC (2012) Types of imaging, Part 2: An overview of Fluorescence Microscopy. Anat Rec 295:1621–1627

38. Jin WW, Lamb JC, Vega JM, Dawe RK, Birchler JA, Jiang J (2005) Molecular and functional dissection of the maize B centromere. Plant Cell 17:1412–1423

39. Jin W, Lamb JC, Zhang W, Kolano B, Birchler JA, Jiang J (2008) Histone modifications associated with both A and B chromosomes of maize. Chromosome Res 16:1203–1214

40. Johnson M (2013) Detergents: Triton X-100, Tween-20, and more. Mater Methods 3:163

41. Knox RB (1970) Freeze-sectioning of plant tissues. Stain Technol 45:265–272

42. Kouzarides T (2007) Chromatin modifications and their function. Cell 128:693–705

43. Kurzbauer M-T, Uanschou C, Chen D, Schlögelhofer P (2012) The Recombinases DMC1 and RAD51 Are Functionally and Spatially Separated during Meiosis in *Arabidopsis*. Plant Cell 24:2058–2070

44. Luger K, Mader AW, Richmond RK, Sargent DF, Richmond TJ (1997) Crystal structure of the nucleosome core particle at 2.8A° resolution. Nature 389:251–260

45. Manzanero S, Arana P, Puertas MJ, Houben A (2000) The chromosomal distribution of phosphorylated histone H3 differs between plants and animals at meiosis. Chromosoma 109:308–317

46. Marttila S, Santén K (2007) Practical aspects of Immunomicroscopy on plant material. In: Méndez-Vilas A, Díaz J (eds) Modern Research and Educational Topics in Microscopy. Formatex, Badajoz, pp 1015–1021

47. McNamara G (2006) Introduction to immunofluorescence microscopy. In: Spector D, Goldman RD (eds) Basic methods in microscopy. Cold Spring Harbor, New York, pp 145–154

48. Moghaddam AMB, Fucs J, Czauderna T, Houben A, Mette MF (2010) Intraspecific hybrids of *Arabidopsis thaliana* revealed no gross alterations in endopolyploidy, DNA methylation, histone modifications and transcript levels. Theor Appl Genet 120:215–226

49. Moser SC, Swedlow JR (2011) How to be a mitotic chromosome. Chromosome Res 19:307–319

50. Paula CMP, Techio VH, Souza Sobrinho F, Freitas AS (2013) Distribution pattern of histone H3 phosphorylation at serine 10 during mitosis and meiosis in *Brachiaria* species. J Genet 92:259–266

51. Pereira RC, Davide LC, Techio VH, Timbó ALO (2012) Duplicação cromossômica de gramíneas forrageiras: uma alternativa para programas de melhoramento genético. Cienc Rural 42:1278–1285

52. Pradillo M, López E, Linacero R, Romero C, Cuñado N, Sánchez-Morán E, Santos JL (2012) Together yes, but not coupled: new insights into the roles of RAD51 and DMC1 in plant meiotic recombination. Plant J 69:921–933

53. Qiao H, Lohmiller LD, Anderson LK (2011) Cohesin proteins load sequentially during prophase I in tomato primary microsporocytes. Chromosome Res 19:193–207

54. Seijo G, Lavia GI, Robledo G, Fernández A, Neffa VGS (2010) La citogenética molecular e inmunocitogenética em el estúdio de los genomas vegetales. In: Levitus G, Echenique V, Rubinstein C, Hopp E, Mroginski L (eds) Biotecnología y mejoramiento vegetal II. Instituto Nacional de tecnologia Agropecuária, Argentina, pp 34–46

55. Stepinski D (2009) Immunodetection of nucleolar proteins and ultrastructure of nucleoli of soybean root meristematic cells treated with chilling stress and after recovery. Protoplasma 235:77–89

56. Steward MW, Steensgaard J (1983) Antibody affinity: Termodynamic aspects and biological significance 1ed. CRC Press, Boca Raton

57. Terasawa M, Shinohara A, Hotta Y, Ogawa H, Ogawa T (1995) Localization of RecA-like recombination proteins on chromosomes of the lily at various meiotic stages. Genes Dev 9:925–934

58. Wang C-JR (2013) Analyzing maize meiotic chromosomes with super-resolution structured illumination microscopy. In: Pawlowski WP, Grelon M, Armstrong S (eds) Plant Meiosis: Methods and Protocols. Springer Science+Business Media, New York, pp 109–118

59. Yang X, Yuan L, Makaroff CA (2013) Immunolocalization protocols for visualizing meiotic proteins in *Arabidopsis thaliana*: Method 3. In: Pawlowski WP, Grelon M, Armstrong S (eds) Plant Meiosis: Methods and Protocols. Springer Science+Business Media, New York, pp 109–118

60. Zhang X, Li X, Marshall JB, Zhong CX, Dawe RK (2005) Phosphoserines on maize Centromeric histone H3 and Histone H3 demarcate the centromere and pericentromere during chromosome segregation. Plant Cell 17:572–583

61. Zhang WL, Friebe B, Gill BS, Jiang JM (2010) Centromere inactivation and epigenetic modifications of a plant chromosome with three functional centromeres. Chromosoma 119:553–563

Chapter 6

CHROMOSOMAL LOCATION OF HWA1 AND HWA2, COMPLEMENTARY HYBRID WEAKNESS GENES IN RICE

Katsuyuki Ichitani[1], Satoru Taura[2], Takahiro Tezuka[3], Yuuya Okiyama[4], and Tsutomu Kuboyama[4]

[1]Faculty of Agriculture, Kagoshima University, 1-21-24 Korimoto, Kagoshima, Kagoshima 890–0065, Japan

[2]Institute of Gene Research, Kagoshima University, 1-21-24 Korimoto, Kagoshima, Kagoshima 890–0065, Japan

[3]Graduate School of Life and Environmental Sciences, Osaka Prefecture University, 1–1 Gakuen-cho, Nakaku, Sakai, Osaka 599–8531, Japan

[4]College of Agriculture, Ibaraki University, 3-21-1 Chuo, Ami, Ibaraki 300–0393, Japan

ABSTRACT

Hybrid weakness phenomena in rice reportedly have two causes: those of *HWC1* and *HWC2* genes and those of *HWA1* and *HWA2* genes. No detailed study of the latter has been reported. For this study, we first produced crosses among cultivars carrying the weakness-causing allele on the *HWA1* and *HWA2* loci to confirm the phenotype of the hybrid weakness and the genotypes of the cultivars on the two loci, as reported earlier. We then confirmed that these cultivars belong to Indica. Subsequent linkage analysis of *HWA1* and *HWA2* genes conducted using DNA markers revealed that both genes are located in the 1,637-kb region, surrounded by the same DNA markers on the long arm of chromosome 11. The possibility of allelic interaction inducing hybrid weakness is discussed.

INTRODUCTION

Many postzygotic reproductive barrier forms have been reported in rice (*Oryza sativa*), such as hybrid weakness (e.g., Oka 1957), hybrid pollen sterility (e.g., Long et al. 2008), and hybrid sterility causing female gamete abortion (e.g.,

Chen et al. 2008). Among them, hybrid weakness is definable as weak growth occurring in F_1 hybrids derived from crosses between two normal strains. According to its degree or symptom, it is also called hybrid lethality, hybrid abnormality, or hybrid necrosis. Hybrid weakness is also apparent in many other plant species including *Arabidopsis thaliana* (Bomblies et al. 2007), *Phaseolus vulgaris* (Shii et al. 1980), interspecific crosses among *Gossypium*(Lee 1981), and interspecific crosses among *Nicotiana* (Tezuka et al. 2007).

In rice, two hybrid weakness phenomena from different intraspecific cross combinations have been reported: one by Oka (1957) and the other by Amemiya and Akemine (1963). The hybrid weakness reported by Oka (1957) resulted from use of a set of complementary dominant genes: L_1 and L_2. They were renamed *L-1-a* and *L-1-b* by Kinoshita (1984); then *Hwa-1* and *Hwa-2* by Sato et al. (1987). According to the new gene nomenclature system for rice (McCouch and Committee on Gene Symbolization, Nomenclature and Linkage, Rice Genetics Cooperative (CGSNL) 2008), the new gene symbols *HWA1* and *HWA2* are used respectively for reference to *Hwa-1* and *Hwa-2*, as presented in Table 1. The latter phenomenon results from the use of a set of complementary dominant genes: *Hybrid weakness C1* (*HWC1*) and *Hybrid weakness C2* (*HWC2*). We have detected the chromosomal locations of these genes and have performed fine mapping (Ichitani et al. 2001; Ichitani et al. 2007; Kuboyama et al. 2009). However, the chromosomal locations of *HWA1* and *HWA2* have remained unknown.

Table 1. Gene symbols frequently used in this study according to the new gene nomenclature system for rice (McCouch and CGSNL 2008)

	Gene symbol					
	Oka (1957)	Kinoshita (1984)	Sato et al. (1987)	This study	Gene full name	Carrier of weakness-causing gene
Locus/gene				*HWA1*	*HYBRID WEAKNESS A1*	
Dominant allele	L_1	*L-1-a*	*Hwa-1*	*Hwa1-1*	*Hybrid weakness a1-1*	P.T.B.10, A.D.T.4, A.D.T.14, M.T.U.9, P.T.B.8
Recessive allele	+	+	*Hwa-1$^+$*	*hwa1-2*	*hybrid weakness a1-2*	
Locus/gene				*HWA2*	*HYBRID WEAKNESS A2*	

Dominant allele	L_2	L-1-b	Hwa-2	Hwa2-1	Hybrid weakness a2-1	P.T.B.7, accession 418
Recessive allele	+	+	Hwa-2 $^+$	hwa2-2	hybrid weakness a2-2	

According to Oka (1957), hybrids carrying both *Hwa1-1* and *Hwa2-1* showed the following symptoms: germination and growth in the seedling stage were quite normal until the seedlings had developed three to four leaves, at which time growth stopped and the leaves yellowed. The phenotype differs considerably from that of hybrids carrying both *Hwc1-1* and *Hwc2-1*, of which the symptoms are characterized by root growth inhibition appearing just 5 days after germination (Ichitani et al. 2001; Saito et al. 2007).

The distribution of *Hwa1-1* and *Hwa2-1* genes was limited to some Indian cultivars (Oka 1957). Based on results of three-way cross combinations, Oka (1957) reported that P.T.B.10 (accession 414) and P.T.B.7 (accession 419) carry *Hwa1-1* and *Hwa2-1*, respectively, whereas Pei-ku (accession 108), Padi ase banda (accession 647), and Kissin (accession 521) carry neither *Hwa1-1* nor *Hwa2-1*. Oka (1957) examined many different cultivars for the distribution of these genes. The four Indian cultivars—A.D.T.4 (accession 415), A.D.T.14 (accession 417), M.T.U.9 (accession 420), and P.T.B.8 (accession 421)—showed weakness in hybrids with P.T.B.7, indicating that these four cultivars also carry *Hwa1-1*. Actually, accession 418, an Indian cultivar whose name is unknown, showed weakness in the hybrid with P.T.B.10, indicating that accession 418 carries *Hwa2-1*. All cultivars relating to this hybrid weakness belong to Indica. The other cultivars that were examined carry neither *Hwa1-1* nor *Hwa2-1*. Information about the gene carriers of *Hwa1-1* and *Hwa2-1* is presented in Table 1.

The molecular mechanism of the hybrid weakness has remained unknown. Elucidating the mechanism necessitates clarification of the causal genes and their gene products. We selected a map-based cloning strategy to identify the two causal genes. As a starting point of map-based cloning, this report describes the chromosomal locations of *HWA1* and *HWA2* genes. It is particularly interesting that both genes are located in the 1,637-kb region, surrounded by the same DNA markers on the long arm of chromosome 11. The possibility of allelic interaction inducing hybrid weakness is discussed in this report.

RESULTS

Confirmation of the Experiment by Oka (1957)

Indian cultivars carrying *Hwa1-1* or *Hwa2-1* gene, P.T.B.10, P.T.B.7, and A.D.T.14, were provided by the Genebank of the National Institute of Agrobiological Sciences, Japan. We produced crosses among these three cultivars and confirmed that the F_1 hybrids from the cross between P.T.B.7 and P.T.B.10, and those from the cross between P.T.B.7 and A.D.T.14 showed the weakness symptom, as reported by Oka (1957) (Fig. 1). The F_1 hybrids from the cross between P.T.B.10 and A.D.T.14 showed normal growth. These results indicate that these Indian cultivars carry one of the hybrid-weakness-causing genes reported by Oka (1957).

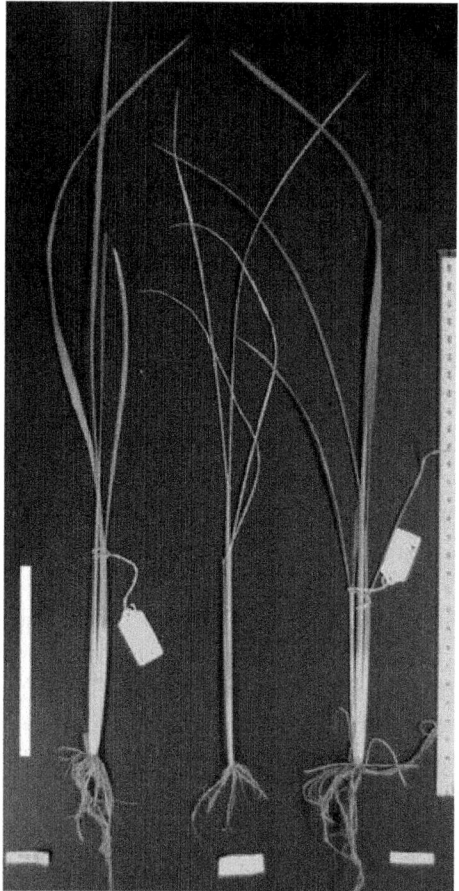

Figure 1. Hybrid weakness caused by two complementary genes: *Hwa1-1* and *Hwa2-1*. Seedlings are shown 40 days after sowing date: *from left to right*, P.T.B.10, F_1 de-

rived from the cross between P.T.B.10 and P.T.B.7, and P.T.B.7. The F$_1$'s developed leaves have turned yellow from the tip. *Bar* shows 10 cm.

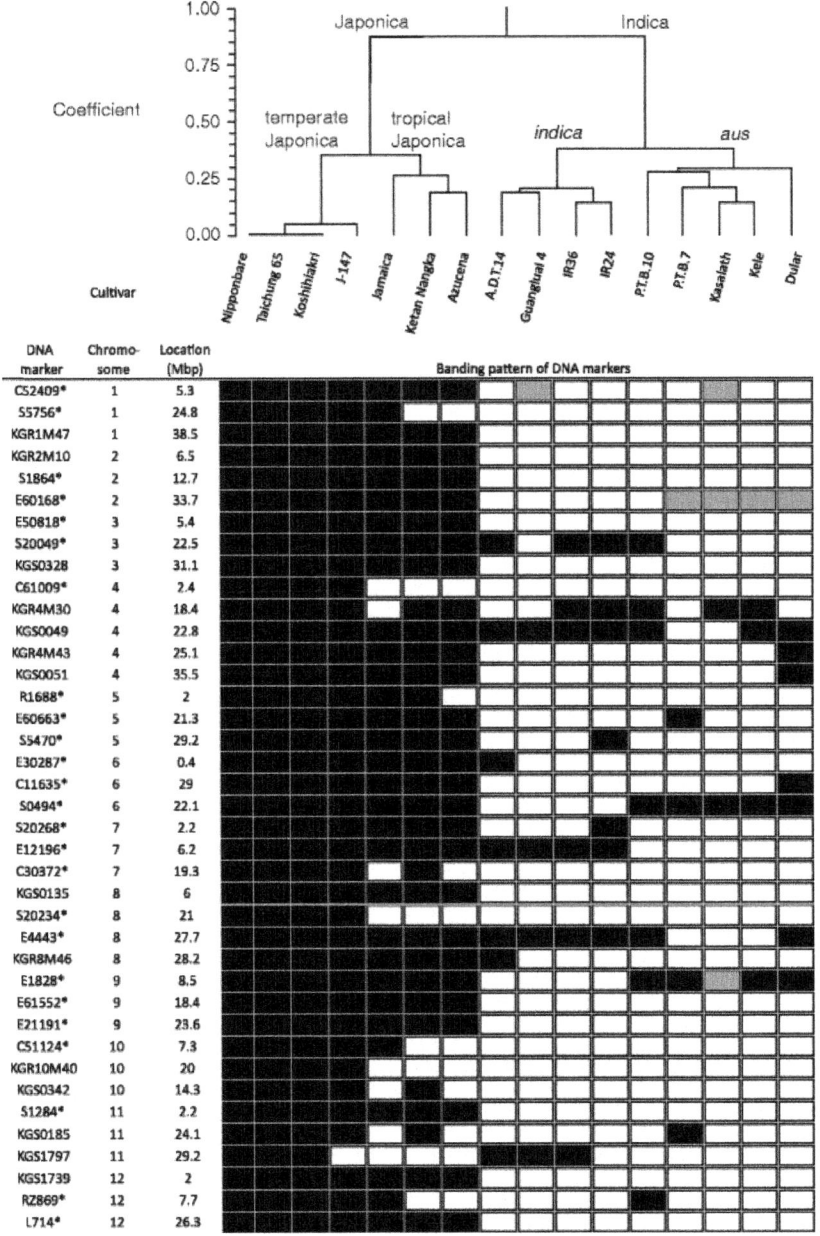

Figure 2. Classification of the Indian cultivars carrying *Hwa1-1* or *Hwa2-1* gene and 13 reference cultivars into four varietal groups based on banding patterns of 39 DNA

markers covering all 12 rice chromosomes. DNA markers with an *asterisk* (*) were developed by RGP (http: //rgp.dna.affrc.go.jp/E/publicdata/caps/index.html); those *without an asterisk* were developed in this study (Table 2). The same banding patterns as Nipponbare are indicated as *solid rectangles*. Other patterns are shown as *open* and *shaded rectangles*. Shown*above* are genetic relations among the 16 cultivars obtained using UPGMA cluster analysis. Each fragment size of the 39 markers was treated as a unique characteristic and scored as present (1) or absent (0). The data matrix was used to calculate genetic similarities using the Jaccard (1908) coefficient.

Oka (1957) reported that all cultivars related to this hybrid weakness belong to Indica. If that is true, then the mapping population of *HWA1* and*HWA2* should be constructed from the cross between these cultivars and Japonica cultivars because DNA polymorphism between Indica cultivars and Japonica cultivars is frequently observed. To confirm the classification by Oka (1957), the three Indian cultivars were compared with 13 reference cultivars (see Materials and methods for details) with respect to the banding patterns of polyacrylamide gel electrophoresis of 39 PCR-based insertion/deletion (indel) markers covering all 12 rice chromosomes (Fig. 2, Table 2).

Table 2. Primer sequences and locations of the indel markers designed for this study

Marker name		Primer sequences (5'–3')	Location on IRGSP pseudomolecules Build05			Source of DNA polymorphism information
			Chromosome	Position		
				From	To	
KGR1M47	F	CACAAATAGAATTACTGATGAAACCTT	1	38529588	38529747	Shen et al. 2004
	R	CGTTACCGCTTATGTAGAGTCATC				
KGR2M10	F	CATACATGGATGTCTAGTCGAAGA	2	6465517	6465683	Shen et al. 2004
	R	CAGTTCCAGTAAGTACATGGGTTT				
KGS0328	F	ATCTAGCAAAATTATTCGAGCAGAA	3	31067890	31068058	Monna et al. 2006
	R	ACTTTACAGTAACAAGGGGTGCAA				
KGR4M30	F	CAAAATAGGGAGGCAGATAGACA	4	18395635	18395794	Shen et al. 2004
	R	CTCCTGGTTGTATGCTCGTAAAT				
KGS0049	F	AAATTATACTTCCAATCAAGCATCAAG	4	22809557	22809823	Monna et al. 2006
	R	AATTGTATTGGATTGGAGCTGGT				
KGR4M43	F	GAGGTTATCCTCCCTAACACCAG	4	25116441	25116556	Shen et al. 2004
	R	TGCCAAATACAATATGACCACAA				
KGS0051	F	CGAAAACAGTTAGGTGTTTGTTAGG	4	35483948	35484104	Monna et al. 2006
	R	CTTACATGTAGTACAAACAACCCACA				
KGS0494	F	GATGGCTAGCTTGACTCCTGAATA	6	22083785	22084001	Monna et al. 2006
	R	CTCTAGGATACAATCATGGCAAACT				
KGS0135	F	ACCTCATTTTATTTCAACATTGCAG	8	6028192	6028300	Monna et al. 2006
	R	CTTCCTGACCAAGTTAAACCCCTA				
KGR8M46	F	CGAATAATTTGTAGCCGAGAAAA	8	28228891	28229086	Shen et al. 2004
	R	GCAGAGTCCAGAGAAGATCCAT				
KGR10M40	F	CGTTTAATTTACGTGCGAATAGG	10	19966648	19966807	Shen et al. 2004
	R	AACCTGAGGCACTAGTTCGTTATC				
KGS0342	F	ATACACACAGCAGACATAAGGTGAT	10	14305320	14305584	Monna et al. 2006
	R	AATAACCGTTTGATTGGACTAAAAA				
KGS0185	F	TGACTACAGAAATAGTGCAGCTTCT	11	24091407	24091582	Monna et al. 2006
	R	CCCCCTCCTTGACTTTGG				
KGC11M1	F	GGCAGGAGAGGAGAAACTGA	11	25263891	25263963	This study
	R	GAAGAAAGTGACCATGGATGAA				
KGS1797	F	AGTGGTGAGCTGCTAACAAATCTCT	11	29167452	29167572	Monna et al. 2006
	R	GCCGAGCGAGCTGAGTATC				
KGS1739	F	AGAGACGCAGGAGCTGCTTA	12	1999528	1999818	Monna et al. 2006
	R	CATGACCCTTCTATGGCAATTAT				

The relations among the 16 cultivars were revealed by UPGMA cluster analysis (Fig. 2). The 16 cultivars were classified clearly into two groups corresponding to Japonica and Indica. All cultivars related to this hybrid weakness belong to Indica, which is consistent with the report by Oka (1957). The Japonica group was divided into temperate Japonica and tropical Japonica. The four cultivars in temperate Japonica showed the same banding patterns with 39 DNA markers, except that J-147 showed a different pattern with KGS1797. The three cultivars in tropical Japonica showed the same banding patterns as temperate Japonica does with 28 DNA markers. The Indica group was divided into two subgroups, with one subgroup comprising IR36, IR24, Guangluai 4, and A.D.T.14, and the other comprising Kasalath, Kele, Dular, P.T.B.10, and P.T.B.7. Garris et al. (2005) classified 234 rice cultivars into five groups: *indica*, *aus*, *aromatic*, *temperate japonica*, and *tropical japonica*. Then IR36 was categorized into *indica*, and Kasalath into *aus*. In the present study, according to Garris et al. (2005), the subgroup comprising Kasalath and the other subgroup comprising IR36 will be designated respectively hereinafter as *aus* and *indica* (Fig. 2). Regarding the three cultivars related to this hybrid weakness, A.D.T.14 was found to belong to *indica*, whereas P.T.B.7 and P.T.B.10 belong to *aus*.

Mapping of *HWA1* and *HWA2*

The three cultivars proved to belong to Indica, as reported by Oka (1957). They showed polymorphism with Japonica cultivars with high frequency (Fig. 2). We selected a Taiwanese temperate Japonica cultivar Taichung 65 (T65) and produced hybrids among the three Indian cultivars and T65 because much polymorphism has been noted among the three Indian cultivars and T65. Furthermore, because it is weakly photoperiod sensitive, hybridization between T65 and other cultivars can be accomplished more easily. All hybrids from crosses between T65 and the three cultivars showed normal growth, indicating that T65 carries neither *Hwa1-1* nor *Hwa2-1*.

Then, linkage analysis of *HWA1* and *HWA2* was conducted using a three-way cross population. For the location of *HWA1*, the F_1 plants derived from the cross between T65 and P.T.B.10 were crossed with P.T.B.7. The three-way cross population [P.T.B.7 × (T65 × P.T.B.10)] ($n=176$) was divided into 105 weak plants carrying both *Hwa1-1* and *Hwa2-1*, and 71 normal plants carrying only the *Hwa2-1* gene. The observed ratio did not fit the expected ratio 1: 1 ($x^2=6.57$, $P=0.01038$). We inferred that the deviation from the expected ratio was probably caused by the linkage between *HWA1* gene and gametophytic reproductive barrier gene(s), which is often found in crosses between Indica and Japonica (see Discussion). Six typical weak plants and six typical normal plants were selected from the three-way cross population. They were subjected to preliminary linkage analysis with 83 DNA markers distributed on the whole rice genome (for details, see Materials and methods). A clear linkage was detected between *HWA1* and DNA markers on the long arm of chromosome 11, RM5349 (McCouch et al. 2002), and RM224 (Chen et al. 1997). Linkage of *HWA1* with RM5349 and RM224 was confirmed by adding the 82 three-way cross plants (Fig. 3).

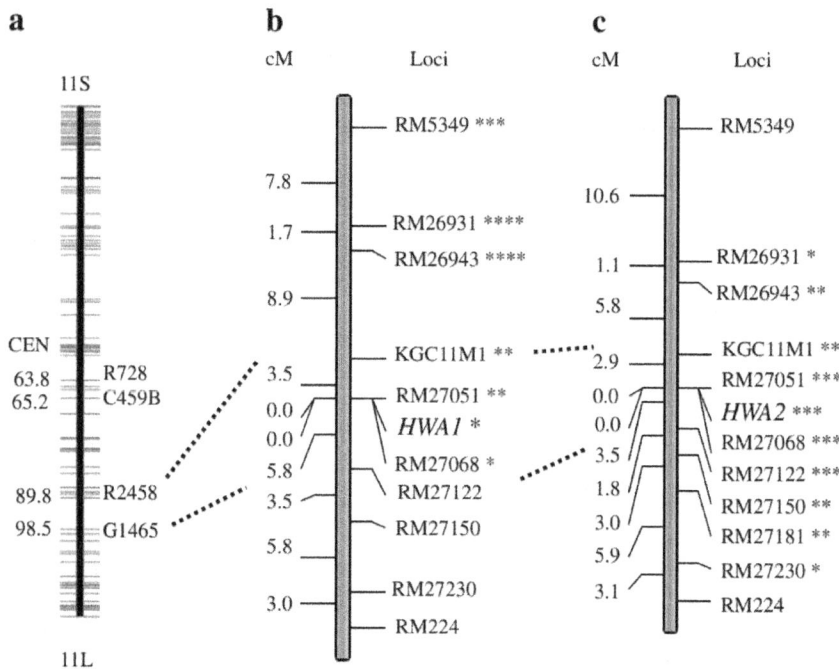

Figure 3. Linkage map showing the location of *HWA1* and *HWA2* on rice chromosome 11. **a** RFLP framework map of chromosome 11 quoted from Harushima et al. (1998). **b** Linkage map of *HWA1* constructed from the three-way cross population [P.T.B.7 × (T65 × P.T.B.10)] (*n* = 176). **c** Linkage map of *HWA2* constructed from the three-way cross population [A.D.T.14 × (T65 × P.T.B.7)] (*n* = 175). Genotypes of RM5349 and RM224 were analyzed for 94 plants in *HWA1* mapping and for 96 plants in *HWA2* mapping to know if these genes are located between the two markers. *, **, ***, and **** mean that the gene segregation was significantly deviated from the expected ratio (1: 1) at 0.05, 0.01, 0.001, and 0.0001 levels, respectively.

Along with the mapping of *HWA1*, linkage analysis of *HWA2* was performed using the same strategy. The F_1 plants derived from the cross between T65 and P.T.B.7 were crossed with A.D.T.14. The three-way cross population [A.D.T.14 × (T65 × P.T.B.7)] (*n* = 175) was divided into 110 weak plants carrying both *Hwa1-1* and *Hwa2-1*, and 65 normal plants carrying only the *Hwa1-1* gene. The observed ratio did not fit the expected ratio 1: 1 (^{2}=11.57, P=0.00067), probably because of the reproductive barrier, as described above. Six typical weak plants and six typical normal plants that had been selected from the three-way cross population were subjected to preliminary linkage analysis with 84 DNA markers distributed throughout the rice genome (see Materials and methods), indicating that *HWA2* was also linked with RM5349 and RM224. Linkage of *HWA2* with RM5349 and RM224 was confirmed by adding the 84 three-way cross plants (Fig. 3).

We selected seven and eight polymorphic simple sequence repeats (SSR) markers reported by the International Rice Genome Sequencing Project (IRGSP) (2005), respectively, for mapping of *HWA1* and *HWA2*. Linkage analysis using these markers revealed a large gap separating RM26943 and RM27051 (Fig. 3). Although we also examined several published SSR markers in IRGSP (2005), they were not applicable to the linkage analysis. To fill the gap, we detected a four-base-pair indel between Nipponbare and Kasalath by comparing the genome sequence of Nipponbare and the BAC-end sequence of Kasalath in the chromosomal region. The PCR products from the primer pair encompassing the indel showed clear banding patterns. Polymorphism between Nipponbare and Kasalath was conserved between T65 and the three Indian cultivars. The DNA marker using this indel was named KGC11M1 (Table 2), which we added to the linkage analysis. The linkage analysis of *HWA1* and eight DNA markers using 176 three-way cross plants showed that *HWA1* was located between KGC11M1 and RM27122, and that it cosegregated with RM27051 and RM27068 (Fig. 3). Linkage analysis of *HWA2* and nine DNA markers using 175 three-way cross plants showed that *HWA2* was also located between KGC11M1 and RM27122, and that it cosegregated with RM27051 and RM27068 (Fig. 3). To sum up, both *HWA1* and *HWA2* genes cosegregated with RM27051 and RM27068, located on the 1,637-kb region surrounded by KGC11M1 and RM27122 on the long arm of chromosome 11.

DISCUSSION

In this study, both *HWA1* and *HWA2* were located between the two DNA markers on chromosome 11: KGC11M1 and RM27122. Harushima et al. (1998) constructed an often-cited restriction fragment length polymorphism (RFLP) marker-based high-density linkage map for rice, in which some RFLP markers have been sequenced. Based on Nipponbare genome sequence, KGC11M1 and RM27122 were close to the two RFLP markers R2458 and G1465, which were reported respectively at positions 89.8 and 98.4 in the map by Harushima et al.

(1998). That location indicates that *HWA1* and *HWA2* are in the chromosomal region surrounded by these markers (Fig. 3).

In the three-way cross-mapping populations, gene segregation was not fitted to the expected ratio, but it was skewed toward weakness-causing alleles for both *HWA1* and *HWA2* loci. Sawamura and Sano (1997) reported that P.T.B.10 carries a gamete eliminator of S_{11} (t), which is linked to *la* on chromosome 11. The P.T.B.10 allele S_{11} (t) induces abortion of gametes carrying the T65 allele $S_{11}{}^a$ (t) only in heterozygotes. The recombination value between S_{11} and *la* (lazy habit) was estimated to be 0.06 ± 0.03 to 0.23 ± 0.23. Miura et al. (2003) conducted linkage analysis of *la* with RFLP markers. The *la* gene was located between R728 and C459B, of which the positions were 63.8 and 65.2, respectively, in the map by Harushima et al. (1998). These experimentally obtained results suggest that distorted segregation at the*HWA1* locus in the three-way cross [P.T.B.7 × (T65 × P.T.B.10)] is attributable to S_{11} (t) linked to *Hwa1-1*. Distorted segregation at the *HWA2*locus in the three-way cross [A.D.T.14 × (T65 × P.T.B.7)] reports that P.T.B.7 might carry the S_{11} (t) allele at S_{11} (t) locus. The heterozygote (S_{11} (t)$S_{11}{}^a$ (t)) induces female and male infertility (Sawamura and Sano 1997). The fact that F_1 plants from the cross (T65 × P.T.B.10) and those from the cross (T65 × P.T.B.7) both showed semi-sterile panicles (Ichitani et al. unpublished results) also suggests that a gamete eliminator such as S_{11} (t) is attributable to the segregating distortion in both cross combinations. However, the peaks of distortion differed between the two populations, implying that a gene(s) other than S_{11} (t) was responsible for the distortion in the three-way cross [A.D.T.14 × (T65 × P.T.B.7)].

Two genetic models can explain hybrid weakness. According to one model, hybrid weakness results from interaction among non-allelic genes: the causal genes of another hybrid weakness in rice, *HWC1* and *HWC2*, are located on the different chromosomes (Ichitani et al.2007; Kuboyama et al. 2009). This case is true for the *A. thaliana* case (Bomblies et al. 2007). The candidate genes in the mapped regions of*HWC1* and *HWC2* do not mutually overlap; they belong to distinct gene families. The other model suggests allelic interaction: a hybrid necrosis symptom observed in interspecific crosses among *Gossypium* was explained by interlocus and intralocus interaction, shown respectively by the dominant *Le*$_2{}^{dav}$ gene from *Gossypium davidsonii* at *Le*$_2$ locus and *Le*$_1$ and/or *Le*$_2$, dominant alleles from *Gossypium hirsutum* at the *Le*$_1$ and *Le*$_2$ loci (Lee 1981). These two loci are thought to be on mutually homoeologous chromosomal segments on the *G. hirsutum* allotetraploid genome (Samora et al. 1994). Increased gene dosage of *Le*$_1$ and *Le*$_2$ with *Le*$_2{}^{dav}$ reportedly hastens necrosis (Rooney and Stelly 1990). In the *Gossypium* case, no candidate gene information was reported. Regarding the reproductive barrier in rice, causal genes were cloned in an allelic interaction case: heterozygotes of Indica allele *S5-i* and Japonica allele *S5-j* at the *S5* locus induce embryo-sac semi-sterility caused by partial abortion of female gametes carrying the Japonica alleles. Chen et al. (2008) cloned the *S5* gene, demonstrating that *S5* encodes an aspartic protease conditioning embryo-sac fertility, the allelic difference being very small at the sequence level. Chen et al. (2008) did not explain the molecular mechanism caused by the heterozygous form at the *S5* locus. In another case, causal genes of reproductive barrier in rice, which had been thought to be allelic before, proved to be a complex of two very tightly linked genes. At the hybrid male sterility locus *Sa*, heterozygotes of Indica allele Sa^i and Japonica allele Sa^j exhibited male semi-sterility. Long et al. (2008) cloned the *Sa* gene, finding that it comprises two adjacently located genes, *SaM* and *SaF*, which respectively encode a small ubiquitin-like modifier E3 ligase-like protein and an F-box protein. Most Indica cultivars contain a haplotype SaM^+SaF^+, whereas all Japonica cultivars have SaM^-SaF^-. Interaction of SaM^- and SaF^+ in the presence of SaM^+ is necessary for this male sterility.

In this study, the physical distance between the KGC11M1 and RM27122 markers was found to be 1,637 kb. According to the Rice Genome Annotation Project (Ouyang et al. 2007, http: //rice.plantbiology.msu.edu/), more than 100 genes are located in the region. We were unable to determine whether *Hwa1-1* and *Hwa2-1* are allelic or not, but the possibility exists that heterozygotes induce the hybrid weakness symptom. To elucidate their relations, whether by tight linkage or allelism, larger mapping populations must be analyzed.

A causal gene controlling hybrid necrosis in *Arabidopsis* was cloned. Reportedly, it encodes an NB-LRR protein (Bomblies et al. 2007). We fine-mapped *HWC2* gene in rice, and narrowed down the area of interest to 19 kb, identifying five candidate genes, one of which encodes an NB-LRR protein (Kuboyama et al. 2009). Alcázar et al. (2009) fine mapped a causal gene of incompatibility or hybrid breakdown to a cluster of TIR-NB-LRR genes in *Arabidopsis*. Yamamoto et al. (2010) reported that *hbd3*, a causal gene controlling hybrid breakdown in rice, is located on the cluster of NB-LRR genes. These reports suggest that plant immune systems conditioned by NB-LRR protein(s) induce hybrid weakness, as reported by Bomblies and Weigel (2007). According to the Rice Genome Annotation Project (Ouyang et al. 2007, http: //rice.plantbiology.msu.edu/), two NB-LRR disease-resistance gene homologs, LOC_Os11g39160 and LOC_Os11g39260, and ten disease resistance-related genes containing NB-ARC domain are located on the target region of *HWA1* and *HWA2*, implying that these genes are causal genes. However, many other genes are also located in the target region. No NB-LRR disease resistance gene homolog is located in the *HWC1* target region (Ichitani et al. 2007). To clarify the fundamental mechanism causing hybrid weakness, many cross combinations causing hybrid weakness in various species should be studied at the molecular level. The cloning of *HWA1* and *HWA2* might provide new information related to hybrid weakness.

Results of cluster analysis show that A.D.T.14 carrying *Hwa1-1* belongs to *indica*, whereas P.T.B.10 carrying *Hwa1-1* and P.T.B.7 carrying*Hwa2-1* belong to *aus*, which suggests a lack of relation between varietal differentiation and the genotypes at the *HWA1* and *HWA2*. However, the three cultivars are insufficient to support discussion of the distribution of these genes. Having learned recently that all the plant materials used by Oka (1957) have been maintained in the National Institute of Genetics, Japan, we plan to use these cultivars to find *Hwa1-1-* or*Hwa2-1*-specific conserved haplotypes of DNA markers. Regarding *HWC2* genes, we found that the DNA marker banding patterns of 14 markers surrounding the *HWC2* locus were conserved completely among the 13 *Hwc2-1* carriers (Kuboyama et al. 2009). We will find potential carriers with the aid of DNA markers from our rice germplasm collection and make test crosses to determine whether they really carry the genes if such a haplotype is found in the *Hwa1-1* or *Hwa2-1* carriers. We plan to add these gene carriers in the cluster analysis to confirm the relation between varietal differentiation and these genes. The increase in the number of gene carriers might disclose conserved haplotypes in the *HWA1* or *HWA2* locus, which help to identify the causal genes.

To identify the causal genes and to confirm whether they are allelic or not, we are undertaking high-resolution mapping and linkage disequilibrium analysis of both *HWA1* and *HWA2* genes. We have also started physiological and close morphological analyses of the hybrid weakness. In three-way cross populations, there existed a few intermediate type plants between normal and weak ones, partly because of diverse genetic backgrounds in the populations (see Materials and methods). We are introducing the *Hwa1-1* gene from P.T.B.10 and A.D.T.14, and the *Hwa2-1* gene from P.T.B.7 in T65 genetic background by backcrossing to develop near-isogenic lines of these genes. Using these near-isogenic lines, we will evaluate the effects of these genes more exactly.

MATERIALS AND METHODS

Plant Materials

The three cultivars related to this hybrid weakness, P.T.B.7, P.T.B.10, and A.D.T.14, were compared with the 13 reference cultivars with regard to the banding patterns of polyacrylamide gel electrophoresis of 39 PCR-based indel markers (Fig. 2). Among the cultivars, Nipponbare and Koshihikari are improved Japanese cultivars, and T65 is an improved Taiwanese cultivar. These three cultivars are generally classified as temperate Japonica. J-147 is a native cultivar in Japan (Sato and Morishima 1988). Ketan Nangka, Azucena, and Jamaica are native cultivars respectively originating in Malaysia, the Philippines, and Peru. They are generally classified as tropical Japonica. Three cultivars generally classified as Indica are Guangluai 4, an improved Chinese cultivar, and IR36 and IR24 developed by the International Rice Research Institute. Kele, Dular, and Kasalath cultivars are native to India. Kele and Dular are categorized as *aus* in Wan and Ikehashi (1997). Kasalath has been categorized as Indica in many studies; DNA marker polymorphism has been observed with high frequency between Kasalath and Japonica cultivars. According to classification by Garris et al. (2005) based on SSR markers or Zhao et al. (2010) based on single nucleotide polymorphism—which divided rice cultivars into *temperate japonica*, *tropical japonica*, *indica*, *aus*, and one more group—Nipponbare and Koshihikari were classified as *temperate japonica*, Azucena as *tropical japonica*, IR36 as *indica*, Kasalath as *aus*, and Dular as admixed.

For hybridization, flowering times of plant materials were mutually synchronized by the use of many sowing dates. Seeds were sown in nurseries with plant spacing of 3×3 cm. One month after sowing, seedlings were transplanted in a paddy field in the experimental farm of Kagoshima University.

The three-way cross populations in which *HWA1* or *HWA2* gene segregated were grown in nurseries along with parental cultivars and F_1 plants inducing the weakness. Plant spacing was 3×6 cm.

DNA Marker Analysis

The DNA of the cultivars used in the cluster analysis was extracted according to Dellaporta et al. (1983) with some modifications. The DNA of the mapping populations was extracted according to the experimental protocols of Rice Genome Research Program (RGP) (http: //rgp.dna.affrc.go.jp/E/rgp/protocols/index.html, written in Japanese) with some modifications. Briefly, each young leaf tip, 2 cm long from a single plant, was put on a well in a 96-deep well plate. Then 100 µl of extraction buffer (100 mM Tris–HCl (pH 8.0), 1 M KCl, and 10 mM EDTA) were added with a 5-mm-diameter stainless steel ball in a well. After covering with a hard lid, the plate was shaken hard (ShakeMaster ver. 1.2; BioMedical Science Inc.) for a minute to grind the leaves. After centrifuging, a plate was incubated at 70°C for an hour. Then 9 µl of supernatant was recovered and 7 µl of 2-propanol was added. After centrifuging, the supernatant was discarded and the DNA pellet was rinsed with 50 μl of 70% ethanol. The DNA pellet was dried and dissolved in 50 μl of sterilized distilled water.

The PCR conditions for indel and SSR markers used for this study were 95°C for 10 min, 40 cycles of 94°C for 30 s, 55°C for 30 s, and 72°C for 30 s with subsequent final extension of 72°C for 1 min. The PCR mixture (5 μl) contained 1 ml of template DNA, 200 mM of each dNTP, 0.2 μM of primers, 0.25 units of Taq polymerase (AmpliTaq Gold; Applied BioSystems), and 1× buffer containing $MgCl_2$. The PCR products were analyzed using electrophoresis in 10% (29: 1) polyacrylamide gel with subsequent ethidium bromide staining and were viewed under ultraviolet light irradiation. Most PCR–based DNA markers used for this study have already been published: Markers with an asterisk in Fig. 2were designed by RGP (http: //rgp. dna.affrc. go.jp/E/publicdata/caps/index.html). Information related to SSR markers was obtained from Panaud et al. (1996), Chen et al. (1997), Temnykh et al. (2001), McCouch et al. (2002), and IRGSP (2005). Some primer pairs did not perform well. Therefore, we redesigned them or developed new DNA markers.

Most of the new markers were based on the indel between Nipponbare and an indica cultivar 93–11 (Shen et al. 2004) or those between Nipponbare and Kasalath or an indica cultivar Guangluai 4 (Monna et al. 2006). Markers were named by adding KG to the original indel information ID, R*** (Shen et al. 2004) or S**** (Monna et al. 2006). In redesigning or developing DNA markers, we made alignments of Nipponbare genome sequence (IRGSP 2005), 93–11 genome sequence (Yu et al. 2002) and/or BAC-end sequence of Kasalath (http: //rgp.dna.affrc.go.jp/blast/runblast.html, Katagiri et al. 2004) using DNAsis Pro (Hitachi Software Engineering Co.). Primer design was performed using Primer 3 (Rozen and Skaletsky 2000). The sequences of primers designed in this study are presented in Table 2.

Cluster Analysis

Genetic relations among the 16 cultivars (Fig. 2) were evaluated using the 39 DNA markers. Each fragment size was treated as a unique characteristic, and scored as present (1) or absent (0). The data matrix was used to calculate genetic similarities using the Jaccard coefficient (Jaccard 1908). A phenogram was constructed using the unweighted pair-group method with the arithmetic mean (UPGMA) with software (NTSYS-pc ver. 2.2; Exeter Software).

Mapping Strategy

About 1 month after the sowing date, the seedlings in the three-way cross populations, [P.T.B.7 × (T65 × P.T.B.10)] ($n=176$) and [A.D.T.14 × (T65 × P.T.B.7)] ($n=175$), developed three to four leaves. Plants of two types appeared. One type was characterized by the yellowing of leaf tips and weak growth as F_1 plants (Fig. 1). It was categorized as a weak plant carrying both *Hwa1-1* and *Hwa2-1* genes. Another type showed normal growth as parental cultivars. It was categorized as a normal plant carrying only *Hwa1-1* or *Hwa2-1*. The tips of lower leaves of normal plants sometimes yellowed also, but the tips of upper leaves remained green. In contrast, in weak plants, all leaf tips except those of the youngest (still expanding) ones yellowed. Therefore, the difference between weak plants and normal plants was clear in most cases. A few intermediate type plants could not be categorized into the above two types. They were excluded from analyses.

For *HWA1* mapping, a total of 83 DNA markers distributed on the whole rice genome were used: 24 polymorphic indel markers in Fig. 2, an indel marker named S21074 designed by RGP, another indel marker named Knindel1 (Matsubara et al. 2007), a sequence characterized amplified region (SCAR) marker named d (Ueda et al. 2005), and 56 published SSR markers. For *HWA2* mapping, a total of 84 DNA markers distributed on the whole rice genome were used: the 23 polymorphic indel markers presented in Fig. 2, 59 SSR markers, and two indel markers designed by RGP—S21074 and S5865.

In each mapping population, six typical weak plants and six typical normal plants were selected and subjected to preliminary linkage analysis with DNA markers, with the result that the linkage of *HWA1* and *HWA2* with RM5349 (McCouch et al. 2002) and RM224 (Chen et al. 1997) was visible on the long arm of chromosome 11. Then linkage analysis using all the plants in the mapping populations and polymorphic DNA markers surrounded by RM5349 and RM224 was conducted using a computer program (MapDisto ver. 1.7; Lorieux 2007). Map distances were estimated using the Kosambi function (Kosambi 1944).

ACKNOWLEDGMENTS

We gratefully acknowledge Dr. Atsushi Yoshimura of Kyushu University and Dr. Lisa Monna of the Plant Genome Center for valuable information related to DNA polymorphism in rice. We are also grateful to Dr. Yo-Ichiro Sato of the Research Institute for Humanity and Nature, the Genebank of the National Institute of Agrobiological Sciences for the kind provision of rice cultivars. This work was funded by the Ministry of Education, Culture, Sports, Science and Technology.

REFERENCES

1. Alcázar R, García AV, Parker JE, Reymond M. Incremental steps toward incompatibility revealed by Arabidopsis epistatic interactions modulating salicylic acid pathway activation. Proc Natl Acad Sci USA. 2009; 106: 334–9.

2. Amemiya A, Akemine H. Biochemical genetic studies on the root growth inhibiting complementary lethals in rice plant (Studies on the embryo culture in rice plant. 3). Bull Nat Inst Agric Sci D. 1963; 10: 139–226. in Japanese with English summary.

3. Bomblies K, Lempe J, Epple P, Warthmann N, Lanz C, Dangl JL, et al. Autoimmune response as a mechanism for a Dobzhansky-Muller-type incompatibility syndrome in plants. PLoS Biol. 2007; 9: 1962–72.

4. Bomblies K, Weigel D. Hybrid necrosis: autoimmunity as a potential gene-flow barrier in plant species. Nat Rev Genet. 2007; 8: 382–93.

5. Chen J, Ding J, Ouyang Y, Du H, Yang J, Cheng K, et al. A triallelic system of *S5* is a major regulator of the reproductive barrier and compatibility of indica–japonica hybrids in rice. Proc Natl Acad Sci USA. 2008; 105: 11436–41.

6. Chen X, Temnykh S, Xu Y, Cho YG, McCouch SR. Development of a microsatellite framework map providing genome-wide coverage in rice (*Oryza sativa* L.). Theor Appl Genet. 1997; 95: 553–67.

7. Dellaporta SL, Wood J, Hicks JB. A plant DNA minipreparation: version II. Plant Mol Biol Rep. 1983; 1: 19–21.

8. Garris AJ, Tai TH, Coburn J, Kresovich S, McCouch S. Genetic structure and diversity in *Oryza sativa* L. Genetics. 2005; 169: 1631–8.

9. Harushima Y, Yano M, Shomura A, Sato M, Shimano T, Kuboki Y, et al. A high-density rice genetic linkage map with 2275 markers using a single F_2 population. Genetics. 1998; 148: 479–94.

10. Ichitani K, Fukuta Y, Taura S, Sato M. Chromosomal location of *Hwc2*, one of the complementary hybrid weakness genes, in rice. Plant Breeding. 2001; 120: 523–5.

11. Ichitani K, Namigoshi K, Sato M, Taura S, Aoki M, Matsumoto Y, et al. Fine mapping and allelic dosage effect of *Hwc1*, a complementary hybrid weakness gene in rice. Theor Appl Genet. 2007; 114: 1407–15.

12. International Rice Genome Sequencing Project (IRGSP). The map-based sequence of the rice genome. Nature. 2005; 436: 793–800.

13. Jaccard P. Nouvelles recherches sur la distribution florale. Bull Soc Vaudoise Sci Nat. 1908; 44: 223–70.

14. Katagiri S, Wu J, Ito Y, Karasawa W, Shibata M, Kanamori H, et al. End sequencing and chromosomal *in silico* mapping of BAC clones derived from an *indica* rice cultivar, Kasalath. Breed Sci. 2004; 54: 273–9.

15. Kinoshita T. Proposal for rules of gene symbolization. Rice Genet Newsl. 1984; 1: 4–15.

16. Kosambi D. The estimation of map distance from recombination values. Ann Eugen. 1944; 12: 172–5.

17. Kuboyama T, Saito T, Matsumoto T, Wu J, Kanamori H, Taura S, et al. Fine Mapping of *HWC2*, a complementary hybrid weakness gene, and haplotype analysis around the locus in rice. Rice. 2009; 2: 93–103.

18. Lee JA. Genetics of D_3 complementary lethality in *Gossypium hirsutum* and *G. barbadense*. J Hered. 1981; 72: 299–300.

19. Long Y, Zhao L, Niu B, Su J, Wu H, Chen Y, et al. Hybrid male sterility in rice controlled by interaction between divergent alleles of two adjacent genes. Proc Natl Acad Sci USA. 2008; 105: 18871–6.

20. Lorieux M. 2007. MapDisto, a free user-friendly program for computing genetic maps. Computer demonstration (p958) given at the plant and animal genome XV conference, Jan 13–17 2007, San Diego, CA. URL: http: //mapdisto.free.fr/.

21. Matsubara K, Ito S, Nonoue Y, Ando T, Yano M. A novel gene responsible for hybrid breakdown found in a cross between *japonica* and*indica* cultivars in rice. Rice Genet Newsl. 2007; 23: 11–3.

22. McCouch SR, Teytelman L, Xu Y, Lobos KB, Clare K, Walton M, et al. Development and mapping of 2240 new SSR markers for rice (*Oryza sativa* L.). DNA Res. 2002; 9: 199–207.

23. McCouch SR, Committee on Gene Symbolization, Nomenclature and Linkage, Rice Genetics Cooperative (CGSNL). Gene nomenclature system for rice. Rice. 2008; 1: 72–84.

24. Miura K, Ashikari M, Matsuoka M, Hsing YIC. Mapping of the lazy gene, *la*. Rice Genet Newsl. 2003; 20: 29–30.

25. Monna L, Ohta R, Masuda H, Koike A, Minobe Y. Genome-wide searching of single-nucleotide polymorphisms among eight distantly and closely related rice cultivars (*Oryza sativa* L.) and a wild accession (*Oryza rufipogon* Griff.). DNA Res. 2006; 13: 43–51.

26. Oka HI. Phylogenetic differentiation of cultivated rice. XV. Complementary lethal genes in rice. Jpn J Genet. 1957; 32: 83–7.

27. Ouyang S, Zhu W, Hamilton J, Lin H, Campbell M, Childs K, et al. The TIGR rice genome annotation resource: improvements and new features. Nucleic Acid Res. 2007; 35: D883–7.

28. Panaud O, Chen X, McCouch SR. Development of microsatellite markers and characterization of simple sequence length polymorphism (SSLP) in rice (*Oryza sativa* L.). Mol Gen Genet. 1996; 252: 597–607.

29. Rooney WL, Stelly DM. Genetic effects on the timing of Le_2^{dav} induced necrosis of cotton. Crop Sci. 1990; 30: 70–4.

30. Rozen S, Skaletsky HJ. Primer3 on the WWW for general users and for biologist programmers. In: Krawetz S, Misener S, editors. Bioinformatics methods and protocols: methods in molecular biology. Totowa: Humana Press; 2000. p. 365–86.

31. Saito T, Ichitani K, Suzuki T, Marubashi W, Kuboyama T. Developmental observation and high temperature rescue from hybrid weakness in a cross between Japanese rice cultivars and Peruvian rice cultivar 'Jamaica'. Breed Sci. 2007; 57: 281–8.

32. Samora PJ, Stelly DM, Kohel RJ. Localization and mapping of the Le_1 and Gl_2 loci of cotton (*Gossypium hirsutum* L.). J Hered. 1994; 85: 152–7.

33. Sato YI, Sano Y, Nakagahra M. Gene symbols for gametic effect, sterility and weakness. Rice Genet Newsl. 1987; 4: 46–51.

34. Sato YI, Morishima H. Distribution of the genes causing F_2 chlorosis in rice cultivars of the Indica and Japonica types. Theor Appl Genet. 1988; 75: 723–727.

35. Sawamura N, Sano Y. Chromosomal location of gamete eliminator, S_{11} (t), found in an Indica-Japonica hybrid. Rice Genet Newsl. 1996; 13: 70–2.

36. Shen YJ, Jiang H, Jin JP, Zhang ZB, Xi B, He YY, et al. Development of genome-wide DNA polymorphism database for map-based cloning of rice genes. Plant Physiol. 2004; 135: 1198–205.

37. Shii CT, Mok MC, Temple SR, Mok DWS. Expression of developmental abnormalities of *Phaseolus vulgaris* L. Interaction between temperature and allelic dosage. J Hered. 1980; 71: 219–22.

38. Temnykh S, DeClerck G, Lukashova A, Lipovich L, Cartinhour S, McCouch S. Computational and experimental analysis of microsatellites in rice (*Oryza sativa* L.): frequency, length variation, transposon associations, and genetic marker potential. Genome Res. 2001; 11: 1441–52.

39. Tezuka T, Kuboyama T, Matsuda T, Marubashi W. Possible involvement of genes on the Q chromosome of *Nicotiana tabacum* in expression of hybrid lethality and programmed cell death during interspecific hybridization to *Nicotiana debneyi*. Planta. 2007; 226: 753–64.

40. Ueda T, Sato T, Hidema J, Hirouchi T, Yamamoto K, Kumagai T, et al. *qUVR-10*, a major quantitative locus for ultraviolet-B resistance in rice, encodes cyclobutane pyrimidine dimer photolyase. Genetics. 2005; 171: 1941–50.

41. Wan J, Ikehashi H. Identification of two types of differentiation in cultivated rice (*Oryza sativa* L.) detected by polymorphism of isozymes and hybrid sterility. Euphytica. 1997; 94: 151–61.

42. Yamamoto E, Takashi T, Morinaka Y, Lin S, Wu J, Matsumoto T, et al. Gain of deleterious function causes an autoimmune response and Bateson–Dobzhansky–Muller incompatibility in rice. Mol Genet Genomics. 2010; 283: 305–15.

43. Yu J, Hu S, Wang J, Wong GK, Li S, Liu B, et al. A draft sequence of the rice genome (*Oryza sativa* L. spp. *indica*). Science. 2002; 296: 79–92.

44. Zhao K, Wright M, Kimball J, Eizenga G, McClung A, Kovach M, et al. Genomic diversity and introgression in *O. sativa* reveal the impact of domestication and breeding on the rice genome. PLoS One. 2010; 5: e10780. doi: 10.1371/journal.pone.0010780.

Chapter 7

STRUCTURE AND STABILITY OF TELOCENTRIC CHROMOSOMES IN WHEAT

Dal-Hoe Koo[1], Sunish K. Sehgal[2], Bernd Friebe[1], and Bikram S. Gill[1]

[1]Department of Plant Pathology, Wheat Genetics Resource Center, Throckmorton Plant Sciences Center, Kansas State University, Manhattan, KS, 66506–5502, United States of America

[2]Department of Plant Science, South Dakota State University, Brookings, SD, 57007, United States of America

ABSTRACT

In most eukaryotes, centromeres assemble at a single location per chromosome. Naturally occurring telocentric chromosomes (telosomes) with a terminal centromere are rare but do exist. Telosomes arise through misdivision of centromeres in normal chromosomes, and their cytological stability depends on the structure of their kinetochores. The instability of telosomes may be attributed to the relative centromere size and the degree of completeness of their kinetochore. Here we test this hypothesis by analyzing the cytogenetic structure of wheat telosomes. We used a population of 80 telosomes arising from the misdivision of the 21 chromosomes of wheat that have shown stable inheritance over many generations. We analyzed centromere size by probing with the centromere-specific histone H3 variant, CENH3. Comparing the signal intensity for CENH3 between the intact chromosome and derived telosomes showed that the telosomes had approximately half the signal intensity compared to that of normal chromosomes. Immunofluorescence of CENH3 in a wheat stock with 28 telosomes revealed that none of the telosomes received a complete CENH3 domain. Some of the telosomes lacked centromere specific retrotransposons of wheat in the CENH3 domain, indicating that the stability of telosomes depends on the presence of CENH3 chromatin and not on the presence of CRW repeats. In addition to providing evidence for centromere shift, we also observed chromosomal aberrations including inversions and deletions in the short arm telosomes of double ditelosomic 1D and 6D stocks. The role of centromere-flanking, pericentromeric heterochromatin in mitosis is discussed with respect to genome/chromosome integrity.

INTRODUCTION

The centromeres, an essential part of all chromosomes, are responsible for chromosome segregation at mitosis and meiosis. Centromeres usually contain highly repetitive DNA, e.g. satellite DNA, which is associated with proteins in higher eukaryotes [1,2]. Although centromeres are not conserved at the DNA sequence level, many core centromeres analyzed to date contain nucleosomes with a histone H3 variant, CENPA in humans [3] and CENH3 in plants [4]. Thus, these specialized nucleosomes serve as the primary marker of centromere identity. In most eukaryotes, centromere proteins assemble at a single location in each chromosome [5] and are visible as a primary constriction in mitotic metaphase chromosomes. The centromere within a chromosome can move to a new location and form a neocentromere, which is often associated with chromosomal rearrangements in humans [6]. Such neocentromeres also have been reported in plants [7–10].

Each chromosome can be identified based on the centromere position; metacentric, submetacentric, acrocentric, or telocentric [11]. Naturally occurring telocentric chromosomes (hereafter referred as telosomes) are rare in plants. Darlington [12] suggested that the absence of telosomes in plants was caused by their instability. Evidence for the existence of stable telosomes has been provided by Marks [13], Strid [14] and Schubert and Rieger [15] in plants and by Southern [16] and Takagi and Sasaki [17] in animals. According to White [18], experimentally produced telosomes are unstable; and thus, all telosomes are unstable. Experimentally produced telosome stocks in plants, such as wheat (*Triticum aestivum* L.), barley (*Hordeum vulgare* L.), rye (*Secale cereal* L.), and rice (*Oryza sativa* L.), are reported [19,20]. Barley telotrisomics are fairly stable, except for triplo 1L, which shows chimaerism [21]. The question arises, why are some telosomes stable whereas others are not?

Centromeres divide quite regularly at mitosis and meiosis. However, when a chromosome is univalent, centric misdivision may occur during meiosis giving rise to telosomes [12]. Such a chromosome aberration is lethal in diploid organisms but can be tolerated in polyploids. The polyploidy nature of hexaploid wheat, *T. aestivum* ($2n = 6x = 42$), tolerates aneuploidy with either the addition or deletion of chromosomes. Sears and Sears [19] developed several aueuploid stocks in Chinese Spring (CS) wheat including ditelosomic stocks (one chromosome pair is substituted by a pair of either short or long arm telosomes, $2n = 40+2t$; such stocks are nullisomic for one of the arms) and double ditelocentric stocks (one chromosome pair is substituted by a pair of short and long arm telosomes, $2n = 40+4t$; such stocks are euploid). These stocks were used intensively for centromere mapping and allocating genes and markers to specific chromosome arms [22,23]. Moreover, the telosomes in the ditelosomic (hereafter Dt) and double ditelosomic (hereafter dDt) stocks are about half the size of a metacentric chromosome, making them amenable for flow sorting [24–26].

Flow-sorted telosomes are the foundation material for chromosome-arm-based BAC libraries and the wheat physical maps developed under the auspices of the International Wheat Genome Sequencing Consortium project [27]. Recently, individual flow-sorted chromosome arms were used to generate a draft sequence of the 17-Gb wheat genome [28]. Even with the extensive use of telosomic stocks for genetic and genomic studies in wheat, their detailed cytogenetic nature is poorly understood. The cytological stability of a telosome depends on the structure of its kinetochore. Steinitz-Sears [29] reported that the relative instability of a telosome may be attributed to the degree of completeness of its kinetochore. Because wheat telosomic stocks were developed by centric misdivision and stably transmitted to progeny, they are supposed to have either complete or nearly complete kinetochores.

In this study, we first developed a wheat CENH3 antibody (see experimental procedure) and then used it to identify and characterize the functional centromeric region of the telosomes. Second, centromeric-specific retrotransposons of wheat (CRWs) [30,31] were used to study the structure of intact and telosomic chromosomes. D-genome-specific, repetitive DNA (pAs1) [32] and single-copy DNA probes [33] were used to identify chromosomes and characterize chromosomal rearrangements. In addition, chromosome-arm-specific molecular markers, derived from the wheat deletion bin map (http://probes.pw.usda.gov:8080/snpworld/Map) [34], were used to detect chromosomal aberrations. The data provide new insights into the structure and stability of telocentric chromosomes and their centromeres; the implications of these results to genetic and genomic studies of wheat are discussed.

MATERIALS AND METHODS

Plant Material and Chromosome Preparation

The cytogenetic stocks used in this study are listed (Table 1). Wheat cultivars, Chinese Spring (CS), Jagger, and TAM111, also were used in molecular and cytogenetical studies. For chromosome preparations, seeds were germinated in petri dishes on moist filter paper. Root tips (1–2 cm long) were treated overnight in ice water. The root tips were fixed overnight in a 3:1 ethanol:glacial acetic acid and then squashed in a drop of 45% acetic acid. For the immunofluorescence of CENH3, ice-cold-treated root tips were fixed immediately using 4% paraformaldehyde in PHEM (60 mM Pipes, 25 mM

Hepes, 10 mM EGTA, 2 mM $MgCl_2$, and 0.3 mM sorbitol, pH 6.8) for 40 min. After washing with 1x PBS (10 mM sodium phosphate, pH 7.0, and 140 mM NaCl), the root tips were treated with 2% cellulase, 1% pectinase (Sigma, St. Louis, MO) and 1% pectolyase in PHEM for 1 h and then squashed on poly-l-lysine coated slides (Sigma). All preparations were stored at -70°C until use.

Table 1. Wheat cytogenetic stocks used in this study.

Cytogenetic stock	Source/Plant ID
Dt1DS	WGRC, TA3087
Dt1DL	WGRC, TA3131
dDt1D	WGRC, TA3158
Dt2DS	WGRC, TA3123
Dt2DL	WGRC, TA3124
dDt2D	WGRC, TA3146
Dt3DS	WGRC, TA3193
Dt3DL	WGRC, TA3192
dDt3D	WGRC, TA3147
Dt4DS	WGRC, TA3125
Dt4DL	WGRC, TA3126
dDt4D	WGRC, TA3148
Dt5DS-Mt5DL	U.C Riverside
Dt5DL	WGRC, TA3127
dDt5D	WGRC, TA3149
Dt6DS	WGRC, TA3128
Dt6DL	WGRC, TA3129
dDt6D	WGRC, TA3150
Dt7DS	WGRC, TA3130
Dt7DL	WGRC, TA3071
dDt7D	WGRC, TA3151
dDt1A	WGRC, TA3132
dDt2A	WGRC, TA3133
dDt3A	WGRC, TA3134
dDt5A	WGRC, TA3136
dDt6A	WGRC, TA3137
dDt7A	WGRC, TA3138
CS dDt1B-dDt2B-dDt3B-dDt4A-dDt5B-dDt6B-dDt7B	WGRC, TA3356

Dt: ditelosome, one chromosome is represented by a pair of either short or long arm telosomes (2n = 40 +2t). dDt: double ditelosome, one chromosome is represented by one pair each of short (S) and long (L) arm telosomes (2n = 40+4t).

doi:10.1371/journal.pone.0137747.t001

Immuno-Detection of CENH3 and Fluorescence *in situ* Hybridization (FISH)

Wheat *CENH3* genes were described previously [31]. A peptide antigen, 'RTKHPAVRKTKALPKK', was synthesized and used to immunize rabbits at Thermo Fisher Scientific (www.thermofisher.com). The raised antisera were purified using an affinity sepharose column comprising the aforementioned peptide. The specificity of the antibody was checked by immunostaining of root tip and pollen mother cells of wheat (data not shown). Immuno-detection of CENH3 and FISH procedures followed previously published protocols [35–37]. The rabbit antibodies to CENH3 were diluted to 1:1000 in TNB buffer (0.1 M Tris-HCl, pH 7.5, 0.15 M NaCl, and 0.5% blocking reagent). Approximately 100 µL of the diluted antibodies was added to each slide, and the slides were incubated in a humid chamber at 37°C for 2–3 h. After three washes in 1x PBS, 100 µL of rhodamine-conjugated goat anti-rabbit antibody (Jackson ImmunoResearch, West Crove, PA) (1:100 in TNB buffer) was added to the slides. Incubation and washes were the same as for the primary antibody. DNA probes of the CRWs, pAs1, pSc119, and the other single-gene probes were labeled with digoxigenin-11-dUTP, biotin-16-dUTP, and/or DNP-11-dUTP, depending on whether two or three probes were used in the FISH experiment. The cDNA clones used in this study were supplied by the National BioResource Project-Wheat, Japan. After post-hybridization washes, the probes were detected with Alexafluor 488 streptavidin for biotin-labeled probes, and rhodamine-conjugated anti-digoxigenin for dig-labeled probe. The DNP-labeled probe was detected with rabbit anti-DNP, followed by amplification with a chicken anti-rabbit Alexafluor 647 antibody.

Multicolor immuno-FISH detection was described previously [36]. Chromosomes were counterstained with 4',6-diamidino-2-phenylindole (DAPI) in Vectashield antifade solution (Vector Laboratories, Burlingame, CA). The images were captured with a Zeiss Axioplan 2 microscope (Carl Zeiss Microscopy LLC, Thornwood, NY) using a cooled CCD camera CoolSNAP HQ2 (Photometrics, Tucson, AZ) and AxioVision 4.8 software. The final contrast of the images was processed using Adobe Photoshop CS5 software.

Sequential Detection of CENH3, CRWs, pSc119 and pAs1

For sequential detection the slides were first incubated with anti-CENH3 overnight at 4°C in a wet chamber. After washes in 1x PBS, the slides were incubated with the appropriate secondary antibody at 37°C for 50 min. Then slides were re-fixed with 4% paraformaldehyde at RT for 30 min. The slides were then denatured in 70% formamide in 2x SSC, 80°C for 2 min, washed in ice-cold 1x PBS for 5 min, and then DNA probe, CRWs, was applied to the slides. Post-hybridization wash and signal detection were the same as FISH procedure. After recording the both CENH3 and CRWs signals, the slides were washed in 4T (4x SSC/0.05% Tween 20) buffer for 1 h at 37°C and re-fixed with 4% paraformaldehyde and dehydrated in an ethanol series. The slides were re-probed with pAs1 and pSc119 to detect additional sequences on the same chromosome.

Genome-Specific Markers and PCR

The genome-specific primers used are listed in Table 2. PCR was performed with 15 μL of the reaction mixture containing 1x PCR buffer (Bioline USA Inc., Taunton, MA), 2 mM MgCl2, 0.25 mM dNTPs, 5 pmol forward primer and reverse primer, respectively, 0.02 U/μL Taq DNA polymerase (Bioline USA Inc., Taunton, MA), and 20 ng genomic DNA. PCR amplification was according to Liu et al. [38]. PCR products were resolved on 2.5% agarose gels and visualized by ethidium bromide staining under UV light.

Table 2. Wheat genome specific EST markers used in this study.

EST marker	Forward primer	Reverse primer	Deletion bin
BE405518	GTCTCAGGTATTGATTGATCCC	GCTGATGCTCCTTGATCTCC	1DS0.70–1.00
BE637971	TGCCTGATGTTTGATGCTCC	CAAAGCGAAGTGACTGTCCA	1DS0.70–1.00
BE444846	TCTTCGCCACAGGAGTACCTA	GGCTCGTAGCGGGTATACAA	1DS0.00–0.48
BE591601	GTTAGTGGCACTCCTACCTG	GATGTCCAACCATAATGCCC	1DS0.00–0.48
BE637864	TCCTCATTTTGTAATCCTTCTCTC	TTTTGTTCCCACCATCAGGT	1DS0.00–0.48
BF202643	GAATAGCAACAGTGCTCATGAAT	GAAGAACAGCAGGGCGTTAC	1DS0.00–0.48
BF474569	CGTACCAACTCAACCCCTC	TGAAGGGTGAGAGAACTCCG	1DS0.00–0.48
BF478737	CTCTTCACAGTTACAACATCAGC	TGAGGCTCAATGATGACCAG	1DS0.00–0.48
BE424523	CAGTAAGGAAATATGGCCGAT	TTGATGCAGAAAAAGTTGGAT	6DS0.79–0.99
BE490604	AAGCGGTTCCATCTCTCC	CTGCCATTGCTTGTCGTAGA	6DS0.79–0.99
BE500768	ACCTCGACCACTCACTCCA	TCAGCGGTCTCAGTTTGTTG	6DS0.79–0.99
BE517858	CCGGTGATGACCGAACTGAT	CCGGATGATCTCGCTGCTCTC	6DS0.79–0.99
BE444631	CTCCAGTTTCAGGGAGCAAG	GTTCTCTGGCAAGTACTTCAAATCC	6DS0.45–0.79
BE445201	AATGAATTGCTACCATTATTCTCA	ACAGCCGTGAACGTTAGTAAGT	6DS0.45–0.79
BF478958	TCACCTGTACAACAACATGATTTCAA	TGTGCTCATATGTTTTAACT	6DS0.45–0.79
BF483025	ATTCTGTAAGCATGACGGC	AAGGAACTAAGGCCAAGCAATT	6DS0.45–0.79
BE405809	AACGATGCAAGGCTAAAATCTGTGT	GAAGCTGCTGGTTTCTTTGG	6DS0.00–0.45
BE426591	CAGACAATCTTCTTGCCGCT	GTTAGAAATACCGTAAAGCTTTTACCATTAC	6DS0.00–0.45

doi:10.1371/journal.pone.0137747.t002

RESULTS

Immunofluorescence of CENH3, coupled with centromeric DNAs (CRWs), was used to identify the centromeric regions of intact chromosomes and telosomes of wheat. The localization pattern of CENH3 and the CRWs on mitotic metaphase chromosome of CS and their derived telosomes is shown in Fig 1. Consistent with previous reports [30,31], the two probes co-localized in most chromosomes of CS wheat. However, in one chromosome pair of CS, identified as 4D, the two probes were clearly separate from each other (Fig 1a), suggesting that not all CRWs are located within the functional centromere.

Interestingly, in the two winter wheat cultivars Jagger and TAM111, the two signals co-localized on chromosome 4D, indicating that the 4D centromeres in CS underwent repositioning (S1 Fig).

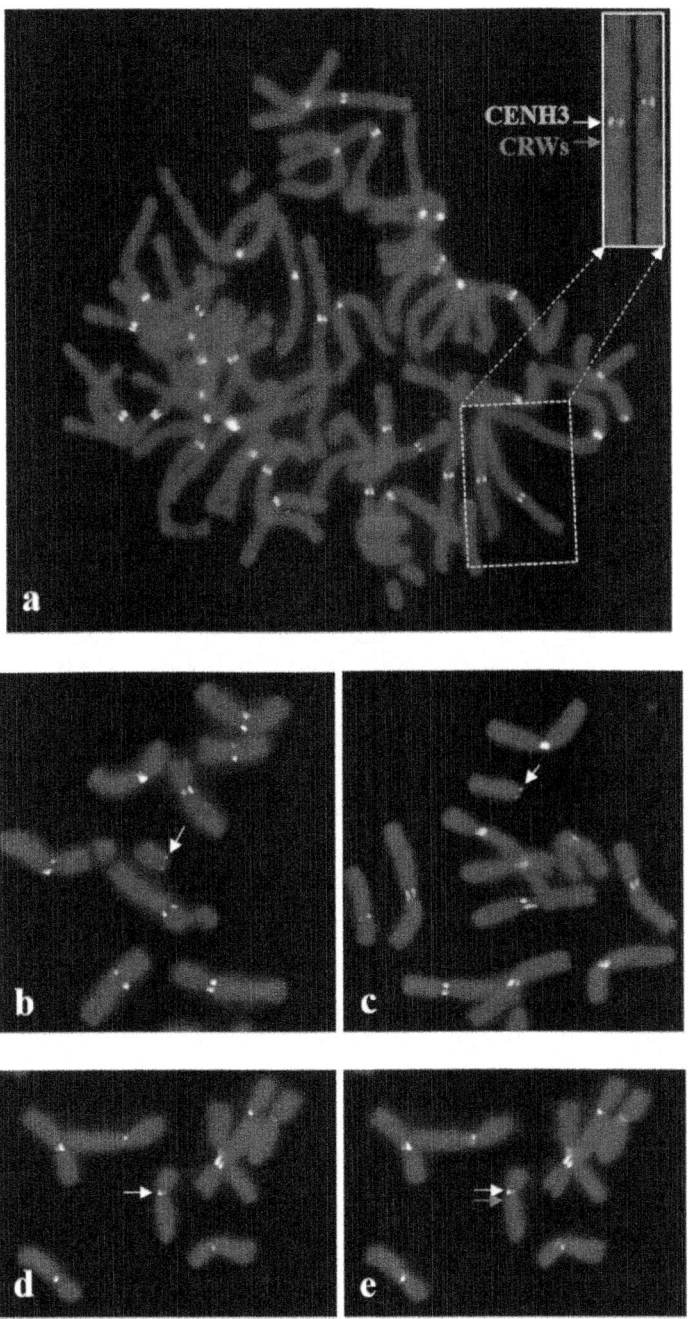

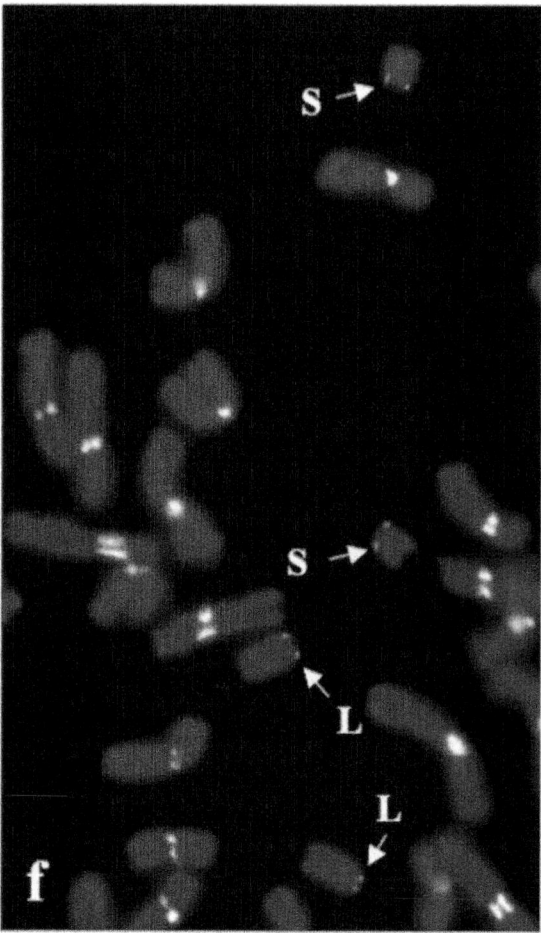

Figure 1. Detection of CENH3 and CRWs on mitotic chromosome of Chinese Spring (CS) wheat and the derived 4D telosomic stocks; a: the insert shows chromosome 4D probed with CENH3 (green) and CRWs (red), straightening was performed using the 'straighten-curved-objects' command of the Image J software, the signals are clearly separated from each other; Relative CENH3 signal intensity of b: t4DS; c: t4DL; f. dDt4D; Chromosome 4D with d: CENH3 (white arrow) and e: CRWs (red arrow) signals.

Next, we applied the wheat CENH3 antibodies to study the structure of t4DS (telosome for 4DS arm) and t4DL, which arose independently from centric misdivison of chromosome 4D. The results showed that CENH3 was detected at the extreme ends of t4DS and t4DL, indicating that they are true telosomes (Fig 1b and 1c). The immunofluorescence of CENH3 on the telosomes was weaker compared to that of the other chromosomes. In order to compare the signal intensity between the intact chromosome 4D and t4DS and to minimize measurement error, root tips from euploid CS and the t4DS stock were squashed on the same slide, and chromosome images probed with CENH3 were captured from the same preparation. The measurement data showed that t4DS had a signal intensity of 43±4.8% (n = 4), compared with that of an intact 4D chromosome. To compare the signal intensity between t4DS and t4DL, we used the dDt4D stock, which contains a pair each of t4DS and t4DL (Fig 1f). The result showed that both telosomes had a similar amount of signal intensity, approximately a 1:1 ratio (79.6±3.8% in t4DS: 80.2±1.7% in t4DL, n = 4) (S2 Fig). Although the dDt4D stock was developed by intercrossing the t4DS and t4DL telosomes, which arose independently from centric misdivison, they showed approximately half of the signal intensity compared with that of the intact 4D chromosome.

We also studied the telosomic derivatives of the seven, D-genome chromosomes of wheat (Figs 2 and 3). Data on the detection of CENH3, the CRWs, pAs1, and single-copy DNA probes on Dt1DS and dDt1DS are shown (Fig 2 and S3 Fig). Signals for CENH3 (arrow in Fig 2a) and the CRWs (arrow in Fig 2b) on Dt1DS were located at the end of the chromosome, indicating that it is a true telosome. In the t1DS of the reconstituted dDt1D stock, however, CENH3 was localized interstitially, forming a small acrocentric chromosome (arrows in Fig 2f). To further discern the chromosomal rearrangement, we performed FISH using pAs1, 1S-1, and 1S-3 as probes. Hybridization signals for pAs1 (arrows in Fig 2c and 2d) and 1S-3 (red dots in Fig 2d) were detected on the terminal region of Dt1DS. However, the pAs1 FISH pattern in dDt1DS showed multiple localizations (arrows in Fig 2g) to both telomeric regions and the interstitial region of the chromosome arm. Probe 1S-3 (arrows in Fig 2i) was detected in the middle of the arm, instead of in the telomeric region, indicating the presence of a paracentric inversion (Fig 2l). Moreover, the FISH signal for 1S-1 (arrow in Fig 2h) was not detected in the pericentromeric region of dDt1DS, implying that dDt1DS has lost the original centromere and now has a *de novo* centromere in a new position (Fig 2l). Supporting evidence came from PCR analysis using six genome-specific markers derived from the 1D proximal bin, which failed to amplify in the dDt1D stock but did amplify in CS and Dt1DS (S4 Fig). Thus, dDt1DS contains multiple chromosomal rearrangements, including a centromere shift, a paracentric inversion, and a deletion. The labeling patterns of CENH3, the CRWs, and pAs1 on dDt1DL were similar with those of Dt1DL (Fig 3).

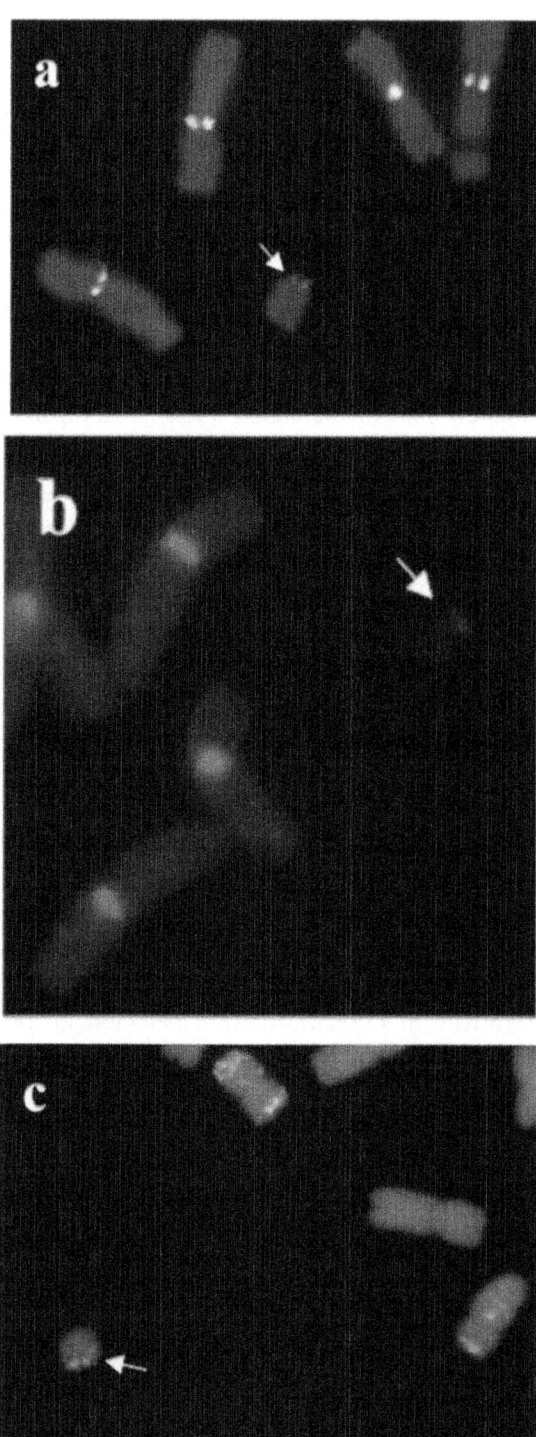

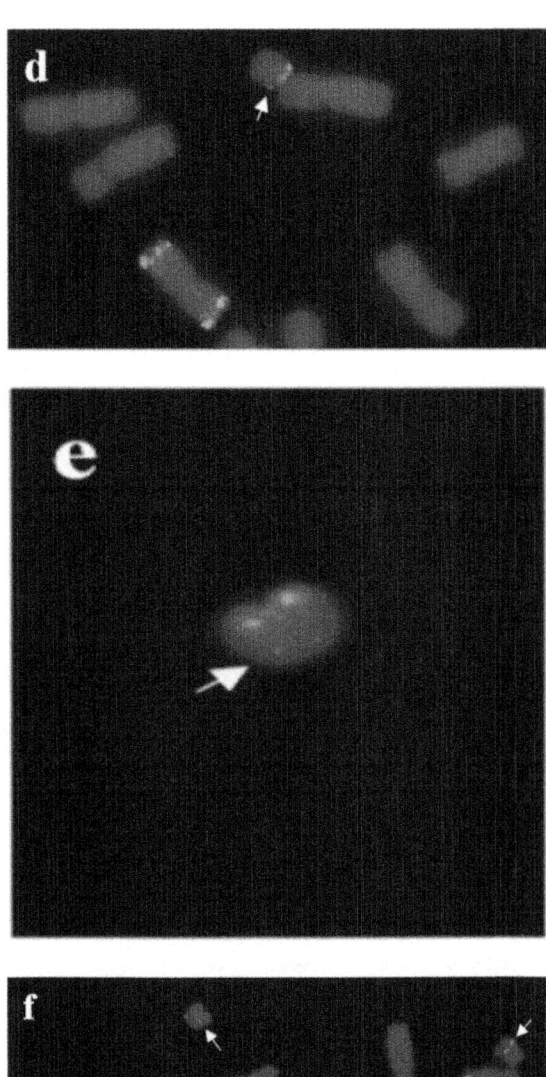

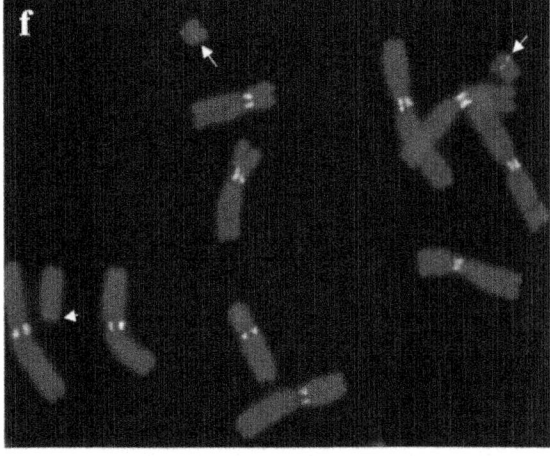

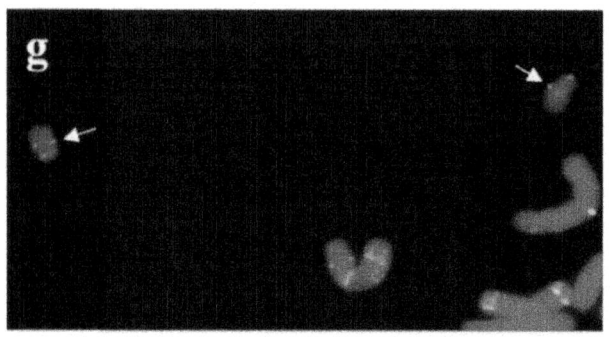

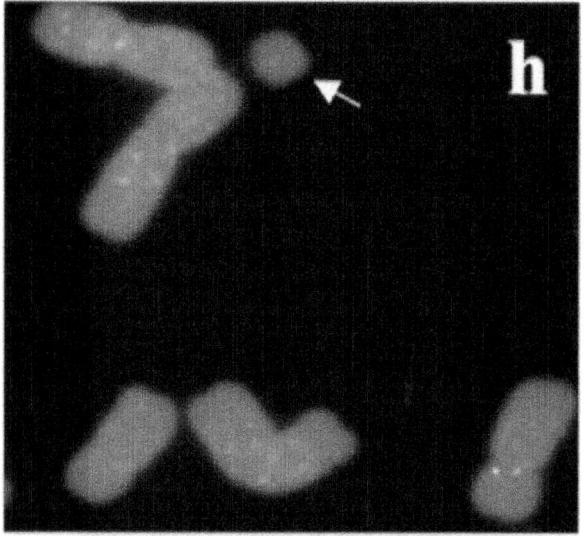

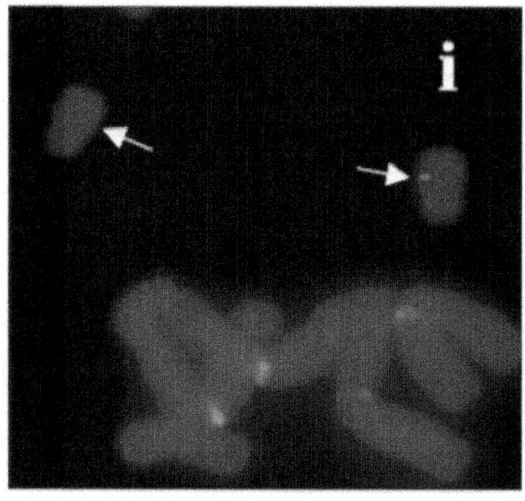

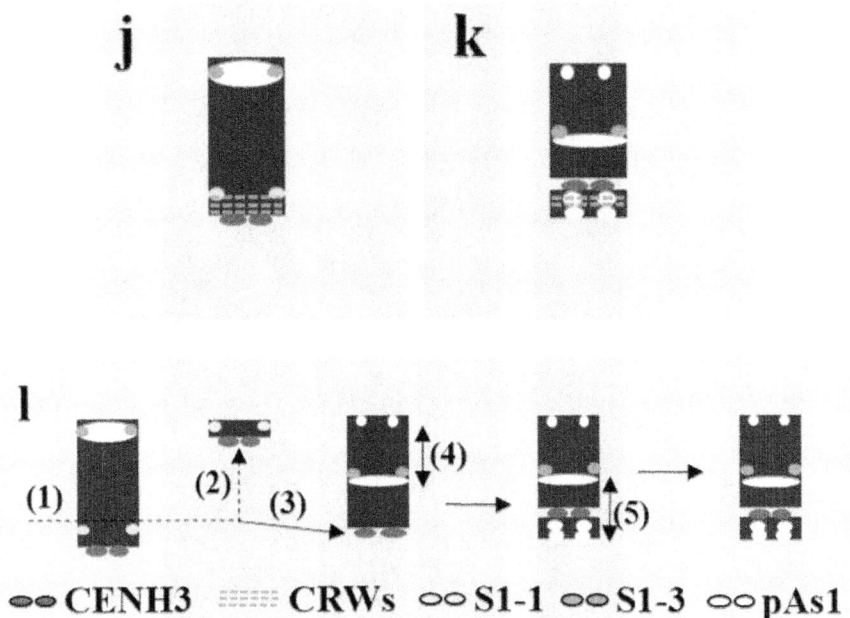

Figure 2. Probing of t1DS telosomes present in the Dt1DS stock with a: CENH3, b: CRWs, c: pAs1, d: 1S-3 (red dots) and pAs1 (green), e: 1S-1 (red dots) and pAs1 (green); Probing of 1DS telosomes present in the dDt1DS stock with f: CENH3, g: pAs1, h: 1S-1 (not detected), and i: 1S-3 (red dots) and CRWs (faint green signals). Simultaneous detection of CENH3, CRWs and pAs1 on dDt1DS also provided in S3 Fig. Ideograms depicting localization of each probe on telosomes, Dt1DS and dDt1DS (j-k, and S3 Fig). Possible scenario of the origin of the chromosomal rearrangements observed in the dDt1DS: (1) chromatin breakage, (2) loss of original centromere, (3) *de novo* formation of a centromere, (4) paracentric inversion, (5) pericentric inversion (s).

The same approach was used to analyze all the D-genome telosomes, and the results are presented (Fig 3). The FISH pattern of pAs1 on chromosome 2D showed multiple localizations, with four FISH sites in the long arm and a single hybridization site on the telomere of the short arm. CENH3 and the CRWs colocalized on the primary constriction (Fig 3). Applying these probes to the Dt and dDt stocks revealed largely identical hybridization patterns with those of an intact 2D chromosome, indicating that there are no rearrangements in these telosomes. Similar patterns were observed for all the remaining D-genome telosome stocks, except for chromosome 6D.

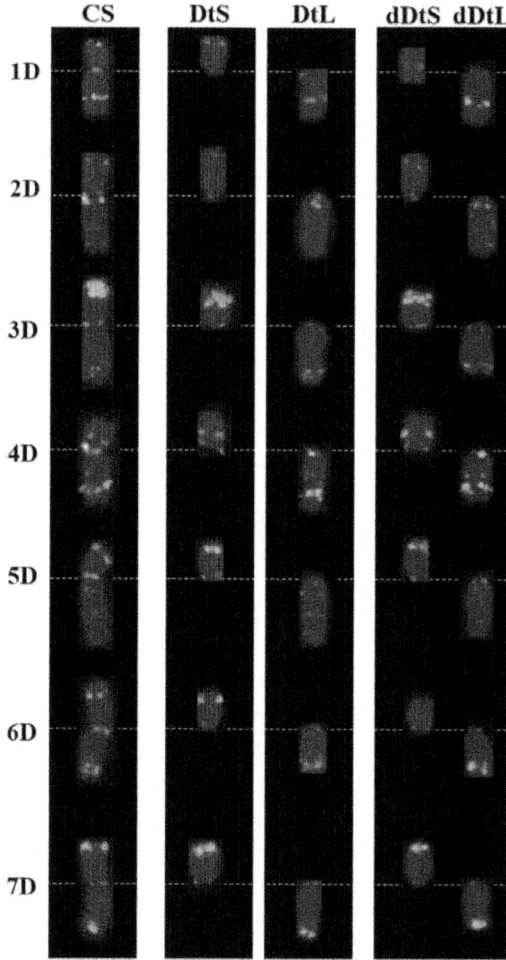

Figure 3. Immuno-FISH based karyotype of D-genome chromosomes of wheat and their derived telosomes using CENH3 (white), CRWs (red) and pAs1 (green) as probes. CRWs (red signals) co-localized with CENH3 (white signals) in most of the chromosome except dDt1DS, 4D, Dt4DS, dDt4DS, Dt5DL and dDt5DL. The centromeric regions of chromosome or chromosome arm were seen as pinkish red colors because the CRWs (red signals) are abundant in centromeric region and much brighter than CENH3 signals except in the above mentioned telosomes. The dDt1DS stock contained multiple chromosome rearrangement including inversion, deletion and centromere shift. Note that the CRWs were not detected in Dt4DS and dDt4DS, instead the pAs1 signal was overlapped with the CENH3 signal in these telosomes. A very faint pAs1 FISH site was detected in the terminal region of dDt6DS, indicating a terminal deletion. Short arm and long arm telosomes present in the ditelosomic stocks are represented as (DtS) and (DtL), respectively and short arm and long arm telosomes present in the double ditelosomic stocks are represented as (dDtS) and (dDtL), respectively.

CENH3 and the CRWs were detected at the end of chromosome arms in Dt6DS (arrows inFig 4a and 4b) and dDt6DS (arrows in Fig 4d and 4e). Whereas a prominent pAs1 signal was located on the subtelomeric region of Dt6DS (arrows in Fig 4c and 4g), a faint hybridization signal was detected on dDt6DS (Fig 4f and 4h) indicating that the chromosome arm of dDt6DS has suffered from a terminal deletion. The 6S-2 FISH signal on Dt6DS was detected at $61.0\pm3.1\%$ (n = 3) from the telomere (Fig 4g). However, in dDt6DS, the 6S-2 FISH signal was detected $39.9\pm3.2\%$ (n = 3) from the telomere (Fig 4h), thus about 20% of the distal region was deleted. In order to verify the deletion at the molecular level, we used genome-specific PCR primers, which were derived from the terminal deletion bin of 6DS (S5 Fig). Four markers derived from terminal bin had no amplification; the other six markers derived from interstitial and proximal bins had amplification (S5 Fig), confirming that about the distal 20% of the telosome was deleted. We made a blastn search of ten EST sequences against the sequence assembly from flow-sorted chromosome arm 6DS [28] and found no hit for markers *XBE424523*, *XBE490604*, *XBE500768*, and *XBE517858* derived from terminal deletion bin. We further analyzed the genome zipper maps of wheat group 6 [28] and observed that this region is deleted in t6DS, whereas the corresponding region is present in t6AS and t6BS. This region corresponds to rice locus *Os02g0116800-Os02g0128800* [39], which is about 595 kb in rice and 702 kb in *Brachypodium Bradi3g01540.1-Bradi3g02817.1* [40]. This region is syntenic with the terminal tip of rice chromosome 2. Nearly 100 genes are annotated in this missing region in the rice genome, 25 of which are syntenic to wheat.

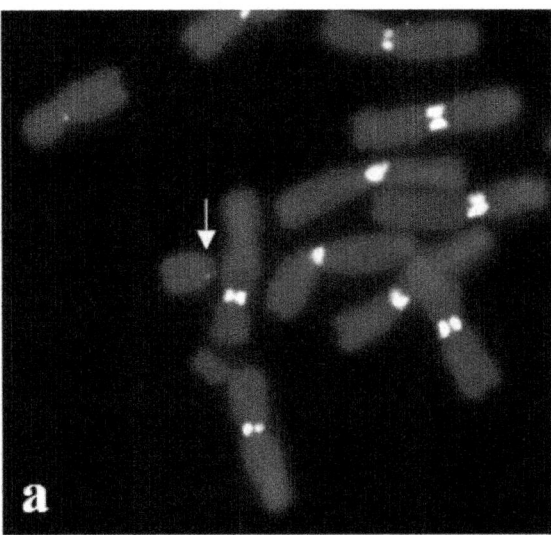

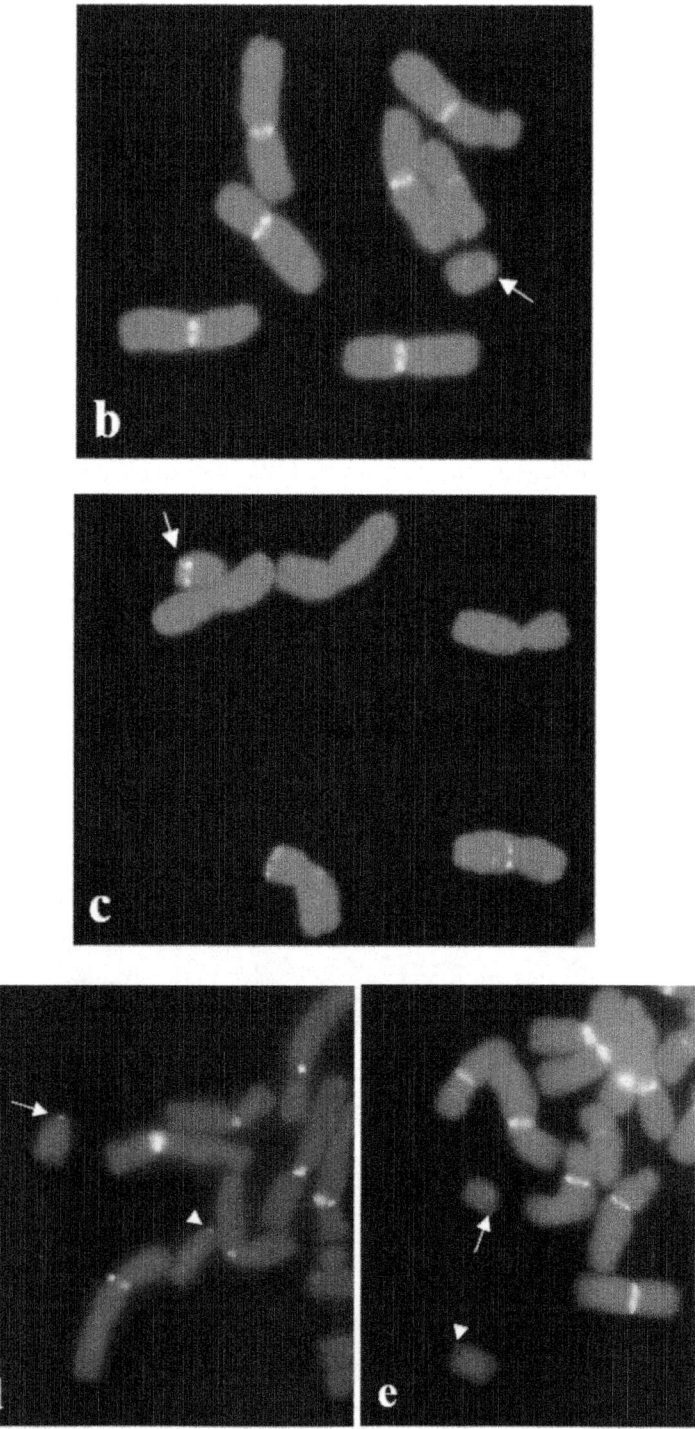

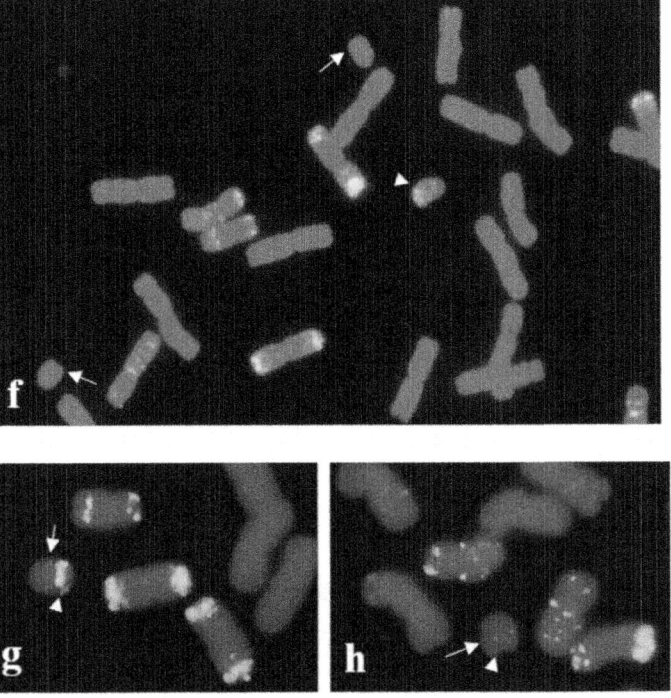

Figure 4. Localization of CENH3 (arrow in a), CRWs (arrow in b) and pAs1 (arrow in c) on Dt6DS; localization of CENH3 (d), CRWs (e) and pAs1 (f) on dDt6D, arrows and arrowhead indicate the 6DS and 6DL telosomes, respectively. Two color detection of single gene probe, 6S-2 and pAs1 on Dt6DS (g) and dDt6DS (h); the hybridization signal for single copy probe 6S-2 (red dots) is indicated by arrows and the pAs1 was labeled with green colors by arrowheads.

Thus, sequence analysis further confirms the loss of a segment from dDt6DS. Because the available dDt1D and dDt6D stocks are rearranged, we re-isolated both stocks, which are now intact and similar to the Dt1DS and Dt6DS stocks. The labeling patterns of CENH3 in the new stocks were similar to those of the DtS lines (data not shown). Cytogenetic analysis is in progress to investigate the mitotic behavior on the newly developed dDt stocks.

To further understand the CENH3 deposition on other telosomes, we used dDt lines that were derived from all the A- and B-genome chromosomes. Our results on immunofluorescence of CENH3 in dDt lines derived from A-genome chromosome showed that the position of CENH3 signals on telosomes is terminal with a signal intensity weaker than that of the other regular chromosomes (S6 Fig). Consistent with dDt4D, signal intensity for CENH3 between the DtS and DtL arms in dDt lines showed similar intensity except in dDt4AS (arrowheads in S6 Fig). The telosome, dDt4AS is known to be acrocentric and, thus, contains a complete centromere [41].

To study the B-genome telosomes, we used a line that has all the B-genome chromosome arms except 4B as telosomes (2n = 28+28t) [42]. This line was obtained by intercrossing the appropriate dDt stocks. Chromosomes t4BS and t4BL in this stock were later identified as t4AS and t4AL telosomes [43]. Likewise, we observed that the CENH3 signal on all B-genome telosomes was smaller compared with that in the complete chromosomes (S6 Fig). In addition, the telosomes, including Dt1BS and Dt6BS in this stock, lack CRWs in the CENH3 region (Fig 5), indicating that the stability of telocentric chromosomes depends on the presence of CENH3 chromatin but not centromeric DNA repeats. Further analysis using the previously known wheat centromeric repeats CCS1 [44], the 192-bp repeat [45], and Quinta [31] showed no signal on these telosomes together with Dt4DS, dDt4DS, Dt5DL, and dDt5DL (data not shown), further indicating that centromeric DNAs are not essential for normal centromere function.

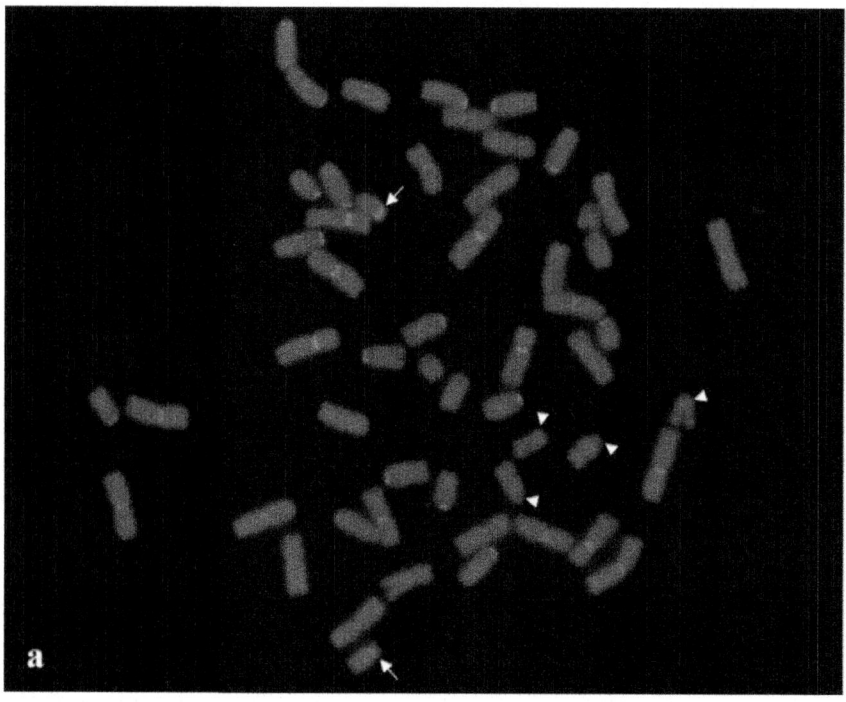

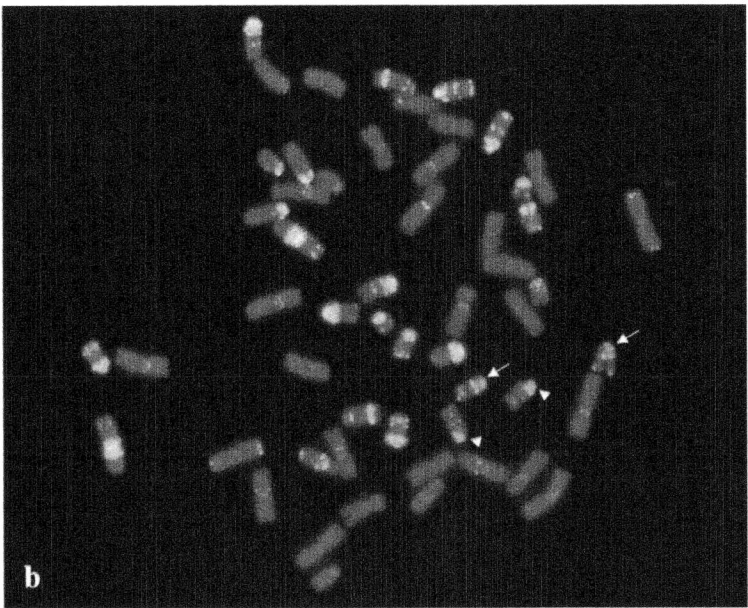

Figure 5. Two color FISH mapping of CRWs (red) and GAAn (green) repeats on mitotic metaphase chromosome of CS containing 28 (24 telosomes from B-genome telosomes and 4 telosomes from t4AS and t4AL). The signals for CRWs were not detected on two pairs of t1BS and t6BS (arrowheads) (a). Instead GAAn repeats were presented in their centromeric region. Arrows indicate the t4AS (a). Arrows and arrowheads indicate the t1BS and t6BS, respectively (b).

DISCUSSION

In many plant species, centromeres consist of complex DNA, including satellite DNA and reterotransposons [2], that are species- and chromosome-specific [46,47]. Some of the centromeric repeats are located within the functional centromere, but centromere-associated repeats also are observed in subtelomeric and interstitial chromosome regions [46,48].

Our study, using immuno-FISH probed with CENH3 and CRWs, shows that CENH3 and CRW elements co-localized on most of the A-, B- and D-genome chromosomes, including their telosomic derivatives. These results indicate that wheat centromeres contain CRW elements that interact with wheat CENH3 [30,31]. However, our results further show that wheat CRWs do not always co-localize with CENH3, as was the case with the complete chromosome 4D and some of the telosomes, including Dt4DS, dDt4DS, Dt5DL, dDt5DL, Dt1BS, and Dt6BS, because these chromosome or chromosome arms lack CRWs in their CENH3.

The CRWs located on chromosome 4D in CS did not overlap with the CENH3 signals but in other wheat cultivars, such as Jagger and TAM111, they co-localized in the same region, providing evidence for centromere repositioning in chromosome 4D of CS (S1 Fig). Comparison of the pAs1 FISH pattern in the pericentromeric region of chromosomes 4D between CS, Jagger, and TAM111 revealed that their pericentromeric localizations were different from each other, implying that the pericentromeric region of 4D in CS underwent a structural rearrangement. The pAs1 FISH site overlapped with CENH3 in CS but not in Jagger or TAM111 (S1 Fig). Lo et al. [49] and Lomiento et al. [50] have reported that neocentromeres form in gene-desert regions containing many repetitive DNAs in many species. These results also indicate that neocentromere function is independent from the presence of original centromeric DNAs [51]. Our immuno-FISH results on telosomes also support their stabilization without centromeric DNAs, because we did not observe CRW signals in Dt1BS and Dt6BS arms. In these telosomes, the GAA repeats co-localized with CENH3 (Fig 5). Thus, wheat centromeres consist of complex DNAs, which may contain CRWs or satellite DNAs, such as pAs1 and GAAn.

Univalent chromosomes at metaphase-I have the tendency to misdivide at the centromere in a transverse manner, which gives rise to telocentrics or potential isochromosomes. Steinitz-Sears [29], suggested that the transverse misdivision split can occur in different regions of the centromere, resulting in telosomes that differ in the completeness of their centromeric regions, and that chromosome arms with incomplete or partial centromeres behave like acentric fragments and are lost during cell division. Because the set of wheat telosomes was produced by centric misdivision and they are stably transmitted to the offspring, they must have received a complete or nearly complete kinetochore. Similarly, barley and rye telocentric chromosomes are cytologically stable. Giemsa N-banding on barley telotrisomics revealed that they contain half of a diamond-shaped kinetochore, whereas complete chromosomes contain an intact, diamond-shaped kinetochore [52]. Rice telocentric chromosomes also contain half of CentO, compared with its normal centromeres [53].

Our immunostaining analysis using the CENH3 antibody suggests that the signals in most Dt and dDt telosomes had approximately half or even less the CENH3 signal intensity compared with that of a complete chromosome. None of the derived telosomes received a complete CENH3 region except dDt4AS, which is an acrocentric chromosome (S6 Fig). These results indicate that telosomes, which only receive half or less than half of the complete CENH3 region, are cytologically stable and transmitted to the offspring. Because most of the wheat telosomes received a partial CENH3 region, it is possible that the transverse misdivision split is not random and may preferentially occur in the middle of the CENH3 chromatin. However, we cannot rule out that breakage also occurs in the entire functional centromere region and that telocentric chromosomes with insufficient CENH3 are mitotically unstable and lost. This is also supported by analyzing the centromere structure of wheat-rye Robertsonian translocations derived from repeated centric breakage-fusion events, which revealed that breakage can occur in different regions of the centromere resulting in wheat-rye hybrid centromeres with different sizes of wheat and rye centromeric repeats [54].

We observed chromosomal aberrations in dDt1DS and dDt6DS. Double labeling of CENH3 and CRWs on the Dt1DS telosomes showed co-localization at the terminal region. In dDt1DS, however, a very faint CRW hybridization signal was observed at the telomere, and CENH3 was localized proximal to it. Single-copy, 1S-1 FISH was not detected in the proximal region of dDt1DS, indicating the presence of a proximal deletion and supporting a previous finding that ESTs mapped in the wheat 1DS deletion bin were absent in dDt1DS [55]. The deletion placed the centromere in a new position. Thus, dDt1DS contains a *de novo* formed centromere (Fig 2l).

The formation of a *de novo* centromere in dDt1DS supports earlier reports in *Drosophila* [56] and chicken (*Gallus gallus*) [57], where neocentromeres formed in regions close to the original centromeres. For instance, when the Z centromere was deleted, neocentromeres most frequently formed near the original Z centromere [57]. CENP-A/CENH3 enrichment in the flanking regions is low but still more enriched compared with that in the rest of the genome [58]. In our study, the formation of *de novo* centromeres or centromere shift was observed in chromosome 4D and in the 1DS telosomes present in the dDt1DS stock. The preference of the formation of the *de novo* centromere near the original centromere is likely caused by the presence of CENH3 in the flanking pericentromeric regions. Chromosome rearrangements after chromatid breaks are a common cause of neocentromere formation in humans [6]. Likewise, neocentromeres reported in plants, maize chromosomes in an oat background and barley chromosomes, also were associated with loss of endogenous centromeres by chromatid breakages [7,8]. We also found a paracentric inversion in telosome dDt1DS using single-copy FISH mapping. Interestingly, the pAs1 signal was observed near the *de novo* centromere region and in an interstitial region of dDt1DS but was absent in Dt1DS, indicating the possibility of another chromosome rearrangement (Fig 2l).

We also identified a deletion in dDt6DS, which comprised about 20% of the terminal deletion bin of chromosome 6D. This deletion was not reported in a recent whole-genome, sequencing analysis [28]. However, our results show that several EST sequences from the terminal deletion bin were missing from the 6DS assembly but present in 6AS and 6BS shotgun sequence assemblies [28]. Thus, when employing these lines for genetic studies, it is important to be aware of the potential presence of chromosomal aberrations in the telocentric chromosome lines to avoid misinterpretation of experimental results. Further relocation of centromeres in chromosomes 4D in CS (the cultivar being used to sequence the wheat genome) compared to other wheat cultivars, Jagger and TAM111, suggests that *de novo* sequencing of more wheat genotypes might provide further insight in the structural organization in the wheat genome.

The wheat dDt stocks were developed by intercrossing the appropriate Dt stocks followed by selection. Therefore, chromosomal rearrangements, including centromere shifts, deletions, and inversions observed in dDt1DS and dDt6DS, might have formed after hybridization of the Dt lines. Rhoades [59] found that telocentric chromosomes in maize undergo structural changes during somatic cell divisions leading to loss or diminution in size. Steinitz-Sears [29] also found that a telocentric chromosome is often unstable and may be lost during plant development in wheat. During mitosis, the structural integrity of the centromeric and flanking pericentric heterochromatic regions is essential for proper assembly of the kinetochore and genome stability [60]. In fission yeast, pericentromeric heterochromatin is an absolute requirement for the establishment of the centromere [61]. In addition to fission yeast, pericentromeric heterochromatin seems to be required for the accurate segregation of chromosome during mitosis in many eukaryotes, including mammals [62]. The implication is that mono-arm oriented pericentromeric heterochromatin in telosomes might be relatively insufficient for maintaining chromosome stability compared to chromosomes with bi-arms oriented pericentromeric heterochromatin. This conclusion is supported by recent findings of Wanner et al. [63] that in monocentrics microtubules attach via CENH3 to both pericentromeres to stabilize the chromosomes during anaphase against the pulling forces.

SUPPORTING INFORMATION

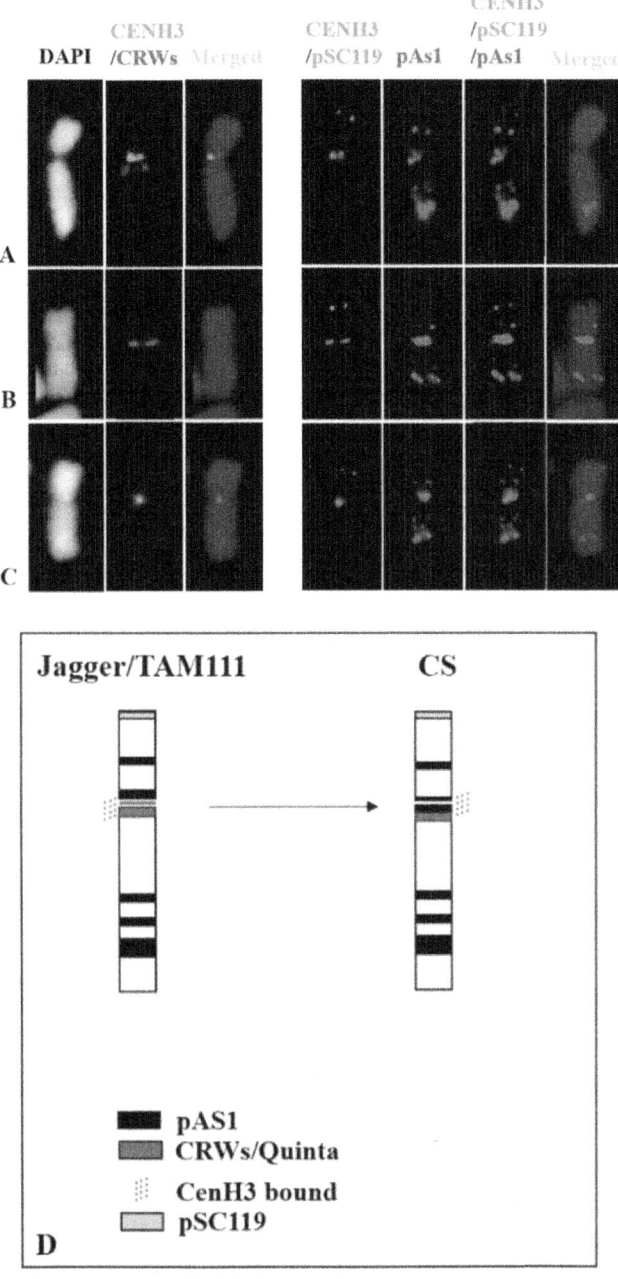

S1 Figure. Sequential detection of CENH3, CRWs, pSc119 and pAs1 on the chromosomes of 4D in CS (A), Jagger (B) and TAM111 (C). The hybridization signals for CENH3 and CRWs were clearly separated from each other in 4D of CS but these were co-localized on the chromosomes of 4D in Jagger and TAM11, indicating the centromere repositioning in 4D of CS. In 4D of CS, pAs1 localization pattern tend to be positioned toward log arm and which is completely overlapped with CENH3. In Jagger and TAM111, however it was positioned toward to the short arm. The pSc119 was used for additional FISH marker to identify the chromosomes 4D in CS, Jagger and TAM111. D, Ideogram depicting distribution of each probe on the chromosomes of 4D in three wheat cultivars.

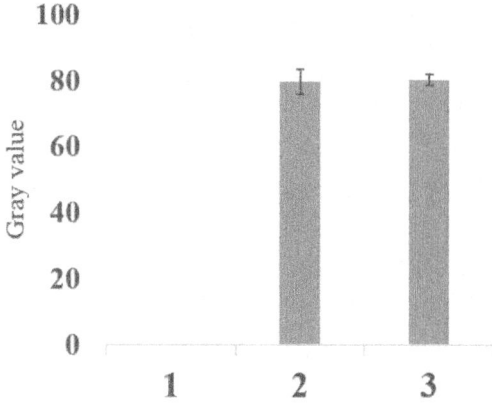

S2 Figure. Graph showing the measurements of the immunofluorescence signal intensity of CENH3. Numbers at y axis represent the gray value (relative signal intensity of antibody to background, background was normalized as zero). 1: background signal, 2: CENH3 signal intensity in dDt4DS, 3: CENH3 signal intensity in dDt4DL. Measurements were done by Image J software. The gray value of CENH3 was 79.6±3.8 (n = 4) and 80.2±1.7 (n = 4) in dDt4DS and dDt4DL, respectively.

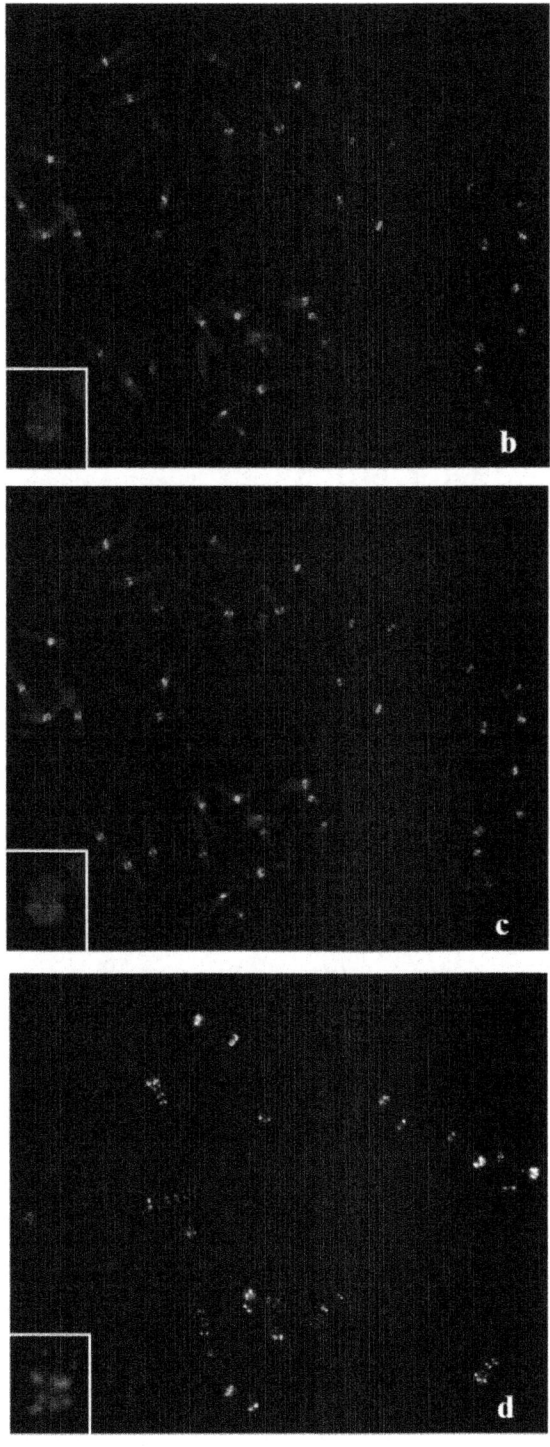

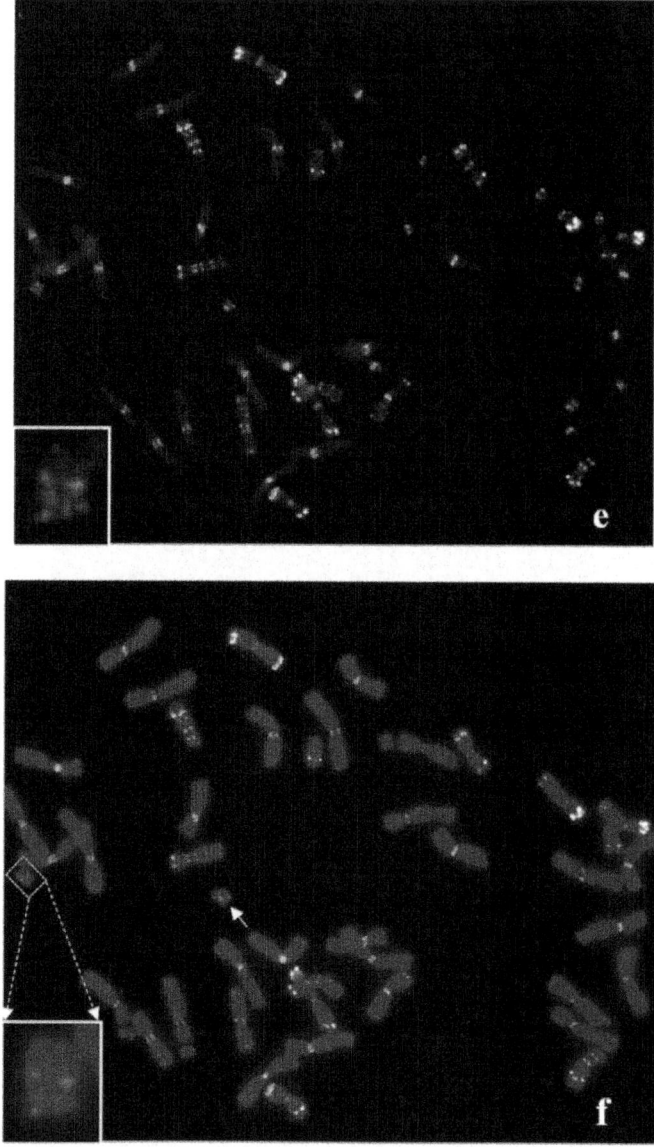

S3 Figure. Multicolor immuno-FISH detection of CENH3 (a), CRWs (b) and pAs1 (d) on telosome, dDt1D. Merged images, CENH3 and CRWs (c), and CENH3, CRWs and pAs1 (e) with DAPI stained metaphase chromosome (f). The inserts show telosome, dDt1DS probed with CENH3 (red), CRWs (green) and pAs1 (white). CENH3 was detected by rhodamine-conjugated anti-rabbit antibodies (red), and the signals were fixed with 4% paraformaldehyde. The same metaphase cell was probed with CRWs (green) and pAs1 (far red, the signals were pseudocolored in white).

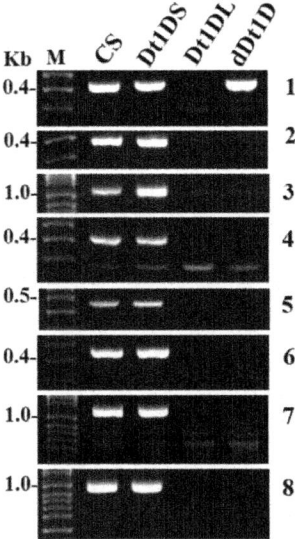

S4 Figure. PCR patterns of CS, Dt1DS, Dt1DL and dDt1D by using genome specific primers: two primers, BE405518 and BE637971, derived from the terminal deletion bin, BE405518 was not amplified (2); six primers, BE444846, BE591601, BE637864, BF202643, BF474569 and BF478737, derived from proximal bin had no amplification (3–8) indicating proximal deletion in dDt1DS.

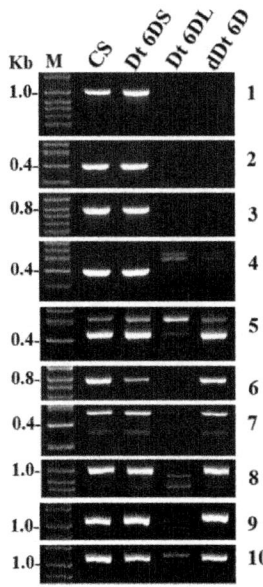

S5 Figure. PCR patterns of CS, Dt6DS, Dt6DL and dDt6D by using genome specific primers: four primers, BE424523, BE490604, BE500768 and BE517858 derived from the terminal deletion bin; four primers, BE444631, BE445201, BF478958 and BF483025 derived from interstitial bin; two primers, BE405809 and BE426591 derived from proximal bin. Four primers derived from terminal deletion bin had no amplification (1–4) while six primers derived from interstitial (5–8) and proximal deletion bin (9–10) had amplification in dDt6DS indicating terminal deletion in dDt6DS.

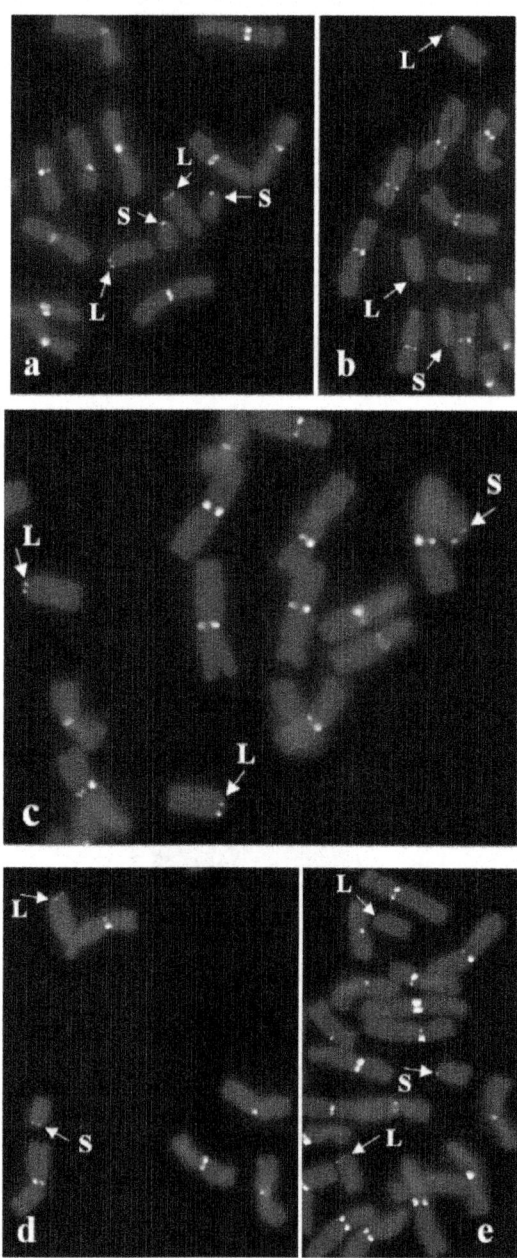

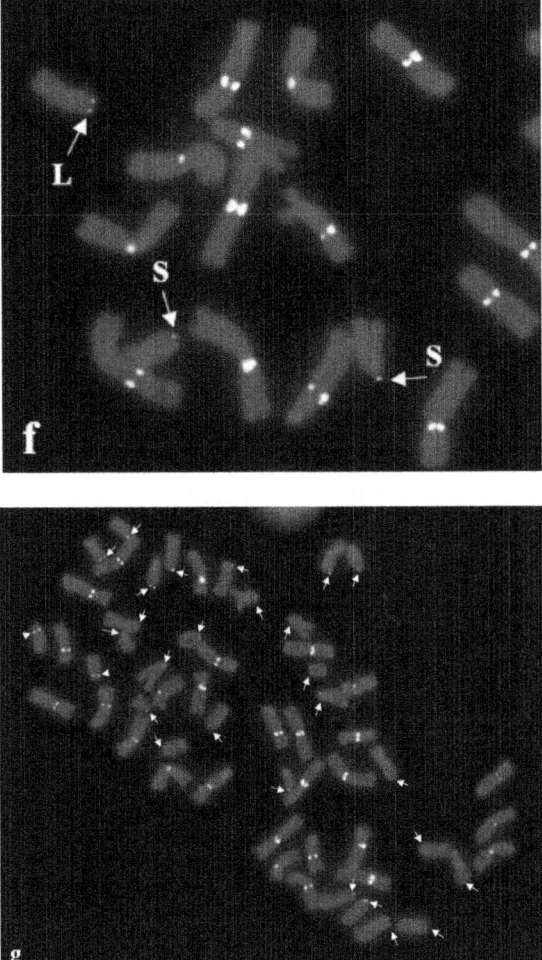

S6 Figure. Partial metaphase cells probed with CENH3 in dDt lines derived from A-genome chromosomes: a, dDt1A; b, dDt2A; c, dDt3A; d, dDt5A; e, dDt6A; f, dDt7A. Telosomes are indicated by arrows. Immunofluorescence of CENH3 on CS containing 28 telosomes (g). The CENH3 fluorescent signals on 24 telosomes (derived from B-genome chromosomes are indicated by arrows) + one pair of t4AL were smaller than those of other intact chromosomes except 4AS (arrowhead) which is an acrocentric chromosome.

ACKNOWLEDGMENTS

We thank W. John Raupp for critical editorial review of the manuscript and Duane Wilson for technical assistance. We are grateful to Dr. Jiming Jiang for his valuable comments on this manuscript. We also thank Adam Lukaszewski

(University of California Riverside) for providing the Dt5DS/Mt5DL line. This is contribution number 15-427-J from the Kansas Agricultural Experiment Station, Kansas State University, Manhattan, KS 66506–5502, U.S.A.

AUTHOR CONTRIBUTIONS

Conceived and designed the experiments: DHK BF BSG. Performed the experiments: DHK SKS. Analyzed the data: DHK BF BSG. Contributed reagents/materials/analysis tools: DHK SKS BF BSG. Wrote the paper: DHK BF BSG.

COMPETING INTERESTS

The authors have the following interests: This research was financially supported by the Kansas Crop Association and by members of the Wheat Genetics Resource Center Industry/University Cooperative Research Center. The following companies and associations make up the membership: Bayer Cropscience, Dow AgroSciences, General Mills, Limagrain, Syngenta, Heartland Plant Innovations, DuPont Pioneer, Kansas Wheat Commission, Kansas Wheat Alliance, and the Kansas Department of Agriculture. The National Science Foundation also provided financial support. The public/ private consortium is aimed at pre-competitive wheat genetics research. Dal-Hoe Koo, Bernd Friebe and Bikram S. Gill are affiliated to the Wheat Genetics Resource Center, Kansas State University. There are no patents, products in development or marketed products to declare. This does not alter the authors' adherence to all the PLoS ONE policies on sharing data and materials.

REFERENCES

1.   Henikoff S, Ahmad K, Malik HS. The centromere paradox: stable inheritance with rapidly evolving DNA. Science. 2001; 293: 1098–1102. pmid:11498581 doi: 10.1126/science.1062939

2.   Jiang JM, Birchler JA, Parrott WA, Dawe RK. A molecular view of plant centromeres. Trends Plant Sci. 2003; 8: 570–575. pmid:14659705 doi: 10.1016/j.tplants.2003.10.011

3.   Earnshaw WC, Rothfield N. Identification of a family of human centromere proteins using autoimmune sera from patients with scleroderma. Chromosoma. 1985; 91: 313–321. pmid:2579778 doi: 10.1007/bf00328227

4.   Talbert PB, Masuelli R, Tyagi AP, Comai L, Henikoff S. Centromeric localization and adaptive evolution of an *Arabidopsis* histone H3

variant. Plant Cell. 2002; 14: 1053–1066. pmid:12034896 doi: 10.1105/tpc.010425

5. Yu H-G, Hiatt EN, Dawe RK. The plant kinetochore. Trends Plant Sci. 2000; 5: 1360–1382. doi: 10.1016/s1360-1385(00)01789-1

6. Marshall OJ, Chueh AC, Wong LH, Choo KHA. Neocentromeres: new insights into centromere structure, disease development, and karyotype evolution. Am J Hum Genet. 2008; 82: 261–282. doi: 10.1016/j.ajhg.2007.11.009. pmid:18252209

7. Nasuda S, Hudakova S, Schubert I, Houben A, Endo TR. Stable barley chromosomes without centromeric repeats. Proc Natl Acad Sci. 2005; 102: 9842–9847. pmid:15998740 doi: 10.1073/pnas.0504235102

8. Topp CN, Okagaki RJ, Melo JR, Kynast RG, Phillips RL, Dawe RK. Identification of a maize neocentromere in an oat-maize addition line. Cytogenet Genome Res. 2009; 124: 228–238. doi: 10.1159/000218128. pmid:19556776

9. Fu SL, Lv ZL, Gao Z, Wu HJ, Pang JL, Zhang B, et al. De novo centromere formation on a chromosome fragment in maize. Proc Natl Acad Sci. 2013; 110: 6033–6036. doi: 10.1073/pnas.1303944110. pmid:23530217

10. Wang K, Wu YF, Zhang WL, Dawe RK, Jiang JM. Maize centromeres expand and adopt a uniform size in the genetic background of oat. Genome Res. 2014; 24: 107–116. doi: 10.1101/gr.160887.113. pmid:24100079

11. Levan A, Fredgra K, Sandberg AA. Nomenclature for centromeric position on chromosomes. Hereditas. 1964; 52: 201–220. doi: 10.1111/j.1601-5223.1964.tb01953.x

12. Darlington CD. Misdivision and the genetics of the centromere. J Genet. 1939; 37: 343–365. doi: 10.1007/bf02982733

13. Marks GE. Telocentric chromosomes. Am Nat. 1957; 91: 223–232. doi: 10.1086/281981

14. Strid A. Stable telocentric chromosomes formed by spontaneous mid-divison in *Nigella doerfleri* (Ranunculaceae). Bot Notiser. 1968; 121: 153–164.

15. Schubert I, Rieger R. Alteration by centric fission of the diploid chromosome number in *Vicia faba* L. Genetica. 1990; 81: 67–69. doi: 10.1007/bf00055238

16. Southern DI. Stable telocentric chromosomes produced following centric misdivision in *Myrmeleotettix maculatus* (Thunb.). Chromosoma. 1969; 26: 140–147. doi: 10.1007/bf00326451

17. Takagi N, Sasaki M. A phylogenetic study of bird karyotypes. Chromosoma. 1974; 46: 91–120. pmid:4134896 doi: 10.1007/bf00332341

18. White MJD. Animal cytology and evolution. 3rd Edition. London: Cambridge University Press. 1973

19. Sears ER, Sears L. The telocentric chromosomes of common wheat. In Proceedings of the 5th International Wheat Genet Symposium (Ramanujam S, ed). New Delhi: Indian Society of Genetics and Plant Breeding. 1978; pp: 389–407.

20. Cheng ZK, Yan H, Yu H, Tang S, Jiang JM, et al. Development and applications of a complete set of rice telotrisomics. Genetics. 2001; 157: 361–368. pmid:11139516

21. Tsuchiya T. Cytogenetics of the telocentric chromosome of the long arm of chromosome 1 in barley. Seiken Ziho. 1972; 23: 47–62.

22. Erayman M, Sandhu D, Sidhu D, Dilbirigi M, Baenziger PS, Gill KS. Demarcating the gene-rich regions of the wheat genome. Nucleic Acids Res. 2004; 12: 3546–3565. doi: 10.1093/nar/gkh639

23. Mutti JS, Sandhu D, Sidhu D, Gill KS. Dynamic nature of a wheat centromere with a functional gene. Mol Breeding. 2010; 26: 177–187. doi: 10.1007/s11032-009-9389-1

24. Gill KS, Arumuganathan K, Lee JH. Isolating individual wheat (*Triticum aestivum*) chromosome arm by flow cytometric analysis of ditelosomic lines. Theor Appl Genet. 1999; 98: 1248–1252. doi: 10.1007/s001220051190

25. Dolezel J, Kubalakova M, Bartos J, Macas J. Flow cytogenetics and plant genome mapping. Chromosome Res. 2004; 12: 77–91. pmid:14984104 doi: 10.1023/b:chro.0000009293.15189.e5

26. Dolezel J, Kubalakova M, Paux E, Bartos J, Feuillet C. Chromosome-based genomics in the cereals. Chromosome Res. 2007; 15: 51–66. pmid:17295126 doi: 10.1007/s10577-006-1106-x

27. Gill BS, Appels R, Botha-Oberholster AM. A workshop report on wheat genome sequencing: international genome research on wheat consortium. Genetics. 2004: 168, 1087–1096. pmid:15514080 doi: 10.1534/genetics.104.034769

28. The international wheat genome sequencing consortium (IWGSC) A chromosome-based draft sequence of the hexaploid bread wheat (Triticum aestivum) genome. Science. 2014; 345: 286. doi: 10.1126/science.1251788

29. Steinitz-Sears LM. Somatic instability of telocentric chromosomes in wheat and the nature of the centromere. Genetics. 1966; 54: 241–248. pmid:17248316

30. Liu Z, Yue W, Li D, Wang RR, Kong X, Lu K, et al. Structure and dynamics of retrotransposons at wheat centromeres and pericentromeres. Chromosoma. 2008; 117: 445–456. doi: 10.1007/s00412-008-0161-9. pmid:18496705

31. Li B, Choulet F, Heng Y, Hao W, Paux E, Liu Z, et al. Wheat centromeric retrotransposons: the new ones take a major role in centromeric structure. Plant J. 2013; 73: 952–965. doi: 10.1111/tpj.12086. pmid:23253213

32. Rayburn AL, Gill BS. Isolation of a D-genome specific repeated DNA sequence from*Aegilops tauschii*. Plant Mol Biol Rep. 1987; 4: 102–109. doi: 10.1007/bf02732107

33. Danilova TV, Friebe F, Gill BS. Development of a wheat single gene FISH map for analyzing homoeologous relationship and chromosomal rearrangements within the Triticeae. Theor Appl Genet. 2014; 127: 715–730. doi: 10.1007/s00122-013-2253-z. pmid:24408375

34. Qi LL, Echalier B, Chao S, Lazo GR, Butler GE, Anderson OD, et al. A chromosome bin map of 10,000 expressed sequence tag loci and distribution of genes among the three genomes of polyploid wheat. Genetics. 2004; 168: 701–712. pmid:15514046 doi: 10.1534/genetics.104.034868

35. Jin WW, Melo JR, Nagaki K, Talbert PB, Henikoff S, Dawe RK, et al. Maize centromeres: organization and functional adaptation in the genetic background of oat. Plant Cell. 2004; 16: 571–581. pmid:14973167 doi: 10.1105/tpc.018937

36. Koo D-H, Jiang JM. Super-stretched pachytene chromosomes for fluorescence *in situ* hybridization mapping and immunodetection of DNA methylation. Plant J. 2009; 59: 509–516. doi: 10.1111/j.1365-313X.2009.03881.x. pmid:19392688

37. Koo D-H, Han F, Birchler JA, Jiang JM. Distinct DNA methylation patterns associated with active and inactive centromeres of the maize B chromosome. Genome Res. 2011; 21: 908–914. doi: 10.1101/gr.116202.110. pmid:21518739

38. Liu W, Rouse M, Friebe B, Jin Y, Gill BS, Pumphrey MO. Discovery and molecular mapping of a new gene conferring resistance to stem rust, *Sr53*, derived from*Aegilops geniculata* and characterization of spontaneous translocation stocks with reduced alien chromatin. Chromosome Res. 2011; 19: 669–682. doi: 10.1007/s10577-011-9226-3. pmid:21728140

39. International Rice Genome Sequencing Project. The map-based sequence of the rice genome. Nature. 2005; 436: 793–800. pmid:16100779 doi: 10.1038/nature03895

40. International Brachypodium Initiative Genome sequencing and analysis of the model grass *Brachypodium distachyon*. Nature. 2010; 463: 763–768. doi: 10.1038/nature08747. pmid:20148030

41. Gill BS, Friebe B, Endo TR. Standard karyotype and nomenclature system for description of chromosome bands and structural aberrations in wheat (*Triticum aestivum*). Genome. 1991; 34: 830–839. doi: 10.1139/g91-128

42. Sears ER, Muramatsu M. A line with all B-genome chromosomes as doubleditelocentric. In Proceedings of the 7th International Wheat Genet Symposium (Miller T.E. and Koebner R.M.D., eds). Institute of Plant Science Research, Cambridge, UK. 1988; pp: 427–431.

43. Naranjo T, Roca A, Goicoechea PG, Giraldez R. Chromosome structure of common wheat: genome reassignment of chromosome 4A and 4B. In Proceedings of the 7thInternational Wheat Genet Symposium (Miller T.E. and Koebner R.M.D., eds). Institute of Plant Science Research, Cambridge, UK. 1988; pp: 115–120.

44. Aragon-Alcaide L, Miller T, Schwarzacher T, Reader S, Graham M. A cereal centromeric sequence. Chromosoma. 1996; 105: 261–268. pmid:8939818 doi: 10.1007/bf02524643

45. Ito H, Nasuda S, Endo TR. A direct repeat sequence associated with the centromeric retrotransposons in wheat. Genome. 2004; 47: 747–756. pmid:15284880 doi: 10.1139/g04-034

46. Lee HR, Zhang W, Langdon T, Jin WW, Yan H, et al. Chromatin immunoprecipitation cloning reveals rapid evolutionary patterns of centromeric DNA in *Oryza* species. Proc Natl Acad Sci. 2005; 102: 11793–11798. pmid:16040802 doi: 10.1073/pnas.0503863102

47. Gong ZY, Wu YF, Koblížková A, Torres GA, Wang K, Iovene M, et al. Repeatless and repeat-based centromeres in potato: Implications for centromere evolution. Plant Cell. 2012; 24: 3559–3574. doi: 10.1105/tpc.112.100511. pmid:22968715

48. Neumann P, Naratilova A, Schroeder-Reiter E, et al. Stretching the rules: monocentric chromosomes with multiple centromere domains. PLoS Genet. 2012; 8: e1002777. doi: 10.1371/journal.pgen.1002777. pmid:22737088

49. Lo AW, Craig JM, Saffery R, Kalitsis P, Irvine DV, et al. A 330 kb CENP-A binding domain and altered replication timing at a human

neocentromere. EMBO J. 2001; 20: 2087–2096. pmid:11296241 doi: 10.1093/emboj/20.8.2087

50. Lomiento M, Jiang ZS, D'Addabbo P, Eichler EE, Rocchi M. Evolutionary-new centromeres preferentially emerge within gene deserts. Genome Biol. 2008; 9: R173. doi: 10.1186/gb-2008-9-12-r173. pmid:19087244

51. Ketel C, Wang HSW, McClellan M, Bouchonville K, Selmecki A, et al. Neocentromeres form efficiently at multiple possible loci in *Candida albicans*. PLoS Genet. 2009; 5: e1000400. doi: 10.1371/journal. pgen.1000400. pmid:19266018

52. Singh RJ. Plant cytogenetics, Second Edition. CRC Press, Boca Raton, FL. 2003; pp: 216–217.

53. **53.**Cheng ZK, Dong FG, Langdon T, Ouyang S, Buell CR, Gu MH, et al. Functional rice centromeres are marked by a satellite repeat and a centromere-specific retrotransposon. Plant Cell. 2002; 14: 1691–1704. pmid:12172016 doi: 10.1105/tpc.003079

54. Zhang P, Friebe B, Lukaszewski AJ, Gill BS. The centromere structure in Robertsonian wheat-rye translocation chromosomes indicates that centric breakage-fusion can occur at different positions within the primary constriction. Chromosoma. 2001; 110: 335–344. pmid:11685533 doi: 10.1007/s004120100159

55. Wicker T, Mayer KFX, Gundlach H, Martis M, Steuernagel B, Scholz U, et al. Frequent gene movement and pseudogene evolution is common to the large and complex genomes of wheat, barley, and their relatives. Plant Cell. 2011; 23: 1706–1718. doi: 10.1105/tpc.111.086629. pmid:21622801

56. Maggert KA, Karpen GH. The activation of a neocentromere in *Drosophila* requires proximity to an endogenous centromere. Genetics. 2001; 158: 1615–1628. pmid:11514450

57. Shang W-H, Hori T, Martins NMC, Toyoda A, Misu S, Monma N, et al. Chromosome engineering allows the efficient isolation of vertebrate neocentromeres. Dev Cell. 2013; 24: 635–648. doi: 10.1016/j. devcel.2013.02.009. pmid:23499358

58. Scott KC, Sullivan BA. Neocentromeres: a place for everything and everything in its place. Trends Genet. 2014; 30: 66–74. doi: 10.1016/j. tig.2013.11.003. pmid:24342629

59. Rhoades MM. Studies of a telocentric chromosome in maize with reference to the stability of its centromere. Genetics. 1940; 25: 483–520. pmid:17246983

60. Sharp JA, Kaufman PD. Chromatin proteins are determinants of centromere function. Curr Top Mirobiol Immunol. 2003; 274: 23–52. doi: 10.1007/978-3-642-55747-7_2

61. Folco HD, Pidoux AL, Urano T, Allshire RC. Heterochromatin and RNAi are required to establish CENP-A chromatin at centromeres. Science. 2008; 319: 94–97. doi: 10.1126/science.1150944. pmid:18174443

62. Peters AH, O'Carroll D, Scherthan H, Mechtler K, Sauer S, Schofer C, et al. Loss of the Suv39h histone methyltransferases impairs mammalian heterochromatin and genome stability. Cell. 2001; 107: 323–337 pmid:11701123 doi: 10.1016/s0092-8674(01)00542-6

63. Wanner G, Schroeder-Reiter E, MA W, Houben A, Schubert V. The ultrastructure of mono- and holocentric plant centromeres: an immunological investigation by structured illumination microscopy and scanning electron microscopy. Chromosoma. 2015; doi: 10.1007/s00412-015-0521-1.

Chapter 8

A MAJOR LOCUS FOR CHLORIDE ACCUMULATION ON CHROMOSOME 5A IN BREAD WHEAT

Yusuf Genc[1,2], Julian Taylor[1], Jay Rongala[2], and Klaus Oldach[1,2]

[1]School of Agriculture, Food and Wine, University of Adelaide, Waite Campus, Glen Osmond, South Australia, Australia

[2]South Australian Research and Development Institute, Plant Genomics Centre, Waite Campus, Glen Osmond, South Australia, Australia

ABSTRACT

Chloride (Cl^-) is an essential micronutrient for plant growth, but can be toxic at high concentrations resulting in reduced growth and yield. Although saline soils are generally dominated by both sodium (Na^+) and Cl^- ions, compared to Na^+ toxicity, very little is known about physiological and genetic control mechanisms of tolerance to Cl^- toxicity. In hydroponics and field studies, a bread wheat mapping population was tested to examine the relationships between physiological traits [Na^+, potassium (K^+) and Cl^- concentration] involved in salinity tolerance (ST) and seedling growth or grain yield, and to elucidate the genetic control mechanism of plant Cl^- accumulation using a quantitative trait loci (QTL) analysis approach. Plant Na^+ or Cl^- concentration were moderately correlated (genetically) with seedling biomass in hydroponics, but showed no correlations with grain yield in the field, indicating little value in selecting for ion concentration to improve ST. In accordance with phenotypic responses, QTL controlling Cl^- accumulation differed entirely between hydroponics and field locations, and few were detected in two or more environments, demonstrating substantial QTL-by-environment interactions. The presence of several QTL for Cl^- concentration indicated that uptake and accumulation was a polygenic trait. A major Cl^- concentration QTL (5A; *barc56/gwm186*) was identified in three field environments, and accounted for 27–32% of the total genetic variance. Alignment between the 5A QTL interval and its corresponding physical genome regions in wheat and other grasses has enabled the search for candidate genes involved in Cl^- transport, which is discussed.

INTRODUCTION

Worldwide, salinity poses a serious threat to agricultural production, with globally salt-affected soils (including saline and sodic soils) totalling around 830 million hectares [1]. Saline soils are generally dominated by sodium and chloride ions, and the ability of crop plants to exclude these ions is often equated to salinity tolerance-ST (i.e. improved growth or yield under salinity stress). Despite intensive research and numerous scientific reports over several decades, there is a lack of consensus on whether Na^+ exclusion is always a useful trait to select for to improve ST, at least in the case of bread wheat. Few studies have shown good phenotypic correlation between Na^+ exclusion and ST [2]–[5], while others have reported weak [6], [7] or no correlation [8]–[10]. In contrast to Na^+ exclusion, there has been very little research on the role of Cl^- exclusion and its contribution to ST [3], [11]–[13]. Similar to Na^+ exclusion, these few studies on Cl^- reported either significant phenotypic correlation or no correlation with ST. It is clear that future research needs to place equal emphasis on the impact of Cl^- in ST as not only Na^+ but also Cl^- is present at toxic concentrations in saline-affected growth media used in the assessment of ST [13], [14].

In screening studies for ST, sodium chloride is the most commonly used salt. Despite both Na^+ and Cl^- ions being present at toxic concentrations in growth media, there has been a lack of interest in Cl^-, which may be attributed to the earlier reports that Na^+ was more toxic than Cl^- [15]. Although these authors cautioned about assumptions made from extrapolations of two lines of bread wheat, their findings were not verified with a large set of genotypes differing in ST. This was later questioned by Martin and Koebner [16] on the basis that while attempting to separate the toxic effects of Na^+ and Cl^-, the authors used phytotoxic concentrations of nitrate in their experiments. Martin and Koebner [16] concluded that Cl^- was more toxic than Na^+, but the full toxic effect was apparent only when Na^+ and Cl^- were present simultaneously. However, more recent studies in barley found that Na^+ and Cl^- had a similar effect upon plant growth [17]. While the issue of whether Na^+ or Cl^- is more detrimental to plant growth remains controversial, it is appropriate to measure the concentration of both ions in the plant and to determine their relevance to ST for reasons mentioned earlier. Munns and Tester [18] argued that toxicity of Na^+ versus Cl^- can be best studied through genetic approaches as the alternative method of using different salts has to date produced equivocal results. The potential of the genetic approach has not been employed extensively.

Chloride is an essential micronutrient and has several functions in plant metabolism; enzyme activation, photosynthesis, a counter ion for cation transport, osmoregulation, and movement of stomata. Like most anions, it is weakly bound to soil particles, mostly in a soluble form in soil solution, and at high concentrations can be toxic to plant growth. Critical toxicity concentration in plants was estimated at 110–200 mmol kg^{-1} DW (4,000–7,000 mg kg^{-1} DW) and 420–1400 mmol kg^{-1} DW (15,000–50,000 mg kg^{-1} DW) for sensitive and tolerant species, respectively [19]. At present there is very little information on genetic control mechanisms of Cl$^-$ exclusion in plant species. A better understanding of genetic control of Cl$^-$ exclusion and identification of molecular markers has the potential to speed up breeding for a complex trait such as ST. To date, rice and barley appear to be the only cereal species in which genetic control of Cl$^-$ exclusion has been investigated. In two studies involving rice F_2 and RIL populations derived from a cross between salt tolerant and salt-sensitive varieties CSR27 and MI48 respectively, few Cl$^-$ concentration QTL were co-located with Na$^+$ concentration QTL, while others mapped to different regions [20], [21]. Similar findings were also reported in barley [22], [23]. If QTL for Na$^+$ and Cl$^-$ concentration map to different locations, this would suggest separate genetic control mechanisms for regulation of these ions. At present, we are not aware of any reports on QTL mapping of Cl$^-$concentration in wheat.

Screening for ST is usually conducted in one of three main environments: hydroponics, soil-based pot assays or field trials. Due to inherent difficulties with field screening such as non-uniform distribution of salinity throughout the experimental area, fluctuations in rainfall and potential nutritional deficiencies, controlled environments are often preferred. However, as reported in recent studies [24], the results of controlled environments can be quite different from those of field environments and, therefore, require verification. The only two genetic studies on Cl$^-$ accumulation to date have been conducted in hydroponics, and how these results correlate to field environments is unknown. It is important to note that the ability of controlled environmental studies to predict yield responses in the field is rarely addressed in the scientific literature, and the validation of controlled environmental studies in the field is important for plant breeding.

In previous studies in hydroponics and field trials [24], [25], we reported QTL for grain yield, grain number m^{-2}, 1000-grain weight, maturity, plant height, seedling biomass, tiller number, chlorophyll content, leaf symptoms, Na$^+$ and K$^+$ concentrations of leaves and shoots for a bread wheat mapping population (Berkut/Krichauff). With the recent renewed interest in Cl$^-$ [13], we revisited this population to (i) elucidate the genetic control mechanisms of Cl$^-$ homeostasis via a QTL approach, and (ii) investigate the relationships among seedling biomass, grain yield and plant Na$^+$, K$^+$ and Cl$^-$ concentrations. Here we report for the first time identification of a major QTL for Cl$^-$ concentration in bread wheat and discuss its importance for marker-assisted selection and fine mapping/discovery of genes involved in Cl⁻ transport in bread wheat. We also demonstrate that Cl⁻ accumulation is a polygenic trait, but does not appear to be a reliable predictor of ST based on grain yield alone.

MATERIALS AND METHODS

Plant Material

A doubled-haploid (DH) population (152 lines) from a cross between bread wheat (*Triticum aestivum* L.) genotypes Berkut [Irene/Babax//Pastor] and Krichauff [Wariquam//Kloka/Pitic62/3/Warimek/Halberd/4/3Ag3/Aroona] was used in this study. The rationale for screening this population for plant Cl⁻ concentration and subsequent QTL detection was that a previous study revealed significantly lower shoot Cl⁻ accumulation in the Krichauff parent than in the Berkut parent [26].

Phenotyping and Trait Analysis

Growth room and field studies were described previously [24], [25]. The data for grain yield, grain number m⁻², 1000-grain weight, maturity, plant height, seedling biomass, tiller number, chlorophyll content, leaf symptoms and Na⁺ and K⁺ concentrations of penultimate leaves and shoots were reported earlier [24], [25]. In the present study shoot and leaf Cl⁻concentrations of DH population grown in hydroponics and field trials (Roseworthy, Balaklava and Georgetown in South Australia) characterized by low, moderate and high salinity [24] were determined. Either single (hydroponics, two replicates) or 15–20 plants per entry (field trials, two replicates) were sampled for elemental analysis. Shoot (hydroponics) and leaf samples (field trials) were dried at 65°C for 48 h and dry weights recorded. The dried plant samples were then ground and analysed for Cl⁻, calcium (Ca^{2+}) and magnesium (Mg^{2+}) concentration using either Inductively Coupled Plasma Optical Emission Spectrometry (ICP-OES) (ARL 3580 B, Appl. Res Lab. SA, Ecublens, Switzerland)[27], [28] or a chloride meter (Model 926, Sherwood, Cambridge, UK). For analysis of Cl⁻ using the ICP-OES method, 0.1 g of ground shoot sample was extracted with hot (95°C) 4% HNO_3 acid in 50 mL capped polypropylene tubes for 90 minutes, whereas for measurements of Cl⁻ using the chloride meter, 0.5 g of ground sample was digested in 40 mL of 1% HNO_3 at 85°C for 5 hours in a 54well HotBlock (Environmental Express, Mt. Pleasant, South Carolina, USA). ICP-OES was used initially for analysis of the Cl⁻concentration of hydroponically-grown plants, but due to prohibitive cost, leaf samples of field-grown plants were analyzed using a chloride meter. As two different analytical methods were used for determination of Cl⁻ concentration in plant tissues, a number of samples were analyzed using both methods, and the high correlation of the measurements between the two methods (hydroponically-grown samples, $r^2=0.98$, n=45; field-grown samples, $r^2=0.97$, n=15) indicated that the methods were comparable. As single measurements were taken from each extraction, duplicate analysis (one sample per batch of 24) was carried out to determine the homogeneity of the samples, and relative standard deviations between the two measurements were below 5% in all cases. Chloride concentration was expressed on a dry mass basis (mmol kg⁻¹ DW).

Linkage Map and Interval Construction

Genotyping and construction of a genetic linkage map for the Berkut/Krichauff population was described earlier [24], [25]. The constructed linkage map initially comprised 557 markers across 21 chromosomes. After omission of co-locating markers this was reduced to 403 markers with an average interval distance of 9.16 cM. For computational purposes the alleles of the Berkut (A)/ Krichauff (B) population were then converted into 1 and -1 respectively and missing marker scores were imputed using the flanking marker method of Martinez and Curnow [29]. A total of 384 inferred interval markers were then constructed using the mid-point interval method of Verbyla et al. [30].

Multi-Environment Analysis

To understand the genetic relationships of the DH lines across the field trials and growth room a multi-environment trials (MET) analysis was conducted for each of the traits Na⁺, K⁺and Cl⁻. For grain yield the MET analysis was restricted to field sites only. The analysis approach follows Smith et al. [31] which involves a linear mixed model including the parsimonious modelling of genetic effects of the DH lines through an appropriate genotype by environment interaction model and also captures non-genetic sources of variation through the use of separate spatial models for the plot errors at each site. The method initially involves the assessment of the spatial or environmental variation occurring at each site through the investigation of the assumption of variance homogeneity, detection of outliers, and the identification of global trends that may exist across the rows or columns of the experiment. For trends existing due to adjacency of plots in a trial, the model included a separable row by column autoregressive correlation structure. Stronger linear trends in either direction are fitted as fixed effects in the model. Design parameters such as genotypic replication or blocking structures were fitted as separate random effects.

The genotype by environment interaction model involved the use of an unstructured heterogeneous correlation matrix that appropriately captures the genetic relationship of the DH lines between trials. If a strong positive genetic correlation exists between any two trials then the relative performance or rankings of the varieties at each of the trials will be similar. As a consequence these trials will most likely share common or co-locating QTL with a common parent being favoured at each locus. Trials that exhibit little or no genetic correlation between them would exhibit different relative rankings for the varieties and most likely have unshared QTL. Each multi-environment analysis was performed using residual maximum likelihood (REML) and the estimated genetic correlations between sites were extracted for interpretation.

Multivariate Analysis

A multivariate linear mixed model analysis of four traits (Na^+, K^+, Cl^- and grain yield) was conducted for each field trial in order to estimate genetic correlations of the DH lines between traits. A multivariate analysis was also conducted for the traits in the growth room and included Na^+, K^+, Cl^-, and seedling biomass. For each of the trials, estimates of the genetic relationships between DH lines were modelled through the use of a trait by genotype interaction with an unstructured heterogeneous correlation matrix. The model also included a separable trait by row by column spatial model for the plot errors with an unstructured heterogeneous correlation matrix for the trait component of the model. This unstructured correlation matrix ensures that traits collected from the same trial are connected phenotypically. This separable structure for the spatial model also assumes that the traits have a common row by column separable autoregressive correlation structure. Similar to the MET models, strong trends for any given trait were captured using the appropriate fixed effects, and random effects were used to model genotypic replication as well as blocking structures existing within each trial. Each multivariate analysis was performed using REML and the estimated genetic correlations between traits were extracted for interpretation.

QTL Analysis

For each of the field sites and the growth room the detection and estimation of QTL for the measured traits was accomplished using the R [32] package wgaim [33]. The package is a computational implementation of Verbyla et al. [30], [34]. In this approach, an initial base linear mixed model is established that contains non-genetic effects that account for extraneous variation. These include an appropriate spatial model for the errors as well as fixed and random effects that are relevant to the trial being examined. The base model also includes a random effect term that captures the genetic variation between the DH lines. Following Verbyla et al. [34] the base model is then extended by including the complete set of inferred interval markers into the base linear mixed model as a contiguous block of random effects with a single variance parameter. The significance of this variance parameter is then checked using a simple residual likelihood ratio test. If found significant, an alternative outlier model is formulated and the inferred interval marker with the largest outlier statistic is chosen as a putative QTL. This inferred interval marker is then removed from the contiguous block of random effects and placed as a separate random covariate in the original base model as well as the extended model that includes the remaining set of inferred interval markers. The forward selection process is then repeated until the variance parameter associated with the remaining inferred interval markers is not significant. The complete set of putative QTL selected appears additively as random covariates and is summarised using the methods of Verbyla et al. [34]. The summary includes the left and right flanking markers of the individual QTL, their effect sizes, approximate LOD scores as well as individual contributions to the overall genetic variance.

For all traits analysed in the present study, the best linear unbiased predictions of the genotypes were extracted from the base linear model and were used to calculate a generalized (broad-sense) heritability using the formula developed by Cullis et al. [35].

Comparative Analysis and Candidate Genes Co-located with the 5A Cl⁻Concentration QTL

To align the genetic position of the 5A Cl⁻ QTL to its physical position in the wheat genome, the sequence of RFLP markers with nearby location to the QTL-flanking SSR markers*gwm304* and *barc141* were identified using the database GrainGenes 2.0 (wheat.pw.usda.gov). The sequences of co-located RFLP markers *bcd21* and *psr128* were 451 bp and 430 bp, respectively, and were both derived from ESTs (GrainGenes 2.0). The sequences were used for homology searches using the BLAST tool athttps://urgi.versailles.inra.fr/blast/blast.php. The search was carried out against the sequences of all bread wheat chromosomes showing the best hits on 5AL, as expected. The gene sequence hits were used for synteny analysis using the Genome Zipper v5 across wheat, rice, *Brachypodium* and sorghum revealing the syntenic regions between 64 non-rendundant wheat ESTs, *Brachypodium* chromosome 4, rice chromosome 9 and sorghum chromosome 2.

To assign potential functions to the wheat genes underlying the 5A Cl⁻ QTL interval, all 64 non-redundant wheat ESTs were used as queries for homology searches at the National Center for Biotechnology Information (NCBI) using BLASTN against the non-redundant nucleotide database.

RESULTS

Responses to Salinity Stress, Distributions and Relationships between Traits

Data on Cl^- concentration provided an opportunity to re-examine the relationships amongst traits associated with ST such Na^+ and K^+ accumulation, seedling biomass and grain yield in low (Balaklava), moderate (Roseworthy) and high (Georgetown and growth room) saline environments [24]. Krichauff had 10–20% lower Na^+ concentration than Berkut in field trials, and these differences diminished in hydroponics, whereas Krichauff had 14–25% lower Cl^- concentration than Berkut in all environments ranging from 362–669 and 482–775 mmol kg^{-1} DW for Krichauff and Berkut, respectively (Table 1), similar to previous studies [26]. It was interesting to observe that Cl^- concentrations were much higher than Na^+ concentrations. Berkut had slightly higher K^+ concentration than Krichauff but only at low to moderately saline field trials at Roseworthy and Balaklava (812–934 and 729–882 mmol kg^{-1} DW for Berkut and Krichauff respectively; Table 1). As for seedling biomass and/ or grain yield production under salinity stress, Berkut produced 18% higher seedling biomass than Krichauff in hydroponics, whilst Krichauff had 6–7% higher grain yield in field trials (Table 1). In all traits, there was evidence of transgressive segregation.

To determine whether selection for Na^+ and/or Cl^- exclusion or even K^+ accumulation would lead to improved ST, multivariate analysis of concentrations of Na^+, Cl^-, K^+, seedling biomass and grain yield was performed for each of the environments (Table 2). The analysis showed that there was generally a good estimated correlation between Cl^- and either Na^+ or K^+ with two exceptions: there were negligible correlations for Na^+ vs Cl^-, and K^+ vs Cl^- at Balaklava and in hydroponics respectively. A negative but moderate correlation was observed between seedling biomass and either Na^+ (r=0.479) or Cl^- (0.527).

Table 1. Parental means, population mean and range for Na$^+$, Cl$^-$, K$^+$, Ca^{2+} and Mg^{2+} concentrations (mmol kg^1 DW) in penultimate leaves (field trials) and whole shoots (hydroponics), seedling biomass [shoot DW (g plant^1)] and grain yield (t ha^1) in Berkut/Krichauff DH population tested for ST in hydroponics and field trials (Roseworthy, Balaklava and Georgetown).

| | | Parental Lines | | DH population | | |
Test environment	Trait	Berkut	Krichauff	mean	range	Heritability h²
Hydroponics	Na⁺ conc.	298	295	270	194–361	0.58
(100 mM NaCl ~10 dS m⁻¹)	Cl⁻ conc.	457	376	378	236–480	0.67
	K⁺ conc.	771	779	749	661–869	0.82
	Shoot DW	1.523	1.296	1.537	1.210–1.835	0.57
	Ca²⁺ conc.	107.2	87.3	97.9	77.4–116.7	0.83
	Mg²⁺ conc.	94.7	77.5	84.1	73.2–95.1	0.77
Roseworthy	Na⁺ conc.	15.2	12.6	12.4	8.8–15.9	0.57
(low salinity, ECe <4 dS m⁻¹)	Cl⁻ conc.	482	362	393	272–510	0.82
	K⁺ conc.	812	729	739	621–847	0.64
	Grain yield	2.237	2.392	2.116	1.451–2.489	0.76
Balaklava	Na⁺ conc.	13.1	11.6	10.9	7.4–14.5	0.78
(Moderate salinity, ECe = 4–8 dS m⁻¹)	Cl⁻ conc.	557	447	474	301–598	0.82
	K⁺ conc.	934	882	865	749–1011	0.77
	Grain yield	2.714	2.888	2.671	1.758–3.035	0.81
	Ca²⁺ conc.	121.3	97.2	110.7	71.9–151.4	0.82
	Mg²⁺ conc.	77.2	62.2	73.5	52.0–95.1	0.79
Georgetown	Na⁺ conc.	20.5	16.3	18.7	10.9–30.4	0.60
(High salinity, ECe >8 dS m⁻¹)	Cl⁻ conc.	775	669	688	571–884	0.84
	K⁺ conc.	1363	1332	1328	1064–1523	0.76
	Grain yield	0.645	0.686	0.573	0.277–0.792	0.72

The means represent predicted values from MET (Na⁺, Cl⁻ , K⁺) and single environment (Ca²⁺ and Mg²⁺) analysis of each trait. Broad-sense heritability is also given for individual traits at each environment.
doi:10.1371/journal.pone.0098845.t001

Table 2. Estimated genetic correlations between shoot DW (hydroponics), grain yield (field) Na$^+$, K$^+$, and Cl$^-$ (field and hydroponics) extracted from the fitted multi-trait model at each environment.

Environment	Na⁺ vs Cl	K⁺ vs Cl	Shoot DW or yield vs Na⁺	Shoot DW or yield vs K⁺	Shoot DW or yield vs Cl
Hydroponics	0.875	0.195	−0.486	−0.146	−0.531
Roseworthy	0.451	0.634	0.200	0.097	0.053
Balaklava	0.122	0.517	−0.200	0.060	0.097
Georgetown	0.319	0.763	0.066	0.148	0.082

doi:10.1371/journal.pone.0098845.t002

To analyse genotype by environment interaction for each of the measured traits, multi-environment analyses and genetic correlations were estimated between environments. Representing low (Balaklava) to moderate (Roseworthy) saline environments, correlations for Na$^+$, Ka$^+$, Cl$^-$ concentrations and grain yield were consistently high between these sites whereas low to moderate correlations were observed for these traits between other higher saline environments (Table 3). Therefore, it is reasonable to assume that selection for similar trait values may not be consistent between higher saline sites, indicating environmental effects controlling ion accumulation and grain yield.

Table 3. Estimated genetic correlations extracted from the fitted multi-environment model for individual traits.

		Roseworthy	Balaklava	Georgetown
Na⁺ concentration	Balaklava	0.861		
	Georgetown	0.016	0.167	
	Growth room	0.182	0.186	0.464
		Roseworthy	Balaklava	Georgetown
K⁺ concentration	Balaklava	0.794		
	Georgetown	0.568	0.598	
	Growth room	0.361	0.322	0.225
		Roseworthy	Balaklava	Georgetown
Cl⁻ concentration	Balaklava	0.897		
	Georgetown	0.582	0.462	
	Growth room	0.276	0.427	0.339
		Roseworthy	Balaklava	
Grain yield	Balaklava	0.706		
	Georgetown	0.130	0.338	

doi:10.1371/journal.pone.0098845.t003

Generalized (Broad-sense) Heritability (h^2)

Estimates of h^2 differed with trait and environment, ranging from moderate to high (0.6–0.8) (Table 1). Amongst the mineral elements, Na⁺ concentration had the lowest h^2, while Cl⁻concentration had the highest h^2. Heritability of shoot DW (0.6) was lower than that of grain yield (0.7–0.8). Heritability of individual traits across environments showed relatively consistent (Ca^{2+}, Mg^{2+} and Cl⁻ concentration) to inconsistent patterns (Na⁺ concentration). As reported earlier [24], consistently higher h^2 values indicate greater ability for selection for the traits, while lower and variable h^2 suggest the presence of substantial environmental effects, difficulty for direct selection, and the need for more replications.

QTL for Cl⁻ Concentration

In the initial analysis in which differences in phenology were not included, there were few Cl⁻ concentration QTL co-locating with QTL for maturity on 5A and 5D (data not shown). Similar to Bonneau et al. [36], after the known differences in the genetic component of the phenology were addressed by fixing the maturity genes in the analyses, 14 QTL were identified (Table 4). However, most QTL were specific to single environments, and only three QTL were detected in two or more environments (3A, 5A, 7D) indicating some genotype by environment interaction. Interestingly, there were no co-located QTL controlling trait variation from hydroponics and field trials. The most significant QTL in hydroponics on chromosome 2A explained 20% of the total

genetic variance and the Krichauff allele was responsible for increased Cl⁻
concentration. QTL on chromosome 3A and 7D were detected from multi-
location trials and explained 4–11% of the total genetic variance, while QTL
on 5A accounted for 27-32% of the total genetic variance (Table 4). Either the
Berkut (3A, 5A) or the Krichauff (7D) allele was associated with increased Cl⁻
concentration at these loci. As the two QTL on 5A appear in tandem, these loci
were further investigated to determine whether there may just be one rather
than two separate QTL. The plot of outlier statistics (Figure 1) shows that
there is in fact just one QTL on 5A expressed at all field locations. The QTL
detected at one location only accounted for a small proportion of the total
genetic variance, varying from 3.6 to 9.9% with either the Berkut or Krichauff
allele being associated with increased Cl⁻ concentration (Table 4).

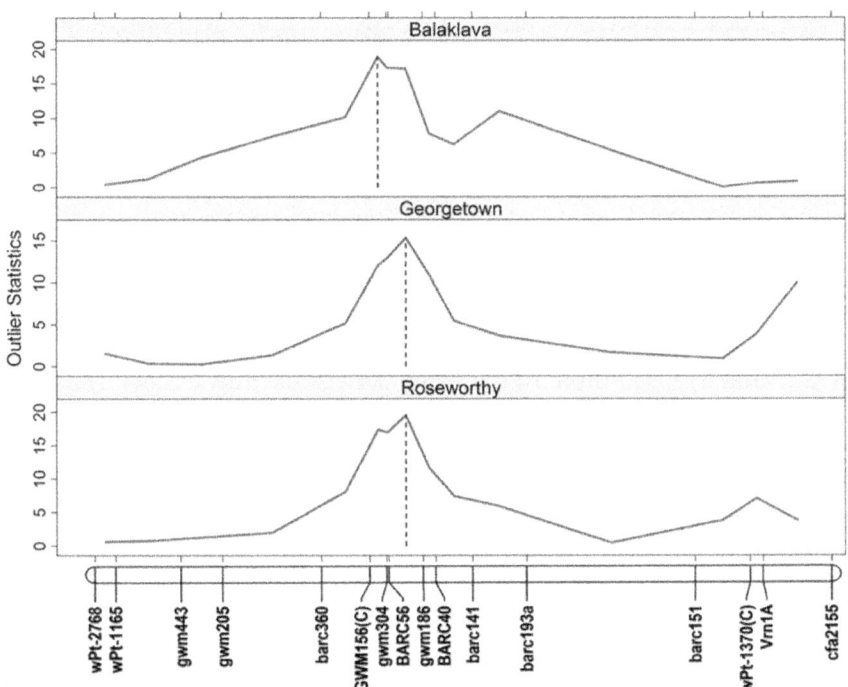

Figure 1. Location of Cl⁻ concentration QTL (*barc56/gwm186*) on chromosome 5A
detected in field trials (Balaklava, Georgetown and Roseworthy) with varying salinity
levels. The outlier statistics represent LOD scores (Table 4).

Table 4. QTL associated with Cl⁻ concentration in hydroponics and field trials (Roseworthy, Balaklava and Georgetown) under varying degrees of salinity stress.

Environment	Ch	Interval	Distance (cM)	Size*	Prob.	% Var.	LOD	**Co-locations with Na^+ or K^+ in field locations and hydroponics reported earlier [24,25]
Roseworthy	2B	cfd011- gwm374(C)	88.0–95.8	−12.5	0.001	4.1	2.4	
	3A	wmc343- cfa2262	75.7–79.7	20.6	0.000	11.1	6.4	
	3B	wPt-9049(C)- wPt-0021	124.4–154.0	13.4	0.002	3.6	2.1	
	5A	barc56/gwm186	90.0–100.5	36.8	0.000	32.4	18.3	Na^+-GT, K^+-RW, K^+-BA, K^+-GT
	6D	cfd005(C)- gpw95010(C)	172.8–174.1	−11.7	0.001	3.9	2.4	
	7D	barc214- gwm437	130.8–133.5	−11.3	0.002	3.6	2.0	Na^+-RW, Na^+- RH, Na^+-WT
	7D	wmc014u- wPt-0695	243.5–246.4	−11.3	0.002	3.6	2.1	
Balaklava	3A	barc324- wmc343	67.0–75.7	9.4	0.002	4.4	2.2	
	3B	gwm247- wPt-4412	195.1–196.1	15.9	0.000	9.9	4.8	
	5A	gwm156(C)- gwm304	84.1–89.3	25.4	0.000	30.8	14.8	
Georgetown	3A	barc56- gwm186	90.0–100.5	30.7	0.000	27	11.1	
	7B	bcg1- wPt-7318	0.0–22.7	−14.5	0.000	5.4	2.9	
	7D	barc214- gwm437	130.8–133.5	13.2	0.001	5.7	2.6	
Hydroponics	1D	wmc216- wPt-1799	80.7–131.2	21.3	0.002	5.8	2.0	
	2A	wPt-3114- wmc170(C)	92.5–109.9	−33.5	0.000	21.1	7.1	Na^+- HYDRO, Na^+-RW, Na^+-BA
	2B	wmc272- barc349	106.5–112.7	−19.4	0.000	8.5	2.7	Na^+-HYDRO
	2B	wPt-7859- wPt-7167	198.9–200.2	18.9	0.000	8.4	2.9	Na^+-HYDRO
	3A	barc324- wmc343	67.0–75.7	17.4	0.001	6.7	2.3	Na^+-RW
	7A	gwm282- wPt-0961(C)	170.2–183.3	−19.3	0.000	7.8	2.7	Na^+-HYDRO

Only those intervals with P values ≤0.01 and LOD>2.0 are presented. Field locations Roseworthy, Balaklava and Georgetown were classified as low, moderate and high salinity, respectively. Please see Genc et al. [24] for soil salinity classification.
*Positive and negative values indicate that Berkut and Krichauff alleles increased the phenotypic values, respectively. **The same co-locations appearing at multiple environments were presented only once. Abbreviations: Hydroponics = HYDRO, Balaklava = BA; Georgetown = GT; Roseworthy = RW. QTL names with letter C indicate several co-locating markers at those loci.
doi:10.1371/journal.pone.0098845.t004

Some QTL for Cl⁻ concentration were co-located with QTL for either Na⁺ or K⁺ concentration from hydroponics and field trials (Table 4). It was interesting to observe that in hydroponics QTL for Cl⁻ concentration were co-located with QTL for Na⁺ concentration (2A, 2B, 2B, 3A, 7A), while in field trials they were co-located with both Na⁺ (3A, 5A, 7D) and K⁺ (5A).

QTL for Calcium (Ca^{2+}) and Magnesium (Mg^{2+}) Concentration

As Ca^{2+} and Mg^{2+} nutrition of the plant can also be affected by salinity [26], [37]–[40], their genetic control under salinity stress was also investigated, in a limited way, using a QTL mapping approach. A total of 13 QTL for Ca^{2+} concentration and 6 QTL for Mg^{2+}concentration were identified in hydroponics and the field (Table 5). At those loci, either the Berkut or Krichauff allele increased the concentration of these cations. The most significant QTL for Ca^{2+} concentration was the QTL in hydroponics accounting for 13% of the genetic variance (Table 5). This QTL was also co-located with the Cl⁻ concentration QTL detected in hydroponic conditions (Table 4). For Mg^{2+} concentration, the QTL on 3A in hydroponics was the most significant, explaining 15% of the genetic variance (Table 5). This QTL was also co-located with a QTL for Cl⁻ concentration detected in environments with low to moderately saline conditions (Roseworthy and Balaklava) and hydroponics e (Table 4). However, there were no common QTL that were detected under hydroponics and the field for either of these two cations.

Table 5. QTL associated with Ca^{2+} and Mg^{2+} concentrations in hydroponics and in a field trial (Balaklava) under varying degrees of salinity stress.

Environment	Trait	Ch	Interval	distance (cM)	Size*	Prob.	% Var.	LOD
Balaklava	Ca²⁺ conc.	3A	cfa2262-wPt-3816	79.7–100.95	5.7	0.000	10.6	3.6
		4A	gwm165a-wmc420	37.8–49.4	5.1	0.000	9.9	3.2
		5B	gwm499-wPt-5851	43.9–46.1	−5.2	0.000	11.4	2.9
		7D	wmc4360-barc214	96.8–130.8	4.6	0.001	6.2	2.2
	Mg²⁺ conc.	2A	gwm294-gdm093(C)	113.5–150.8	2.4	0.002	6.2	2.1
		4B	gwm149(C)-wPt-1505(C)	48.0–48.7	2.3	0.001	9.1	2.3
		5B	wPt-3437-wPt-1250	51.5–54.3	−2.4	0.001	10.1	2.2
Hydroponics	Ca²⁺ conc.	1D	wmc216-wPt-1799	80.7–131.2	4.9	0.000	13.3	10.1
		2B	wPt-5707-wPt-8161	58.4–61.0	−1.9	0.001	3.9	2.5
		3B	gwm389-wPt-8093(C)	0.0–0.7	−1.9	0.000	4.0	3.4
		3B	wPt-4412-wPt-8352(C)	196.1–197.1	1.6	0.002	2.8	2.1
		4A	wmc106-gwm165a	28.4–37.8	2.5	0.000	6.3	4.4
		4D	gpw95001-gwm165b	49.7–50.5	−2.9	0.000	9.0	6.5
		5B	gwm371-gwm499	37.6–43.9	−2.5	0.000	6.3	3.1
		6B	cfd0760-wPt-4924	119.8–124.4	3.4	0.000	11.8	6.6
		7B	wPt-6372(C)-wPt-2833	63.2–66.8	2.2	0.000	3.1	3.4
	Mg²⁺ conc.	2A	gwm294-gdm093(C)	113.5–150.8	1.6	0.000	6.6	2.8
		3A	gwm733a-barc1121(C)	2.2–8.8	−1.4	0.000	7.6	2.9
		3A	barc324-wmc343	67.0–79.7	2.1	0.000	15.4	5.5
		4B	wPt-513(C)-gwm495	43.4–44.7	1.6	0.000	9.9	3.4

Only those intervals with P values ≤0.01 and LOD>2.0 are presented. Balaklava location was classified as moderate salinity. Please see Genc et al. [24] for soil salinity classification.
*Positive and negative values indicate that Berkut and Krichauff alleles increased the phenotypic values, respectively. QTL names with letter C indicate several co-locating markers at those loci.
doi:10.1371/journal.pone.0098845.t005

Known Chloride Transporters/Channel in Grass Genomes and the 5A Cl⁻QTL'

In previous studies, employing transcription analysis under salt stress, cellular localization and transgenesis for functional characterization, CLC (chloride channel) and CCC (cation chloride co-transporter) genes had been identified to play a role in Cl⁻ homoeostasis in plants; examples are CLC1 in tobacco and soybean [13], [41], [42], [43] and CCC in Arabidopsis [44].

To investigate the presence of candidate genes such as CLCs, CCCs and other ion transporters within the QTL interval on chromosome 5A, we physically positioned the 5A Cl⁻concentration QTL in the wheat genome sequence. For this purpose, the gene-based sequences of RFLP markers, *bcd21* and *psr128* with close linkage to the QTL-flanking SSR markers *gwm304* and *barc141* (Figure 2) were used to find wheat genome sequences. As expected, both RFLP sequences had their best hits in bread wheat chromosome 5AL and allowed to retrieve matching contigs of 4.2 kb and 9.1 kb for *bcd21* and *psr128*, respectively. As both RFLPs had originally been derived from ESTs, the corresponding wheat genome contigs (4.2 and 9.1 kb) identified gene hits in rice chromosome 9 (*Os09g0321900* and *Os09g0412200*). These rice genes functioned as borders of the physical interval in the comparative analysis between rice, *Brachypodium* and wheat using the alignment in Genome Zipper v5 (Figure 2). In rice (MSU Release 7 at rice.plantbiology.msu.edu/cgi-bin/gbrowse/rice/#search), the syntenic interval contained 547 genes.

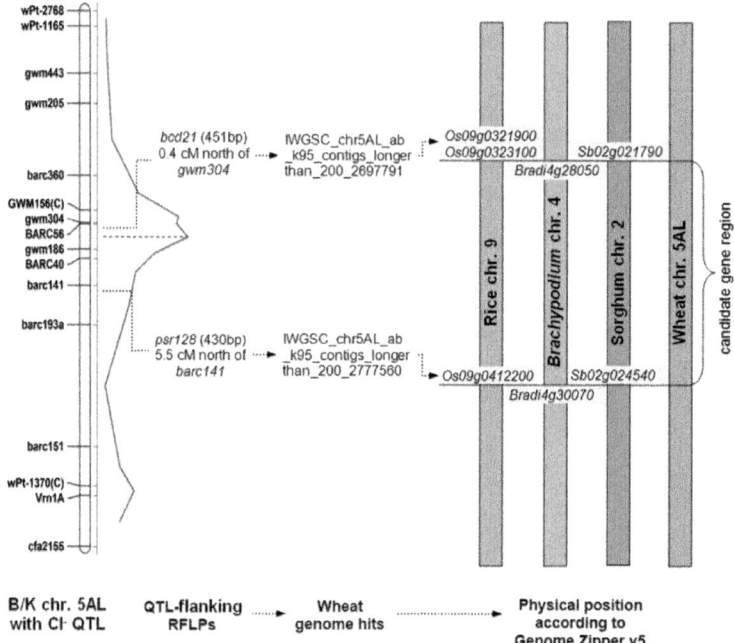

Figure 2. Inferred physical position of the Cl⁻ concentration QTL on 5AL in Berkut/ Krichauff identified at Roseworthy field location onto 5AL in wheat.

Table 6. Candidate genes underlying the physical interval of the Cl⁻ QTL on chromosome 5A.

Brachypodium (v1.2)	Rice (Gene ID at MSU[a]/at IRGSP v2)	Sorghum (v1.4)	Wheat (v5[b])	Predicted protein function[a]
-	-	Sb02g022750	-	ABC transporter
Bradi4g28660	LOC_Os09g19734/OS09G0361400	Sb02g022910	WHE0957_E03_J05ZT; Traes_5AL_99BEC1C3B	Voltage dependent anion channel 1 (VDAC1)
Bradi4g29102	LOC_Os09g20480/Os09g0371000	Sb02g023340	Traes_5AL_F8B48EC59	ABC transporter
Bradi4g29110	LOC_Os09g20490/Os09g0371100	Sb02g023370	-	ABC transporter
Bradi4g29120	"	Sb02g023380	-	"
Bradi4g29110	LOC_Os09g20500/Os09g0371200	Sb02g023370	-	ABC transporter
Bradi4g29120	"	Sb02g023380	-	"
Bradi4g29140	LOC_Os09g20510/Os09g0371300	Sb02g023360	-	ABC transporter
Bradi4g29110	LOC_Os09g20520/Os09g0371400	Sb02g023370	-	ABC transporter
Bradi4g29120	"	Sb02g023380	-	"
Bradi4g29347	LOC_Os09g21000/Os09g0376900	Sb02g023620	Traes_5AL_864648FE6	Potassium transporter family (HKT23-like)
Bradi4g29440	LOC_Os09g21340/Os09g0381100	Sb02g023720	WHE1104_A05_B10ZS; WHE0807_A06_B11ZS; Traes_5AL_3E0C865DF	Nucleobase-ascorbate transporter
Bradi4g29650	LOC_Os09g23110/Os09g0394500	Sb02g024060	Traes_5AL_01A13992D; Traes_5AL_88B668113	ABC transporter
Bradi4g29720	LOC_Os09g23300/Os09g0396900	Sb02g024130	WHE1787_E02_I03ZS; Traes_5AL_F80B422BA	Vacuolar iron transporter 1 (VIT1)
Bradi4g29810	LOC_Os09g23640/Os09g0401100	-	Traes_5AL_678EA44B2	ABC transporter

[a]according to MSU Rice Genome Annotation Project release 7, Ensembl Plants release 22 or NCBI;
[b]Genome Zipper v5.
doi:10.1371/journal.pone.0098845.t006

Within the physical interval underlying the 5A Cl⁻ QTL, there were five different classes of genes encoding different transporters or channels: eleven ABC-transporter genes, and single genes encoding a nucleobase-ascorbate transporter, a vacuolar iron transporter (VIT1), a voltage-dependent anion channel (VDAC1) and a potassium transporter (HKT23-like) (Table 6).

DISCUSSION

Implications of Screening for Cl⁻ Concentration in Hydroponics and Field Environments

Phenotyping for ST is generally conducted under hydroponic conditions, and results are rarely validated in relevant field environments. Our recent studies in bread wheat demonstrated that phenotyping for Na⁺ exclusion or ST in hydroponics had limited value in predicting field responses, and QTL differed vastly between hydroponics and field locations[24]. Given the renewed interest in Cl⁻ exclusion and ST [13], we analysed Cl⁻ concentration of hydroponically- and field-grown plants of Berkut/Krichauff DH population to determine the value of hydroponics for QTL analysis of Cl⁻ concentration, and elucidate genetic control mechanisms of Cl⁻ accumulation. The results demonstrated that plant Cl⁻ accumulation varied significantly between hydroponics and field trials and as a result, different QTL were identified between the two systems. As was the case with Na⁺ concentration [24], [25], this was most probably due to the two systems being vastly different [24]. These results suggest that there may be very little value in hydroponics testing to predict field responses to Cl⁻ concentration in bread wheat, and future studies should consider field testing or soil-based-pot assays as an alternative.

ST and Na⁺ or Cl⁻ Exclusion

In most studies to date Na⁺ exclusion, and to a limited extent Cl⁻ exclusion, are traits contributing to ST. However, studies that also examined relationships between ST (absolute or relative growth) and Na⁺ or Cl⁻ concentration reported inconsistent correlations [3]–[8],[10], [12], [45]–[48]. It is noteworthy that studies reporting high correlations ($r^2 > 0.5$) analysed either a small number of genotypes or were conducted under controlled environmental conditions [4], [7], [12], [46], [49]. In the present study genetic correlations were investigated using multivariate analysis. Moderate correlations in controlled environments (generally phenotypic) are not uncommon [13], [21]–[23], [25], while such correlations involving a large number of genotypes in the field are almost non-existent [24]. The lack of genetic correlation between Na⁺ and/or Cl⁻ exclusion and grain yield in field studies suggests that other biochemical and physiological processes need to be taken into consideration for identifying mechanisms associated with ST. These results also indicate that a reductionist approach, such as selection for Na^+ and/or Cl^- exclusion only, may not substantially improve ST in bread wheat. A more reliable approach would be to select for grain yield, unless specific physiological traits are shown to have significant and consistent correlations with grain yield under saline environments.

Ion Channels and Transporters Involved in Na^+, K^+ and Cl^- Homeostasis

Here we aim to introduce a brief discussion on transport of these ions from the soil solution into the root cell and their movements within the plant with respect to ion channels and transporters. However, a greater focus will be placed on Cl^-, and for Na^+ and K^+, readers are referred to recent reviews [41], [50], [51]. It is well established that the initial entry of Na^+ from the soil solution into the root cell is passive along the concentration gradient [52], and Na^+ uptake occurs primarily via non-selective cation channels and transporters [41], [51],[53]. However, Na^+ efflux (from cytosol into vacuole or removal from xylem) as a tolerance mechanism has to be active as it requires energy [41]. Potassium uptake from the soil solution into the root cell is an active process (i.e. moving across the membrane against its concentration gradient) and largely mediated by genes encoding channels and transporters [54]. Chloride uptake can be both active and passive depending on the external concentration. Under non-saline conditions transport of negatively charged Cl^- across negatively charged plasma membrane requires energy, therefore is an active process and mediated by transporters, while under saline conditions most Cl^- influx across the plasma membrane becomes passive [14]. However, its movement within or out of the plant must be active and aided by transporters [13]. Despite being the most abundant anion in the plant cells, compared to a number of well characterised Na^+ and K^+ transporters and encoding genes [41], very little is known about Cl^- transport mechanisms and the genes involved. To date two groups of gene families have repeatedly been discussed in relation to Cl^- homeostasis: CLCs and CCCs. From limited studies, it appears that CLCs, with their location in endomembranes, are involved in turgor regulation, stomatal movement and NO_3^- transport [55] but not root Cl^- uptake [41], while CCCs are involved in long distance transport of Na^+, K^+ and Cl^- and function as K^+:Cl^-, Na^+:Cl^- or Na^+: K^+:Cl^- co-transporters[44], [56]. However, there may be other gene families involved in Cl^- transport, as discussed by Teakle and Tyerman [13].

Genetic Control Mechanisms of Na⁺, K⁺ and Cl⁻Homeostasis under Salinity Stress

A good understanding of inheritance of homeostasis of these ions is required for a successful breeding strategy. Our present knowledge of mechanisms of their inheritance is a mere reflection of the number of studies conducted on them. For instance, due to greater focus on Na^+, we know more about Na^+ than K^+ or Cl^-. The limited studies in rice and wheat suggest that Na^+ and K^+ homeostasis under salinity stress are under separate genetic control [25], [57]–[59], although studies in barley [22], [23] found that QTL for Na^+ co-located with QTL for K^+, suggesting one or more genes regulating Na^+ and/or K^+ transport such as vacuolar sodium-hydrogen antiporter (NHX) genes [22]. As for Cl^-, to date there have been only four quantitative genetic studies that reported QTL for Cl^- concentration; two in rice[20], [21] and two in barley [22], [23]. In those studies as well as in the present study, moderate correlations between Cl^- and either Na^+ or K^+, and QTL for different ions mapping to the same and/or different regions indicate the presence of common (i.e. CCCs) and specific transporters for the uptake of these ions [i.e. high-affinity potassium (HKT) and (CLCs)] which in turn suggests common and separate genetic control. For instance, in the present study, Cl^- QTL on 2A and 5A co-located with QTL for Na^+ and K^+ concentration, respectively [24]. A physiological explanation for these co-locations and correlations between cations (Na^+ and K^+) and anions (Cl^-) may be the charge balance between these two groups of ions since the net movement of ions must be balanced so that there is charge equivalence with small difference [13]. Whereas there were no CCC and CLC genes physically close to the Cl^- concentration QTL on chromosome 5A there are 15 transporter and channel genes in the physical interval as candidates for the observed Cl^-accumulation. Although we have used the latest release of the comparison between wheat ESTs and contigs with sequenced grass genomes (Genome Zipper v5), it is possible that other genes reside in this region in wheat and the microsynteny is less well preserved as it appears to be so far.

Nguyen et al. [22] recently reported a Cl^- concentration QTL under salt stress in barley but used an incorrect barley chromosome nomenclature so that the actual chromosome 5H was mislabelled as 7H (other chromosomes were also mislabelled). The physical position of the RFLP markers *ABC324* and *ABC302* that flank their barley Cl^- QTL on 5HL [22] suggest a position close to the physical chromosomal region corresponding to the 5AL Cl^- concentration QTL reported here. In fact, the physical position of the northern flanking marker *ABC324* (position chromosome 5H: 399,222,591) slightly overlaps with the position of the southern end of the5A QTL flanked by RFLP marker *psr128* (position chromosome 5H: 412,653,548). It is possible that both QTL are caused by orthologous genes in wheat and barley, although this is far from certain as the QTL intervals contain hundreds of genes and the barley QTL was identified under salt stress in a hydroponics system whereas the QTL in wheat was repeatedly observed under salt stressed field conditions but not in hydroponics. Only further work such as fine mapping and gene expression analysis of the candidate genes will prove unequivocally the identity of the underlying gene for the differential Cl⁻ accumulation between Berkut and Krichauff. The present study provides a compelling case that the 5A QTL contains a K⁺:Cl⁻ co-transporter gene several other candidates capable of moving Cl⁻ ions across membranes. However, as this is the first report in wheat, there is clearly a need for testing other mapping populations and genetics resources to identify other Cl⁻ transporter gene(s) to gain a better understanding of genetic control mechanisms of Cl⁻ homoeostasis in crops.

Genetic Control Mechanisms of Ca^{2+} and Mg^{2+} Accumulation under Salinity Stress

The inheritance of plant Ca^{2+} and Mg^{2+} accumulation was also investigated, given the reports of salinity-induced nutritional deficiencies such as Ca^{2+} and Mg^{2+} [26], [37], [39], [40]and the importance of maintenance of adequate nutrition for these essential elements to plant growth and yield under salinity stress. To our knowledge, there have only been two studies in barley [22], [23] that investigated inheritance of Ca^{2+} and Mg^{2+} uptake or accumulation under salinity stress. However, only in one study [23] QTL for Ca^{2+} and Mg^{2+}were detected under salinity stress; one QTL for Mg^{2+} concentration on 6H, and three QTL for Ca^{2+} concentration on 1H, 6H and 7H. The QTL on 6H was common not only to Ca^{2+} and Mg^{2+} but also to ST. To our knowledge, this is the first time in the literature that a QTL for a nutrient other than Na^+ was co-located with ST, providing evidence for the role of Ca^{2+} and Mg^{2+} nutrition in growth and yield under salinity stress. These results also indicate that $Ca^{2+}$and Mg^{2+} uptake may occur through common as well as independent pathways. In contrast to the barley study, in the present study, none of the Ca^{2+} and Mg^{2+} concentration QTL co-located with each other or ST (measured as seedling biomass or grain yield), suggesting independent genetic control. However, further studies are required to enable better understanding of their genetic control mechanisms.

CONCLUSIONS

As was the case with Na^+ [24], plant Cl⁻ responses and related QTL differed widely between hydroponics and field tests, indicating substantial genotype and QTL interactions with environments. The results also indicated that hydroponics-based seedling assays may be very limited in their ability to predict field responses to salinity, and soil-based assays may be the second best option after field testing. As Cl^- concentration in the plant correlated only moderately with seedling biomass and showed no correlation with grain yield in the field, it does not appear, on its own, to be a reliable physiological parameter to select for in a breeding context, at least in bread wheat. Further research involving other mapping populations/genetic resources is warranted to be definitive. In the short term, selection for grain yield, which is integrative of all tolerance mechanisms, appears a more reliable strategy, while in the long term identification of donors for various physiological traits and subsequently combining them in a genotype (pyramiding) is likely to be the way forward[60]. This latter process can be fast-tracked via marker assisted selection. Finally the presence of several QTL for Cl^- concentration indicates that Cl^- uptake/accumulation is a polygenic trait. The discovery of a major QTL for Cl^- concentration on 5A that co-locates with several candidate genes that could be involved in Cl^- transport in bread wheat provides a starting point for further analysis through fine mapping and functional studies.

ACKNOWLEDGMENTS

We would like to thank Dr Hugh Wallwork (South Australian Research and Development Institute) for access to the Berkut/Krichauff DH population and the coordination of the field trials, the staff at Australian Grain Technologies, the LongReach Plant Breeders, the University of Adelaide Barley Breeding Program and Mr Jim Lewis (South Australian Research and Development Institute) for sowing, harvesting and general maintenance of the field trials, Mr Robin Hosking (the Australian Centre for Plant Functional Genomics) for his construction and creative modification of the supported hydroponic system and technical support throughout this project, Mrs Teresa Fowles and Mr Lyndon Palmer (the University of Adelaide) for their help with ICP-OES analysis, Prof. Mark Tester (Australian Centre for Plant Functional Genomics) for the use of a chloride meter, Drs Graham Lyons and John Harris (the University of Adelaide) for their critical review of the manuscript prior to submission, and the editor and the reviewers for their helpful comments.

AUTHOR CONTRIBUTIONS

Conceived and designed the experiments: YG JT JR KO. Performed the experiments: YG JR KO. Analyzed the data: YG JT KO. Contributed reagents/materials/analysis tools: YG JT KO. Wrote the paper: YG JT KO.

REFERENCES

1.   Martinez-Beltran J, Manzur CL (2005) Overview of salinity problems in the world and FAO strategies to address the problem. In Proceedings of the international salinity forum; Riverside, California. pp. 311–313.

2.   Schachtman DP, Munns R, Whitecross MI (1991) Variation in sodium exclusion and salt tolerance in *Triticum tauschii*. Crop Science 31: 992–997. doi: 10.2135/cropsci1991.0011183x003100040030x

3.   Ashraf M, O'Leary JW (1996) Responses of newly developed salt-tolerant genotype of spring wheat to salt stress: yield components and ion distribution. Journal of Agronomy and Crop Science 176: 91–101. doi: 10.1111/j.1439-037x.1996.tb00451.x

4.   Rashid A, Querishi RH, Hollington PA, Wyn Jones RG (1999) Comparative responses of wheat cultivars to salinity at the seedling stage. Journal of Agronomy and Crop Science 182: 199–207. doi: 10.1046/j.1439-037x.1999.00295.x

5.   Poustini K, Siosemardeh A (2004) Ion distribution in wheat cultivars in response to salinity stress. Field Crops Research 85: 125–133. doi: 10.1016/s0378-4290(03)00157-6

6.   Hollington PA (2000) Technological breakthroughs in screening/breeding wheat varieties for salt tolerance. In: Gupta SK, Sharma SK, Tyagi NK, editors. National Conference on Salinity Management in Agriculture. Central Soil Salinity Research Institute, Karnal, India. pp. 273–289.

7.   Huang Y, Zhang G, Wu F, Chen J, Zhou M (2006) Differences in physiological traits among salt-stressed barley genotypes. Communications in Soil Science and Plant Analysis 37: 567–570. doi: 10.1080/00103620500449419

8.   Ashraf M, McNeilly T (1988) Variability in salt tolerance of nine spring wheat cultivars. Journal of Agronomy and Crop Science 160: 14–21. doi: 10.1111/j.1439-037x.1988.tb01160.x

9.   Bagci SA, Ekiz H, Yilmaz A (2007) Salt tolerance of sixteen wheat genotypes during seedling growth. Turkish Journal of Agriculture and Forestry 31: 363–372.

10. Genc Y, McDonald GK, Tester M (2007) Reassessment of tissue Na^+ concentration as a criterion for salinity tolerance in bread wheat. Plant, Cell and Environment 30: 1486–1498. doi: 10.1111/j.1365-3040.2007.01726.x

11. Royo A, Aragüés R (1999) Salinity-yield response functions on barley genotypes assessed with a triple line source sprinkler system. Plant and

Soil 209: 9–20.

12. El-Hendawy S, Hu Y, Schmidhalter U (2005) Growth, ion content, gas exchange and water relations of wheat genotypes differing in salt tolerances. Australian Journal of Agricultural Research 56: 123–134. doi: 10.1071/ar04019

13. Teakle NL, Tyerman SD (2010) Mechanisms of Cl⁻ transport contributing to salt tolerance. Plant, Cell and Environment 33: 566–589. doi: 10.1111/j.1365-3040.2009.02060.x

14. White PJ, Broadley MR (2001) Chloride in soils and its uptake and movement within the plant: A Review. Annals of Botany 88: 967–988. doi: 10.1006/anbo.2001.1540

15. Kingsbury R, Epstein E (1986) Salt sensitivity in wheat. A case for specific ion toxicity. Plant Physiology 80: 651–654. doi: 10.1104/pp.80.3.651

16. Martin P, Koebner R (1995) Sodium and chloride ions contribute synergistically to salt toxicity in wheat. Biologia Plantarum 37: 265–271. doi: 10.1007/bf02913224

17. Tavakkoli E, Rengasamy P, McDonald GK (2010) The response of barley to salinity stress differs between hydroponics and soil systems. Functional Plant Biology 37: 621–633. doi: 10.1071/fp09202

18. Munns R, Tester M (2008) Mechanisms of salinity tolerance. Annual Review of Plant Biology 59: 651–681. doi: 10.1146/annurev.arplant.59.032607.092911

19. Xu G, Magen H, Tarchitzky J, Kafkafi U (2000) Advances in chloride nutrition. Advances in Agronomy 68: 96–150. doi: 10.1016/s0065-2113(08)60844-5

20. Ammar MHM, Pandit A, Singh RK, Sameena S, Chaucan MS, et al. (2009) Na⁺, K⁺and Cl⁻ ion concentrations in salt tolerant *Indica* rice variety CSR27. Journal of Plant Biochemistry and Biotechnology 18: 139–150. doi: 10.1007/bf03263312

21. Pandit A, Rai V, Bal S, Sinha S, Kumar V, et al. (2010) Combining QTL mapping and transcriptome profiling of bulked RILs for identification of functional polymorphism for salt tolerance genes in rice (*Oryza sativa* L.). Molecular Genetics and Genomics 284: 121–136. doi: 10.1007/s00438-010-0551-6

22. Nguyen VL, Ribot SA, Dolstra O, Niks RE, Visser RGF, et al. (2013) Identification of quantitative trait loci for ion homeostasis and salt tolerance in barley (*Hordeum vulgare* L). Molecular Breeding 31: 137–152. doi: 10.1007/s11032-012-9777-9

23. Nguyen VL, Dolstra O, Malosetti M, Kilian B, Graner A, et al. (2013) Association mapping of salt tolerance in barley (*Hordeum vulgare* L). Theoretical and Applied Genetics. DOI.10.1007/s00122-013-2139-0.

24. Genc Y, Oldach K, Gogel B, Wallwork H, McDonald GK, et al. (2013) Quantitative trait loci for agronomical and physiological traits for a bread wheat population grown in environments with a range of salinity levels. Molecular Breeding 32: 39–59 DOI: 10.1007/s110320139851y.

25. Genc Y, Oldach K, Verbyla A, Lott G, Hassan M, et al. (2010) Sodium exclusion QTL associated with improved seedling growth in bread wheat under salinity stress. Theoretical and Applied Genetics 121: 877–894. doi: 10.1007/s00122-010-1357-y

26. Genc Y, Tester M, McDonald GK (2010) Calcium requirement of wheat in saline and non-saline conditions. Plant and Soil 327: 331–345. doi: 10.1007/s11104-009-0057-3

27. Wheal MS, Palmer LT (2010) Chloride analysis of botanical samples by ICPOES. Journal of Analytical Atomic Spectrometry 25: 1946–1952. doi: 10.1039/c0ja00059k

28. Wheal MS, Fowles TO, Palmer LT (2011) A cost effective acid digestion method using closed polypropylene tubes for inductively coupled plasma optical emission spectrometry (ICP-OES) analysis of plant essential elements. Analytical Methods 3: 2854–2863. doi: 10.1039/c1ay05430a

29. Martinez O, Curnow RN (1992) Estimating the locations and sizes of the effects of quantitative trait loci using flanking markers. Theoretical and Applied Genetics 85: 480–488. doi: 10.1007/bf00222330

30. Verbyla AP, Cullis BR, Thompson R (2007) The analysis of QTL by simultaneous use of the of the full linkage map. Theoretical and Applied Genetics 116: 95–111. doi: 10.1007/s00122-007-0650-x

31. Smith A, Cullis BR, Thompson R (2001) Analysing variety by environment data using multiplicative mixed models. Biometrics 57: 1138–1147. doi: 10.1111/j.0006-341x.2001.01138.x

32. R Development Core Team (2013) R: A language and environment for statistical computing, R Foundation for Statistical Computing. Vienna, Austria.

33. Taylor JD, Diffey S, Verbyla AP, Cullis BC (2013) wgaim: Whole Genome Average Interval Mapping for QTL detection using mixed models, R package version 1.3–0.

34. Verbyla AP, Taylor JD, Verbyla KL (2012) RWGAIM: An efficient high dimensional random whole average (QTL) interval mapping approach. Genetics Research 94: 291–306. doi: 10.1017/s0016672312000493

35. Cullis BR, Smith AB, Coombes NE (2006) On the design of early generation variety trials with correlated data. Journal of Agricultural, Biological and Environmental Statistics 11: 381–393. doi: 10.1198/108571106x154443

36. Bonneau J, Taylor J, Parent B, Bennett D, Reynolds M, et al. (2013) Multi-environment analysis and improved mapping of a yield-related QTL on chromosome 3B of wheat. Theoretical and Applied Genetics 126: 747–761. doi: 10.1007/s00122-012-2015-3

37. Ehret DI, Redmann RE, Harvey BL, Cipywnyk A (1990) Salinity-induced calcium deficiencies in wheat and barley. Plant and Soil 128: 143–151. doi: 10.1007/bf00011103

38. Bergmann W (1992) Nutritional disorders of plants: Developments, visual and analytical diagnosis. Gustav Fisher, Jena, Stuttgart, New York.

39. Cramer GR (2002) Sodium-calcium interactions under salinity stress. In: Lauchli A, Luttge U, editors. Salinity: Environment- plants- molecules. Kluwer, Dordrecht. pp. 205–227.

40. Adcock KG, Gartrell JW, Brennan RF (2001) Calcium deficiency of wheat grown in acidic sandy soil from Southwestern Australia. Journal of Plant Nutrition 24: 1217–1227. doi: 10.1081/pln-100106977

41. Mian AA, Senadheera P, Maathuis FJ (2011) Improving crop salt tolerance: anion and cation transporters as genetic engineering targets. Plant Stress 5: 64–72.

42. Lurin C, Geelen D, Barbier-Brygoo H, Guern J, Maurel C (1996) Cloning and functional expression of a plant voltage-dependent chloride channel. Plant Cell 8: 701–711. doi: 10.2307/3870345

43. Wong TH, Li MW, Yao XQ, Lam HM (2013) The GmCLC1 protein from soybean functions as a chloride ion transporter. Journal of Plant Physiology 170: 101–104. doi: 10.1016/j.jplph.2012.08.003

44. Colemenero-Flores JM, Martinez G, Gamba G, Vazuez N, Iglesias DJ, et al. (2007) Identification and functional characterization of cation-chloride cotransporters in plants. Plant Journal 50: 278–292. doi: 10.1111/j.1365-313x.2007.03048.x

45. Isla R, Royo A, Aragüés R (1997) Field screening of barley cultivars to soil salinity using a sprinkler and a drip irrigation system. Plant and Soil 197: 105–117. doi: 10.1023/a:1004240622652

46. Khan MA, Shirazi MU, Khan MA, Mujtaba SM, Islam E, et al. (2009) Role of proline, K/Na ratio and chlorophyll content in salt tolerance of wheat (*Triticum aestivum* L.). Pakistan Journal of Botany 41: 633–638.

47. Rawson HM, Richards RA, Munns R (1988) An examination of selection criteria for salt tolerance in wheat, barley and triticale genotypes. Australian Journal of Agricultural Research 39: 759–772. doi: 10.1071/ar9880759

48. Salam A, Hollington PA, Gorham J, Wyn Jones RG, Gliddon C (1999) Physiological genetics of salt tolerance in wheat (*Triticum aestivum* L): Performance of wheat varieties, inbred lines and reciprocal F_1 hybrids under saline conditions. Journal of Agronomy and Crop Science 183: 145–156. doi: 10.1046/j.1439-037x.1999.00361.x

49. Munns R, James RA (2003) Screening methods for salinity tolerance: a case study with tetraploid wheat. Plant and Soil 253: 201–218. doi: 10.1023/a:1024553303144

50. Maathuis FJM, Amtmann A (1999) K^+ Nutrition and Na^+ Toxicity: The Basis of Cellular K^+/Na^+ Ratios. Annals of Botany 84: 123–133. doi: 10.1006/anbo.1999.0912

51. Tester M, Davenport R (2003) Na^+ tolerance and Na^+ transport in higher plants. Annals of Botany 91: 503–527. doi: 10.1093/aob/mcg058

52. Cheeseman JM (1982) Pump-leak sodium fluxes in low salt corn roots. Journal of Membrane Biology 70: 157–164. doi: 10.1007/bf01870225

53. Byrt CS, Platten JD, Spielmeyer W, James RA, Lagudah ES, et al. (2007) HKT1;5-like cation transporters linked to Na^+ exclusion loci in wheat, *Nax2* and *Kna1*. Plant Physiology 143: 1918–1928. doi: 10.1104/pp.106.093476

54. Horie T, Brodsky DE, Costa A, Kaneko T, Schiavo FL, et al. (2011) K^+ transport by the OsHKT2;4 transporter from rice (*Oryza sativa*) with atypical Na^+ transport properties and competition in permeation of K^+ over Mg^{2+} and Ca^{2+} ions. Plant Physiology 156: 1493–1507. doi: 10.1104/pp.110.168047

55. Hechenberger M, Schwappach B, Fischer WN, Frommer WB, Jentsch TJ, et al. (1996) A family of putative chloride channels from Arabidopsis and functional complementation of a yeast strain with a CLC gene disruption. Journal of Biological Chemistry 271: 33632–33638. doi: 10.1074/jbc.271.52.33632

56. Kong X-Q, Gao X-H, Sun W, An J, Zhao Y-X, et al. (2011) Cloning and functional characterization of a cation-chloride cotransporter gene OsCCC1. Plant Molecular Biology 75: 567–578. doi: 10.1007/s11103-011-9744-6

57. Garcia A, Rizzo CA, Ud-Din J, Bartos SL, Senadhira D, et al. (1997) Sodium and potassium transport to the xylem are inherited independently

in rice and the mechanisms of sodium: potassium selectivity differs between rice and wheat. Plant, Cell and Environment 20: 1167–1174. doi: 10.1046/j.1365-3040.1997.d01-146.x

58.   Koyama ML, Levesley A, Koebner RMD, Flowers TJ, Yeo AR (2001) Quantitative trait loci for component physiological traits determining salt tolerance in rice. Plant Physiology 125: 406–422. doi: 10.1104/pp.125.1.406

59.   Lin HX, Zhu MZ, Yano M, Gao JP, Liang ZW, et al. (2004) QTLs for Na^+ and K^+uptake of the shoots and roots controlling rice salt tolerance. Theoretical and Applied Genetics 108: 253–260. doi: 10.1007/s00122-003-1421-y

60.   Peng S, Ismail AM (2004) Physiological basis of yield and environmental adaptation in rice. In: Nguyen HT, Blum A, editors. Physiology and biotechnology integration for plant breeding. CRC Press, New York, Basel. pp. 72–118.

Chapter 9

TRANSCRIPTOMIC ANALYSIS OF FIBER STRENGTH IN UPLAND COTTON CHROMOSOME INTROGRESSION LINES CARRYING DIFFERENT GOSSYPIUM BARBADENSE CHROMOSOMAL SEGMENTS

Lei Fang, Ruiping Tian, Jiedan Chen, Sen Wang, Xinghe Li, Peng Wang, and Tianzhen Zhang
National Key Laboratory of Crop Genetics and Germplasm Enhancement, Cotton Hybrid R & D Engineering Center (the Ministry of Education), Nanjing Agricultural University, Nanjing, China

ABSTRACT

Fiber strength is the key trait that determines fiber quality in cotton, and it is closely related to secondary cell wall synthesis. To understand the mechanism underlying fiber strength, we compared fiber transcriptomes from different *G. barbadense* chromosome introgression lines (CSILs) that had higher fiber strengths than their recipient, *G. hirsutum* acc. TM-1. A total of 18,288 differentially expressed genes (DEGs) were detected between CSIL-35431 and CSIL-31010, two CSILs with stronger fiber and TM-1 during secondary cell wall synthesis. Functional classification and enrichment analysis revealed that these DEGs were enriched for secondary cell wall biogenesis, glucuronoxylan biosynthesis, cellulose biosynthesis, sugar-mediated signaling pathways, and fatty acid biosynthesis. Pathway analysis showed that these DEGs participated in starch and sucrose metabolism (328 genes), glycolysis/ gluconeogenesis (122 genes), phenylpropanoid biosynthesis (101 genes), and oxidative phosphorylation (87 genes), etc. Moreover, the expression of MYB- and NAC-type transcription factor genes were also dramatically different between the CSILs and TM-1. Being different to those of CSIL-31134, CSIL-35431 and CSIL-31010, there were many genes for fatty acid degradation and biosynthesis, and also for carbohydrate metabolism that were down-regulated in CSIL-35368. Metabolic pathway analysis in the CSILs showed that different pathways were changed, and some changes at the same developmental stage in some pathways. Our results extended our understanding that carbonhydrate metabolic pathway and secondary cell wall biosynthesis can affect the fiber strength and suggested more genes and/or pathways be related to complex fiber strength formation process.

INTRODUCTION

The cotton fiber is a terminally differentiated single cell derived from the epidermal cell of the developing ovule. After initiation, the fiber cell undergoes 1000- to 3000-fold elongation during its development. The development of cotton fibers involves four partially overlapping stages: initiation (−3 to +3 days post-anthesis; DPA), elongation and primary cell wall formation (3–23 DPA), secondary cell wall formation (16–40 DPA) and maturation (40–50 DPA) [1]–[6]. The most rapid period of fiber cell elongation begins around 10–16 DPA and continues to ~20 DPA. Primary and secondary cell wall synthesis overlaps during the period of 16–25 DPA. During the secondary cell wall formation stage, the speed of cell elongation slows down and even stops.

Fiber strength is an important indicator of cotton fiber quality, and depends on formation of the secondary cell wall. Cellulose synthesis plays a predominant role in fiber cells, and cellulose accounts for >95% of the dry weight of the mature cotton fiber [3], [7]. Genome and EST sequencing have revealed that there are at least ten different CesA genes for cellulose synthase in *Arabidopsis*; CesA-like genes have also been reported in rice and barley[8]–[10]. In cotton (*Gossypium raimondii*), at least 15 cellulose synthase (CESA) sequences are required for cellulose synthesis [11]. A recent investigation in *Arabidopsis thaliana* using microarrays led to the identification of genes that are highly co-expressed with cellulose synthase genes and two mutants, irx8 and irx13, that have irregular xylem phenotypes, were also identified [12]. Sucrose synthase (Susy) is the enzyme that catalyzes the hydrolysis of sucrose to UDP-glucose that is then used as a substrate for cellulose synthesis. In cotton, the expression of Susy is higher at 16–32 DPA, and this enzyme plays a major role in partitioning carbon toward cellulose synthesis in the fiber [13]. SusC is another new sucrose synthase gene with a high level of expression during secondary cell wall synthesis [14]. Peroxide, mainly as H_2O_2, promotes cellulose synthesis as a signal of secondary cell wall synthesis [15], [16].

At present, many ovule- and fiber-specific cDNA libraries have been constructed and sequenced, and more than 268,000 expressed sequence tags (ESTs) from *Gossypium* are deposited in the NCBI database (http://www.ncbi. nlm.nih.gov). For genetic characterization of rapid cell elongation in cotton fibers, approximately 14,000 unique genes were assembled from 46,603 expressed sequence tags (ESTs) from developmentally-staged fiber cDNAs of a cultivated diploid species (*G. arboreum* L.). Eighty-one genes that were significantly up-regulated during secondary cell wall synthesis were found to be involved in cell wall biogenesis and energy/carbohydrate metabolism, which is consistent with the stage of cellulose synthesis during secondary cell wall modification in developing fibers[17]. Transcriptome profiling of the cotton fiber early in development by high-throughput tag-sequencing (Tag-seq) analysis using the Solexa Genome Analyzer reveals significant differential expression of genes in a fuzzless/lintless mutant [18]. High-throughput, genome-wide transcriptomic analysis of cotton under drought stress revealed a significant down-regulation of genes and pathways involved in fiber elongation, and an up-regulation of defense response genes [19]. More research have been processed in fiber initiation and elongation stage [20]–[24]. Saturated very-long-chain fatty acids (VLCFAs; C20:0–C30:0) exogenously applied in ovule culture medium significantly promoted fiber cell elongation in cotton (*G. hirsutum L.*) by activating ethylene biosynthesis [25], [26]. Previous investigations into cotton fiber development mainly focused on the elongation stage, and the number of genes reported from the later stages is quite small. Most of the genes up-regulated during secondary cell wall synthesis were related to cellulose synthesis, cell wall biosynthesis, and carbohydrate metabolism [17], [22], [27].

Chromosome segment introgression lines (CSILs) consist of a battery of near-isogenic lines that have been developed to cover the entire genomes of some crops, including tomato, rice, wheat, and cotton [28]–[31]. With the exception of a single, homozygous chromosome segment transferred from a donor parent, the remaining genome of each CSIL is the same as the recipient parent [31]. We used *G. barbadense* CSILs in the background of the standard genetic line of *G. hirsutum*, cv. TM-1, in order to understand the molecular mechanism behind superior quality fiber formation. Multi-point tests showed that three CSILs produced stronger fibers when compared to the recipient parent TM-1, but one CSIL produced weaker fibers. Using Solexa Genome sequencing, we analyzed transcriptome profiles from the CSILs and TM-1. We found that many genes were either up- or down-regulated at the stage of secondary cell wall synthesis, and that many metabolic pathways were altered in the CSILs.

MATERIALS AND METHODS

Plant Materials

G. hirsutum cv. TM-1, the genetic standard for Upland cotton, was obtained from the Southern Plains Agricultural Research Center, USDA-ARS, College

Station/Texas, USA [32].*G. barbadense* cv. Hai7124, an extra-long staple cotton that is widely grown in China, is descended from a selected individual in a study of inheritance of resistance to *Verticillium dahlia* [33], [34]. In this study, we identified three CSILs with stronger fiber or high fiber strength that carried different *G. barbadense* chromosome segment(s) in the recurrent parent TM-1. The detailed method of developing CSILs has been described previously [31]. We selected three CSILs, CSIL-35431, CSIL-31134, and CSIL-31010, in which the average fiber strength were 35.1, 34.73 and 34.28 cN/tex, respectively, significantly higher than TM-1, and also CSIL-35368 which had poorer fiber strength than TM-1(28.71 cN/tex) (Table S1). The introgressed *G. barbadense* chromosomal segments were different in the four lines[35]. Fiber samples were collected at 15, 20, and 25 DPA, frozen in liquid nitrogen, and stored at $-70°C$.

RNA Isolation and Evaluation

Total RNA was extracted from frozen tissue using an improved CTAB extraction protocol [36]. RNAs were evaluated for quality using RNA Pico Chips on an Agilent 2100 Bioanalyzer (Agilent Technologies, Santa Clara, CA, USA). All RNA samples were quantified and qualified with an RNA Integrity Number (RIN) >8, and 28S/18S rRNA band intensity (2:1).

Library Construction and Sequencing

Digital gene expression libraries were constructed using the Illumina Gene Expression Sample Preparation Kit according to the manufacturer's instructions. We constructed and sequenced 14 libraries derived from immature fibers at 15, 20, and 25 DPA using the Solexa Genome Sequencing Analyzer system provided by BGI (Beijing Genomics Institute at Shenzhen, China), which gave 21 bp tags. The process was described in detail previously [18].

Data Processing, Statistical Evaluation, and Selection of Differentially Expressed Genes (DEGs)

Raw data reads were filtered by the Illumina pipeline to produce clean data. All low-quality data, such as short tags (<21 nt) and singletons, were removed. A database of 21-base-long sequences was produced beginning with CATG using 37,505 reference genes from the diploid species *G. raimondii* (http://www.phytozome.net). The remaining high quality sequences were then mapped to this database; only a single mismatch was allowed, and more than one match was excluded. Gene expression levels were the summation of tags aligned to the different positions of the same gene. Expression levels are expressed as TPM, transcripts per million. To identify DEGs during fiber elongation, we compared pairs of DEG profiles from different libraries. Three fiber development periods for the four CSILs were compared with the same period for TM-1, and 11 comparisons were obtained. P- and Q-values were also calculated for every comparison [37]. DEGs were defined as FDR≤0.001 with an absolute value of |log$_2$Ratio|≥1 to judge the significance of differences in transcript abundance.

Digital Tag Profiling Analysis

DEG clustering in CSILs at different developmental stages were performed with Cluster3.0 (http://bonsai.hgc.jp/~mdehoon/software/cluster/software. htm). We also performed clustering with the 'Self-organizing tree algorithm' (SOTA, Multiple Array Viewer software, MeV 4.9.0)[38].

GO enrichment and KEGG (Kyoto Encyclopedia of Genes and Genomes) pathway analysis was done using BLAST2GO (http://www.blast2go.com/ b2ghome). Mapman was also used to analyze metabolic pathway base on KEGG database [39].

Quantitative RT-PCR

Quantitative RT-PCR assays were performed on a 7500 Real-Time PCR system (Applied Biosystems, San Francisco, CA, USA). Reactions were performed in a final volume of 15 µL and contained 2 µL of diluted cDNA, 7.5 µL of 2× SYBR mix (Roche, Basel, Switzerland), and 200 nM of the forward and reverse primers. Primer lengths were designed to be between 18 and 24 nt using Beacon Designer 7, and PCR amplicon lengths were designed to be between 100 bp and 150 bp (Table S2). The thermal cycling conditions were 40 cycles of 95°C for 15 s, 60°C for 30 s, and 72°C for 30 s. All reactions were run in triplicate, and the cotton *histone3* gene (ACC NO. AF024716) was used as an internal control for normalization of expression levels (F: 5′-GGTGGTGTGAAGAAGCCTCAT-3′, and R: 5′-AATTTCACGAACAAGCCTCTGGAA-3′). The relative gene expression levels were presented as $2^{-\Delta CT}$.

RESULTS

Statistical Analysis of Transcriptome Data

The total number of sequence tags per library ranged from 7.0 to 8.5 million, and the number of distinct sequence tags was between 1.8 and 2.2 million. Approximately 50% of the clean tags were mapped to reference genes, and

60% of the reference genes were mapped with unambiguous tag (Table 1 and Table S3).

Table 1. The distribution of total and distinct tags.

Summary	TM-1			CSIL-35431			CSIL-31010		
	15DPA	20DPA	25DPA	15DPA	20DPA	25DPA	15DPA	20DPA	25DPA
Raw Data	8374304	7231305	14267931	7461562	7389820	7298224	7471212	7007447	7203054
Distinct Raw Tag	389162	394430	494345	375999	371748	317147	283130	337525	386302
Clean Tag	8144920	7013147	14002534	7264426	7213791	7124575	7369012	6840101	7002766
Distinct Clean Tag	169700	177887	251173	182629	199257	148121	182184	172775	188639
Unique Clean Tag Mapping to Gene	3983215	3206328	6829327	3961228	3411249	3463314	3671014	3381955	3552566
Total % of clean tag	48.90%	45.72%	48.77%	54.53%	47.29%	48.61%	49.82%	49.44%	50.73%
Unique Distinct Clean Tag Mapping to Gene	45380	40475	56172	53187	48453	41065	46720	45762	48147
Total % of distinct tag	26.74%	22.75%	22.36%	29.12%	24.32%	27.72%	25.64%	26.49%	25.52%
Unambiguous Tag-mapped Genes	21498	20901	22594	21781	22325	20590	21900	22457	22811
Percentage of reference genes	57.32%	55.73%	60.24%	58.07%	59.53%	54.90%	58.39%	59.88%	60.82%

Clean tags: tags after filtering dirty tags (low quality tags) from raw data.
Distinct tags: different kinds of tags.
Unambiguous tags: the clean tags after removing tags mapped to reference sequences from multiple genes.
doi:10.1371/journal.pone.0094642.t001

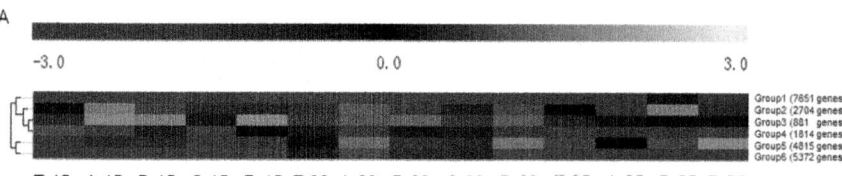

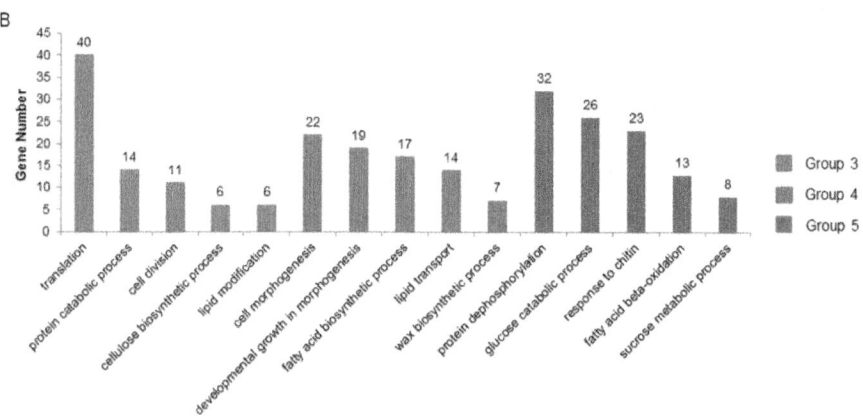

Figure 1. Statistical analysis of transcriptome data. (A) SOTA clustering of the different genes using Log2(TPM). T, TM-1; A, CSIL-35431; B, CSIL-31010; C, CSIL-31134; D, CSIL-35368. 15, 15 DPA; 20, 20 DPA; 25, 25 DPA. (B) Distribution of functions of genes in different clusters. Yellow square indicated group 3, green square indicated group 4 and blue square indicated group 5. X-axis indicated different enriched process and Y-axis indicated number of hit-found genes in these processes.

To see whether the fiber transcriptomes at different developmental stages were different, the 23,237 genes which were expressed in at least three libraries at one stage (15 DPA, 20 DPA, or 25 DPA) were classified into six groups using the Multiple Array Viewer using TPM value (Figure 1A). Genes in Group 3 had higher expression levels at 15 DPA and 20 DPA than at the later stage (25 DPA). Genes in Group 4 had higher expression levels at 15 DPA than at either 20 DPA or 25 DPA. Genes in Group 5 showed the opposite expression pattern, with higher expression levels at 20 DPA and 25 DPA compared to 15 DPA. The other groups also showed distinct expression patterns (Figure 1A).

Classification by gene function revealed that Group 3 is enriched in genes involved in protein catabolism, cell division, and cellulose biosynthesis, Group 4 is enriched in genes for cell morphogenesis, fatty acid biosynthesis, lipid transport, and wax biosynthesis, and Group 5 has more genes involved in glucose catabolism, response to chitin, and sucrose metabolism (Figure 1B). The unbalanced pattern of the expressed-gene functional distribution could possibly reflect some physiological events involved in secondary cell wall biosynthesis.

Cluster Analysis of Differentially Expressed Genes (DEGs) Between and/or Among CSILs

We specifically looked for DEGs in secondary cell wall fibers from 15 to 25 DPA, because previous studies have reported that the different sets of transcripts responsible for fiber secondary cell wall formation may be enriched at these stages of development [17], [22],[27]. Three fiber development periods for the four CSILs were compared with TM-1 at the same period. DEGs were defined as FDR$\leq$0.001 with an absolute value of $|log_2 Ratio|\geq 1$. Analysis of the data indicated that many genes showed differential expression in the 11 comparison groups. The number of DEGs were about 6,000–8,000 in CSILs from 15 DPA to 25 DPA (Figure 2A). But the number of DEGs in CSIL-31010 at 20 DPA, CSIL-31010 at 25 DPA, and CSIL-31134 at 15 DPA, were 4,600, 10,106 and 2,060, respectively. We also found that the DEGs that were up-regulated or down-regulated were different in CSILs. There were ~1,500–3,500 DEGs in common from 15 DPA to 25 DPA between CSIL-35431, CSIL-31010, and CSIL-35368 (Figure 2B).

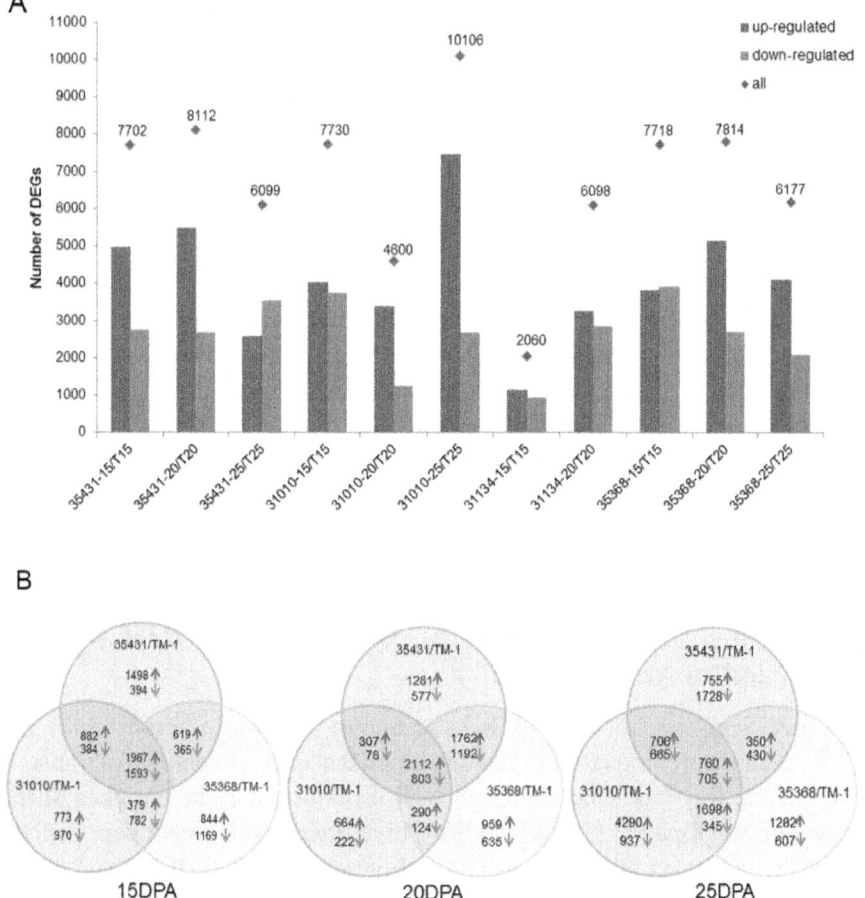

Figure 2. Statistical of DEGs between CSILs and TM-1 at 15, 20 and 25 DPA. (A) Up-regulated and down-regulated genes in different comparison. Red bar, up-regulated genes compared to TM-1; green bar, stand for down-regulated genes compared to TM-1, Blue square, total DEGs. CSILs included CSIL-35431, CSIL-3010, CSIL-31134, CSIL-35368 and TM-1. 15, 15DPA; 20, 20DPA; 25, 25DPA. (B) Common and special DEGs at 15 DPA, 20 DPA and 25 DPA.

To understand the mechanisms behind the changes in fiber strength observed in the CSILs, we also analyzed the common DEGs among CSIL-35431, CSIL-31010 and CSIL-31134. A total of 727 and 1796 common DEGs were selected at 15 and 20 DPA in three stronger fiber CSILs, respectively (Figure 3). More functional enrichment were shown at 15 DPA, including major CHO metabolism (carbohydrate), cell wall biosynthesis, amino acid metabolism and secondary metabolism (Figure 3E). Among these genes,

321 and 998 common upregulated DEGs between the same CSILs at 15 and 20 DPA were indentified, respectively (Figure 3). These common DEGs or processes maybe directly related to the fiber strength. However, these DEGs maybe function as downstream genes altered by the introgressed segments since these CSILs were inserted different *G. barbadense* segments in recipient TM-1.

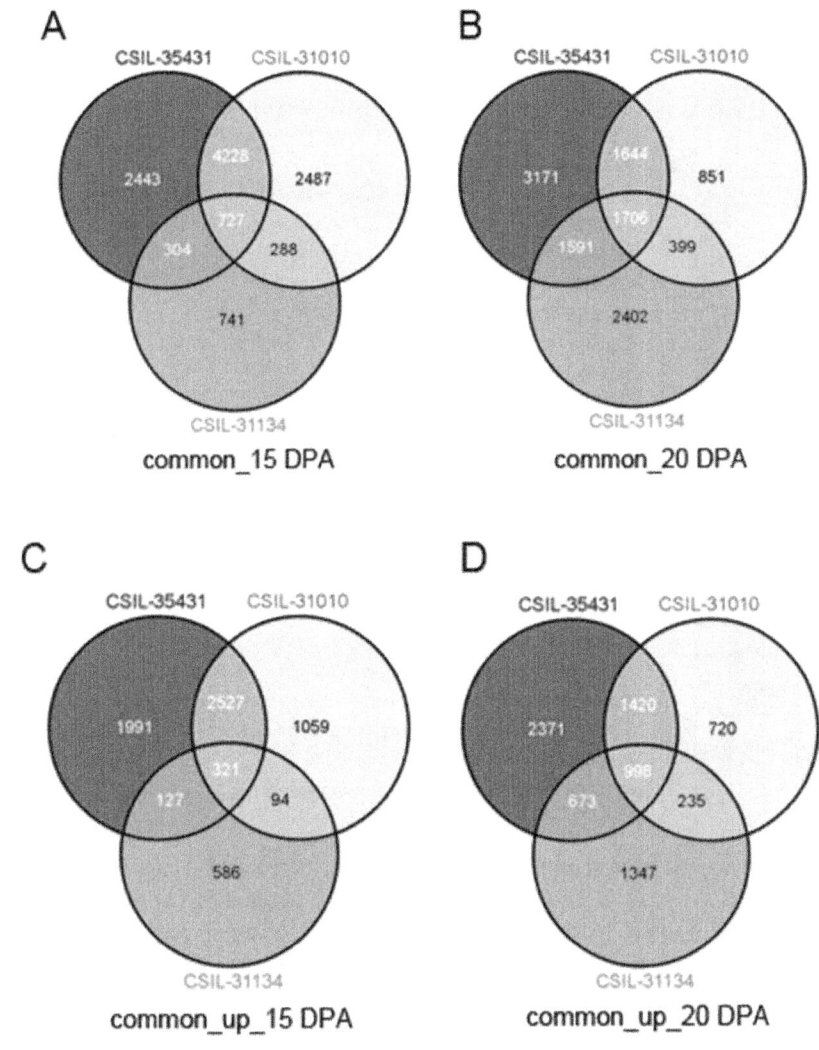

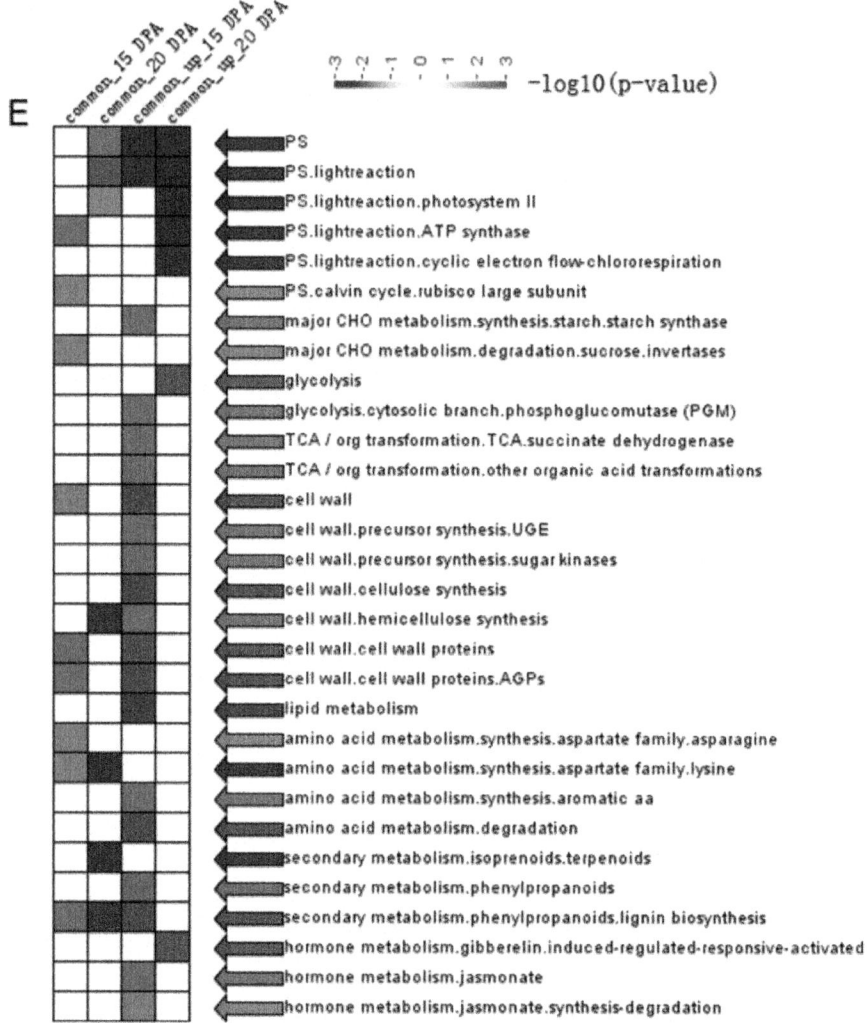

Figure 3. Analysis of common and common upregulated DEGs among three stronger fiber CSILs. (A, B, C, D) Common and common upregulated DEGs among three stronger fiber CSILs at 15 and 20 DPA. Common_up, common regulated DEGs. (B) Functional enrichment analysis of these DEGs using mapman software (Summary statistic type, wlcoxon). Colors from blue to red indicated that functions were enriched more significantly with smaller p-values.

To visualize the expression patterns of DEGs, we performed cluster analysis of 18,288 genes that were differentially expressed between CSIL-35431 and CSIL-31010 (Figure 4). These DEGs could be grouped into six clusters, designated G1–G6, based on their expression patterns. From 15 DPA to 20 DPA, the stages of fast fiber elongation and secondary cell wall deposition overlap, with the latter reaching a peak at around 20–25 DPA. We focused on clusters G1, G4, and G6 to conduct data analysis in order to identify genes that were either up-regulated or down-regulated during the secondary cell wall synthesis stage. Compared to the TM-1 control, 3,658 genes in cluster G1 were highly expressed at 15 and 20 DPA, 4,487 genes in G4 were highly expressed at 15 DPA, 20 DPA, and 25 DPA, 3,033 genes in G6 were highly expressed only at 25 DPA, and the other three groups showed various different expression patterns. Clustering results for 19,742 DEGs from the four CSILs showed five groups, indicating that the gene expression pattern in CSIL-31134 was distinct from the others at 15 DPA and 20 DPA, and that CSIL-35368 was similar to CSIL35431 and CSIL-31010 (Figure S1).

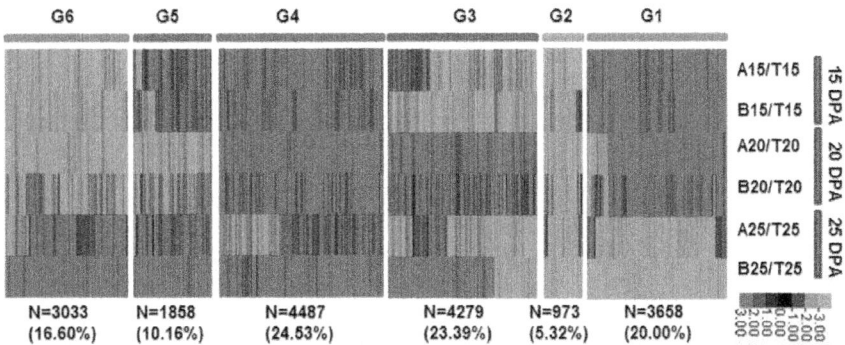

Figure 4. Heat map analysis of the expression of DEGs between CSILs and TM-1. A, B and T indicated CSIL-35431, CSIL-31010 and TM-1, respectively. 15, 15DPA; 20, 20DPA; 25, 25DPA. Red color indicated up-regulated genes and green color indicated down-regulated genes. N=number of DEGs in different group.

Functional Annotation by GO Enrichment and KEGG Analysis

To understand the mechanisms behind the changes in fiber strength observed in the CSILs, we analyzed DEG enrichment in the major functional GO categories of biological process, molecular function, and cellular component between CSIL-35431 and CSIL-31010. Based on the clustering results shown in Figure 4, G1 was enriched in genes for secondary cell wall biogenesis, glucuronoxylan biosynthesis, microtubule-based movement, and cellulose biosynthesis, G4 was enriched in genes for protein phosphorylation, response to chitin, and sugar-mediated signaling pathways, and G6 was enriched in fatty acid biosynthesis genes (Table 2). These data suggest that in the developmental stage of secondary cell wall deposition, DEGs were enriched for carbohydrate synthesis and cell wall formation.

Table 2. Enrichment analysis of gene ontologies from 15 to 25 DPA.

Cluster	GO-ID	GO Ontology (Biological process)
G1	GO:0010417	glucuronoxylan biosynthetic process
15DPA up-regulated	GO:0009834	secondary cell wall biogenesis
20DPA up-regulated	GO:0007018	microtubule-based movement
25DPA down-regulated	GO:0030244	cellulose biosynthetic process
	GO:0009753	response to jasmonic acid stimulus
G2	GO:0015031	protein transport
15DPA down-regulated	GO:0015991	ATP hydrolysis
20DPA down-regulated	GO:0032544	plastid translation
25DPA down-regulated	GO:0016075	rRNA catabolic process
	GO:0006511	ubiquitin catabolic process
G3	GO:0009734	auxin mediated signaling pathway
15DPA down-regulated	GO:0030259	lipid glycosylation
20DPA up-regulated	GO:0018106	peptidyl-histidine phosphorylation
25DPA down-regulated	GO:0009722	detection of cytokinin stimulus
G4	GO:0006468	protein phosphorylation
15DPA up-regulated	GO:0010200	response to chitin
20DPA up-regulated	GO:0006096	glycolysis
25DPA up-regulated	GO:0010182	sugar mediated signaling pathway
	GO:0009966	regulation of signal transduction
G5	GO:0015884	protein folding
15DPA up-regulated	GO:0006886	intracellular protein transport
20DPA down-regulated	GO:0006122	mitochondrial electron transport
25DPA up-regulated	GO:0015914	phospholipid transport
G6	GO:0007267	cell-cell signaling
15DPA down-regulated	GO:0010025	wax biosynthetic process
20DPA down-regulated	GO:0006633	fatty acid biosynthetic process
25DPA up-regulated	GO:0006723	hydrocarbon biosynthetic process

G1–G6 according to Figure 3.
doi:10.1371/journal.pone.0094642.t002

We applied the same GO analysis to the common DEGs at 15 DPA and 20 DPA in CSIL-35431 and CSIL-31010, respectively. These DEGs were enriched in genes for similar functional categories, such as cellular metabolic processes and carbohydrate metabolism, etc. We also found genes for some processes that were enriched only in CSIL35431 or CSIL-31010 (Figure S2).

Further GO analysis for CSIL-35368 and CSIL-31134 indicated that the DEGs in CSIL-35368 at 15 and 20 DPA were enriched in genes for lignin biosynthesis, secondary cell wall biogenesis, and response to chitin, which was similar to the enrichment found in CSIL-35431 and CSIL-31010. But at 15 and 20 DPA in the stronger fiber line CSIL-31134, GO enrichments were different from the other three lines, mainly in genes for ATP synthesis, proton transport, copper ion export, and oxidoreductase activity, but not in cell wall biosynthesis (Table S4).

Based on the results of GO analysis, we know that the secondary cell wall related biological process were impacted in the CSILs, but it is still not very clear how secondary cell wall biosynthesis was affected in the CSILs. Therefore, we performed pathway analysis on 18,288 DEGs in CSIL-35431 and CSIL-31010. The most highly enriched pathways found are listed in Table 3. KEGG analysis showed that the genes were enriched in pathways for starch and sucrose metabolism (328 genes), glycolysis/gluconeogenesis (122 genes), phenylpropanoid biosynthesis (101 genes), and oxidative phosphorylation (87 genes) (Table 3 and Figure S3). The regulation of some enzymes that catalyze sucrose, starch, and cellulose biosynthesis may have a direct or indirect impact on fiber quality. This could be especially true for sucrose and pectin metabolism, and many genes in these pathways were up-regulated. We also found that genes involved in phenylpropanoid and flavonoid biosynthetic processes were enriched in the CSILs.

Table 3. KEGG analysis of DEGs in CSIL-35431 and CSIL-31010.

Pathway	DEGs with pathway annotation (2576)	References genes with pathway annotation(4601)	Ratio	DEGs distribution in each groups					
				G1	G2	G3	G4	G5	G6
Starch and sucrose metabolism	328	563	58.26%	87	15	54	89	31	52
Purine metabolism	251	459	54.68%	35	20	53	65	30	48
Phenylalanine metabolism	140	261	53.64%	29	5	26	47	12	21
Amino sugar and nucleotide sugar metabolism	137	211	64.93%	47	5	23	36	12	14
Pyrimidine metabolism	128	227	56.39%	15	13	26	37	15	22
Glycolysis/Gluconeogenesis	122	188	64.89%	28	4	23	35	13	19
T cell receptor signaling pathway	115	212	54.25%	28	4	23	34	11	15
Pentose and glucuronate interconversions	109	227	48.02%	24	8	20	27	7	23
Glycerolipid metabolism	102	174	58.62%	18	4	29	21	15	15
Pyruvate metabolism	102	182	56.04%	16	8	18	29	12	19
Phenylpropanoid biosynthesis	101	220	45.91%	26	6	17	28	8	16
Galactose metabolism	94	160	58.75%	23	2	20	24	7	18
Cysteine and methionine metabolism	94	145	64.83%	21	2	21	28	10	12
Glycerophospholipid metabolism	94	167	56.29%	14	8	29	21	10	12
Arginine and proline metabolism	93	144	64.58%	16	5	22	19	12	19
Oxidative phosphorylation	87	184	47.28%	19	13	12	11	13	19
Carbon fixation in photosynthetic organisms	84	151	55.63%	19	3	11	34	8	9
Fatty acid degradation	83	133	62.41%	17	5	18	21	7	15

G1–G6 according to Figure 3.
doi:10.1371/journal.pone.0094642.t003

Based on the cluster analysis of the weaker fiber line CSIL-35368, we hypothesized that changes in other biochemical pathways led to reduced fiber strength (Figure S1). Considering only those that were down-regulated in CSIL-35368, we found genes that participated in fatty acid degradation and biosynthesis, and also in carbohydrate metabolic pathways (Figure 2B and Figure S4).

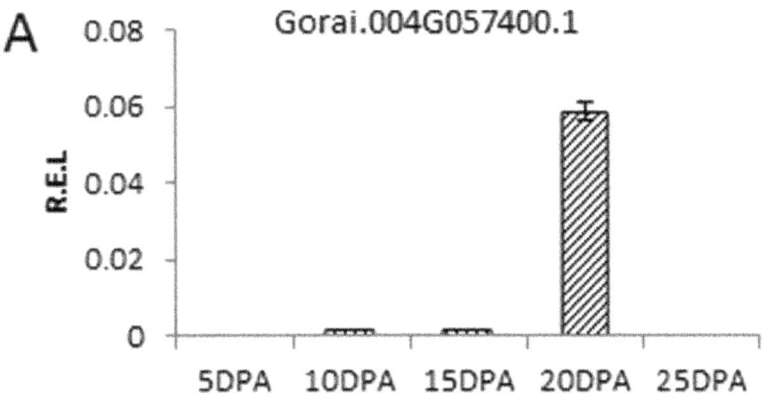

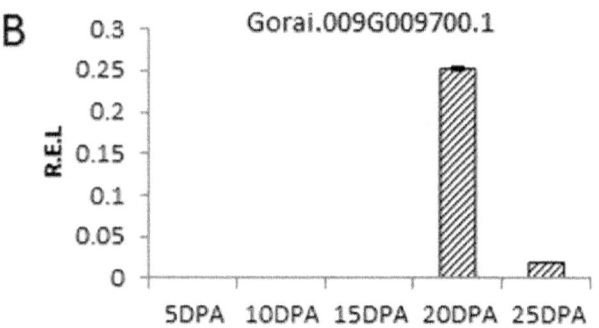

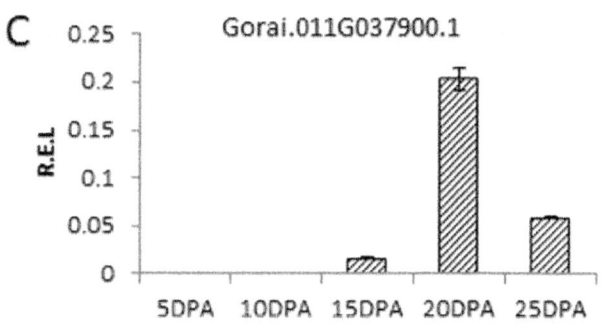

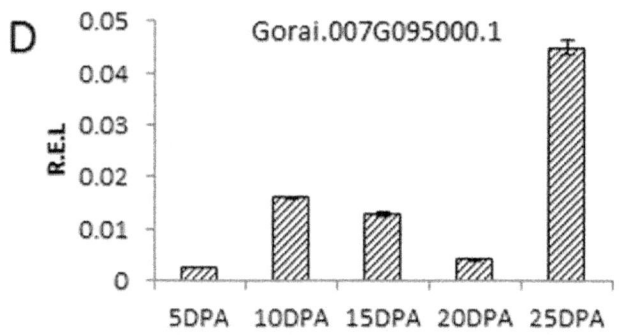

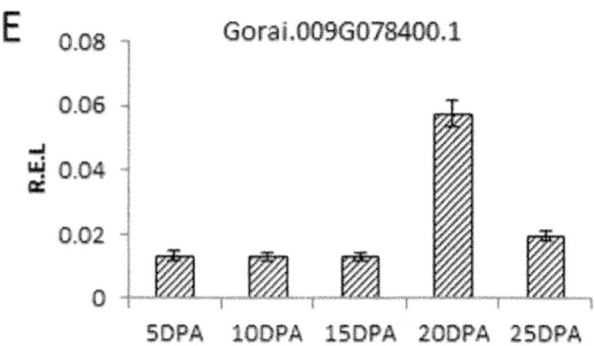

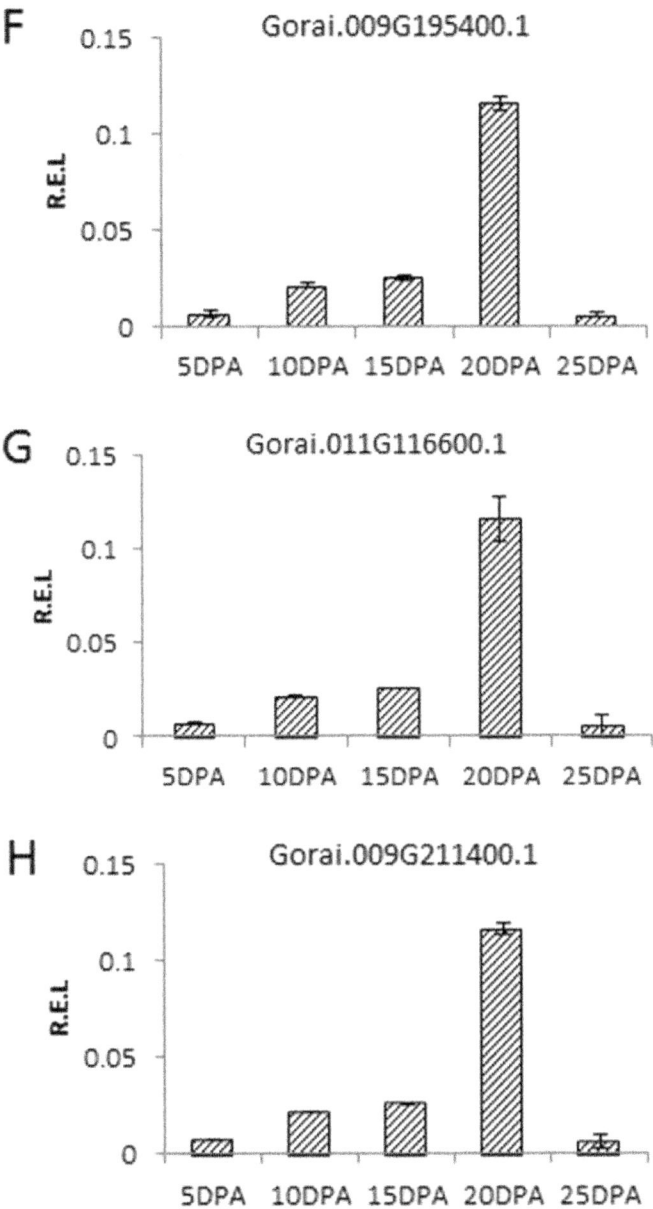

Figure 5. Quantitative RT–PCR validation of tag-mapped genes in TM-1. These genes have been reported before, including 3 CesA genes (A,B,C) (homologous with AtCE-SA4, AtCESA7, AtCESA8, respectively), xyloglucan endotransglucosylase (D), beta -galactosidase (E), glycosyl hydrolase 9B7 (F), xylan alpha-glucuronosyltransferase 1, GUX1 (G), xylan alpha-glucuronosyltransferase 2, GUX2 (H).

Eight genes previously reported in the carbohydrate pathway were selected for quantitative RT-PCR. The expression patterns of these genes were consistent with the DEG data in TM-1 (Figure 5) and in the CSILs as well (Figure 6B and Figure 7B).

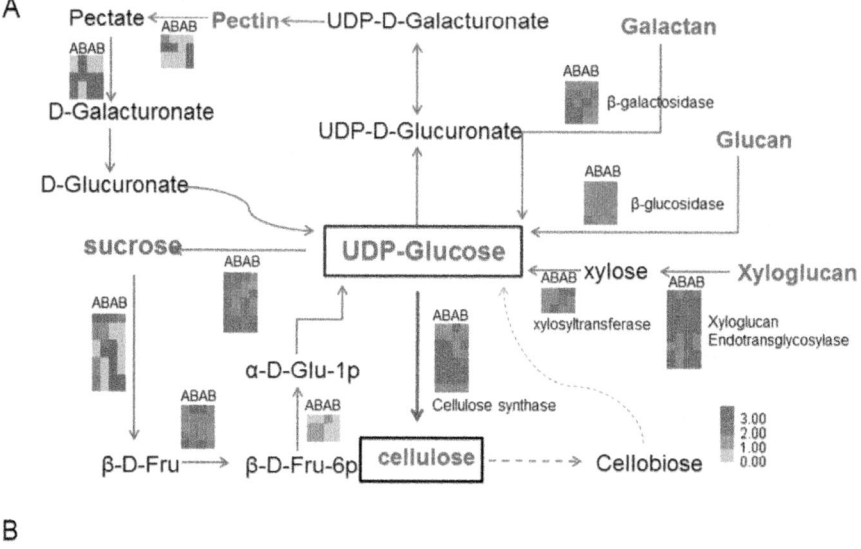

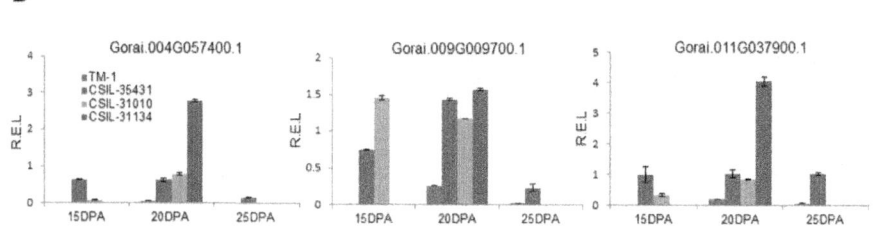

Figure 6. Carbohydrate pathways that are differentially regulated during the secondary cell wall synthesis stage. (A) Carbohydrate pathways. Genes up-regulated in CSIL-315431 and CSIL-31010 were selected to do heat map. ABAB indicated DEGs in CSIL-35431 at 15 DPA, CSIL-35431 at 20DPA, CSIL-3010 at 15DPA and CSIL-31010 at 20DPA, from left to right. Every square stand for one gene and every line stand for the same gene. Genes with red color expressed higher in CSILs than TM-1 and gray color stand for no difference. β-D-Fru, β-D-Fructose; α-D-Glu-1p, α-D-Glucose-1-phosphate; β-D-Fru-6p, β-D-Fructose-6-phosphate. (B) Quantitative RT–PCR validation of four CesA genes in CSILs and TM-1, Gorai.004G057400.1, Gorai.009G009700.1 and Gorai.011G037900.1 homologous with *AtCESA4*, *AtCESA7* and *AtCESA8*, respectively.

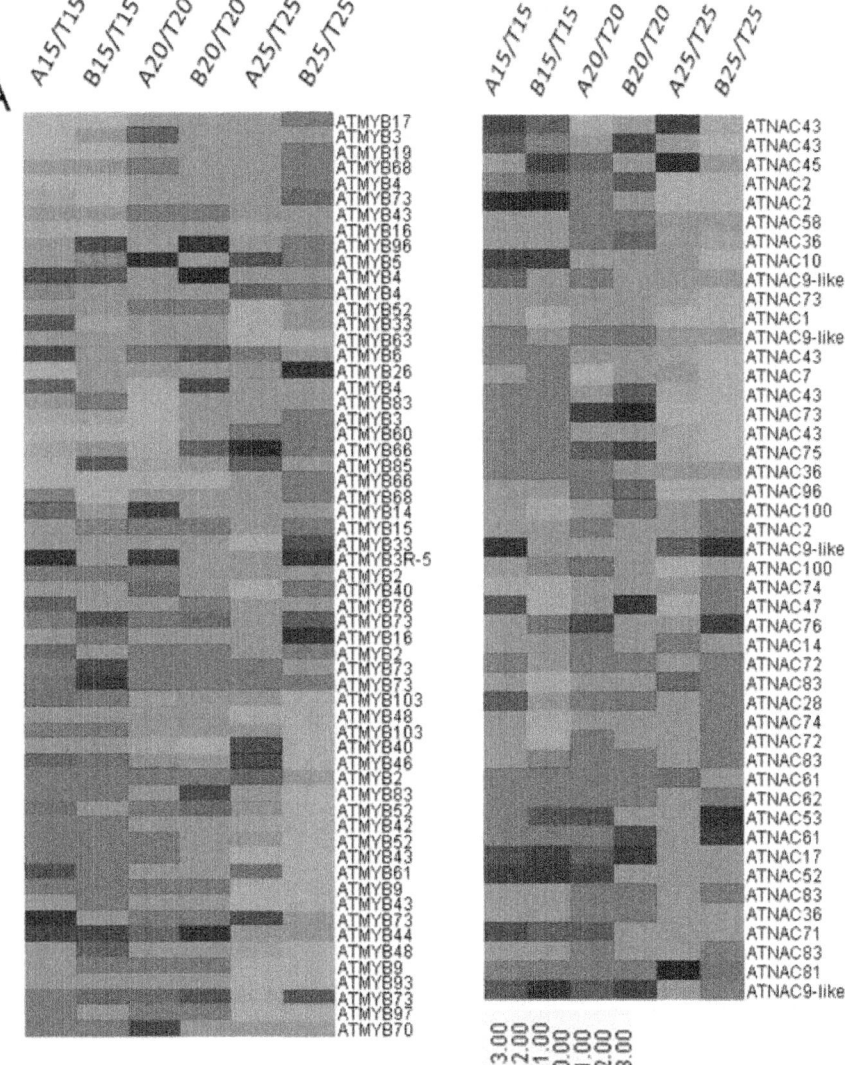

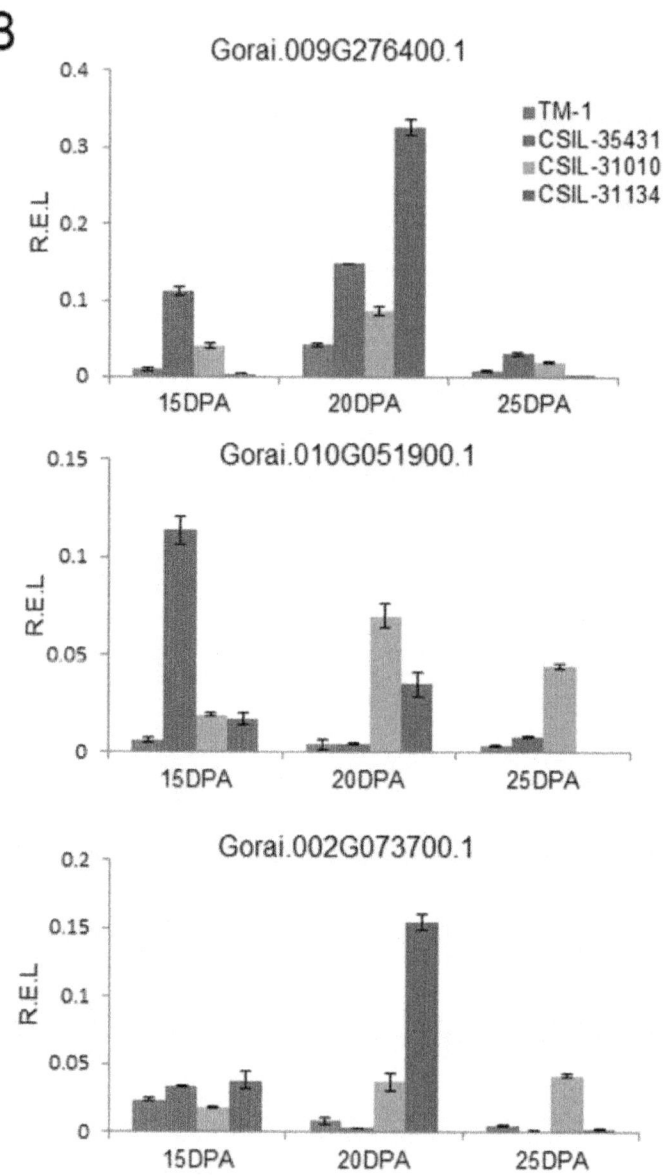

Figure 7. NAC and MYB family genes involved in the regulation of secondary wall biosynthesis. (A) 59 MYB family genes and 47 NAC family genes showed different expression level between CSILs and TM-1 at 15DPA, 20DPA and 25DPA. |Ratio|>2 and FDR<0.001. A, B, T indicated CSIL-35431, CSIL-31010 and TM-1. 15, 15DPA; 20, 20DPA; 25, 25DPA. (B) Quantitative RT–PCR validation of three transcription factors.

Carbohydrate Metabolism in the Secondary Cell Wall Synthesis Stage

Following the start of secondary cell wall formation, protein and carbohydrate metabolism genes involved in cell wall biosynthesis will be up-regulated [27]. We selected 72 DEGs associated with carbohydrate metabolism to investigate the mechanism of fiber development. These genes were related to pectin, sucrose, galactan, glucan, xyloglucan, and cellulose biosynthesis. We were interested in genes that are up-regulated in fiber cells at 15 DPA and 20 DPA, at the start of secondary cell wall formation. A heat map showing the different expression levels for these genes including cellulose synthase, sucrose synthase, pectin lyase, and other polysaccharides degradation in CSIL-35431 and CSIL-31010 is shown in Figure 6A. We found that the cellulose synthase genes were up-regulated in the CSILs at 15 DPA-25 DPA. It has been reported that cellulose biosynthesis predominates, and that many other metabolic pathways are down-regulated during secondary cell wall synthesis [27]. Moreover, we confirmed the expression patterns of cellulose synthase genes, annotated with the *Arabidopsis* genes *AtCESA4*, *AtCESA7* and *AtCESA8*, using quantitative RT-PCR (Figure 4B). Proteins encoded by *AtCESA4*, *7*, and *8* are specifically required to form a functional cellulose synthase complex (CSC) that is essential for secondary cell wall formation [40]–[42].

Transcription Factors Associated with Secondary Cell Wall Synthesis

Recent molecular and genetic studies have identified transcription factors that are involved in regulating secondary cell wall synthesis in *Arabidopsis* [43]–[45]. In our study, 97 MYB-type and 68 NAC-type transcription factors showed changes in expression between the CSILs and TM-1 (Table S5, Table S6). It was interesting that some NACs and MYBs were up-regulated in CSIL-35431 and CSIL-31010 during the secondary cell wall synthesis stage, especially at 15 DPA and 20 DPA. Defined as $|\log_2\text{Ratio}|\geq2$, 59 MYB and 47 NAC transcription factors were selected for heat-map analysis (Figure 7A). Among these transcription factors, genes homologous with *ATMYB2*, *ATMYB43*, *ATMYB73*, *ATNAC52*, and *ATNAC61* were expressed at higher levels in the CSILs. We confirmed that three transcription factors were up-regulated in CSILs from 15 DPA to 25 DPA (Figure 7B). In the MYB family, it has been reported that the expression of genes for MYB85, MYB52, MYB54, MYB69, MYB42, and MYB43 are developmentally associated with cells undergoing secondary wall thickening [45].

Different Metabolic Pathways Associated with Altered Fiber Strength

In order to investigate the mechanisms underlying changes in fiber strength, we analyzed several metabolic pathways including cell wall, lipids, minor CHO (carbohydrate) metabolism, and two secondary metabolite pathways. It is interesting that DEGs involved in cell wall proteins, cell wall pectin esterase, cell wall modification, cell wall cellulose synthesis, cell wall degradation/pectate lyases, lipid metabolism/FA synthesis, and lipid degradation showed distinct expression patterns or differential up/down-regulation at 20 DPA (Figure 8A). We found that up-regulated DEGs were similar to down-regulated DEGs both in CSIL-35431 and CSIL-35368. However, most of DEGs in CSIL-31010 were up-regulated at 20 DPA, while the opposite was true for DEGs in CSIL-31134, especially those genes involved in cell wall modification. In CSIL-31134, we also found a few genes in these metabolic pathways that were changed at 15 DPA except in cell wall modification, and in CSIL-31010, we found DEGs enriched in these metabolic pathways at 25 DPA (Figure S5). From the secondary metabolism results, we identified a few DEGs involved in flavonoid biosynthesis in CSIL-35431 and CSIL-31010 at 15 DPA. In contrast, more genes were up-regulated or down-regulated in CSIL-35368 at 15 DPA. It was obvious that DEGs from the phenylpropanoid pathways at 25 DPA were different from one another, and the expression pattern of DEGs in CSIL-31010 changed dramatically.

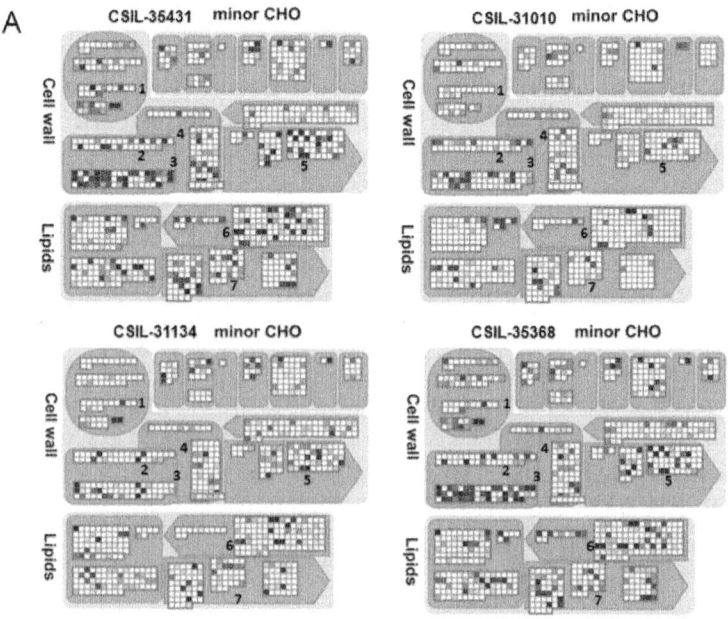

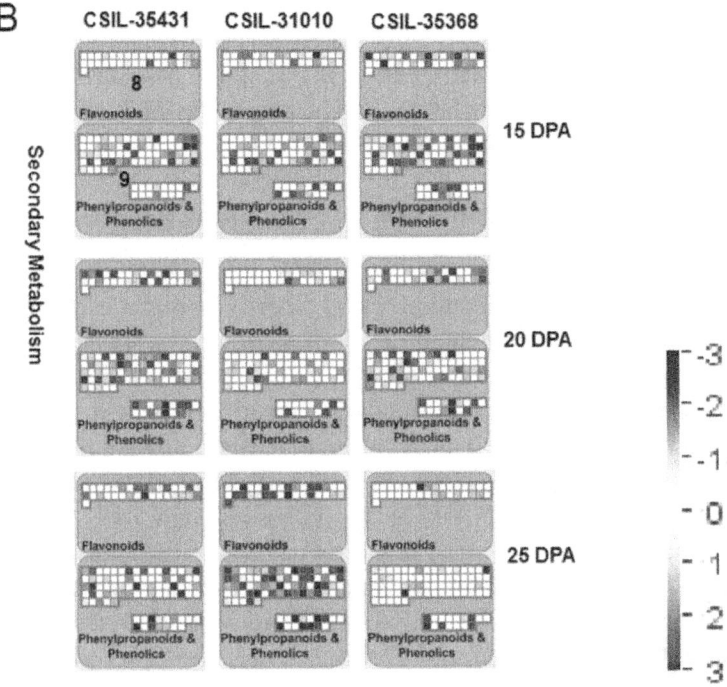

Figure 8. Metabolism analysis of DEGs in CSILs during the secondary cell wall bio-synthesis stage. (A) Motabolism overview in four CSILs at 20 DPA. (B) Secondary motabolism analysis in three CSILs at 15 DPA, 20 DPA and 25 DPA. 1, cell wall protein; 2, cell wall pectin esterases; 3, cell wall modification; 4, cell wall cellulose synthesis; 5, cell wall degradation/pectate lyases; 6, lipid metabolism/FA synthesis; 7, lipid degradation; 8, flavonoids; 9, phenylpropanoids/lignin biosynthesis. Blue square, down-regulated genes; Red square, up-regulated genes.

Moreover, there were few genes that were up-regulated or down-regulated in CSIL-35368 at 25 DPA (Figure 7B). We assume that metabolic pathways in the CSILs at different developmental stages were changed in various ways as a result of the introgressed chromosmal segments from *G. barbadense*.

DISCUSSION

G. hirsutum produces a high yield of cotton with moderate fiber strength. *G. barbadense* is characterized by a low yield, but with increased fiber fineness and strength. As a breeding target, we tried to combine the high yield of *G. hirsutum* with the superior fiber qualities of *G. barbadense*, and we also wanted to elucidate the molecular mechanism behind the formation of superior quality fibers. Fiber strength is an important indicator of the cotton fiber quality, and depends on the formation of the secondary cell wall. Genome-wide transcriptome profiling is effective at revealing significant genes and pathways involved in secondary cell wall formation. Transcriptome analysis showed that gene expression patterns and functional distribution were different during secondary cell wall biosynthesis.

Carbohydrate Metabolism Plays an Important Role in Secondary Cell Wall Synthesis

It is well known that the mature cotton fiber is composed of nearly pure cellulose, and that such a high level of cellulose synthesis requires an abundant supply of UDP-glucose [46],[47]. This means that a large amount of cellulose is required during the secondary cell wall synthesis stage. Functional classification and enrichment analysis showed that following the initiation of secondary cell wall synthesis, DEGs were enriched for secondary cell wall biogenesis, glucuronoxylan biological processes, and other carbohydrate metabolic pathways in the CSILs (Table 2). Focusing on carbohydrate metabolic pathways, it is obvious that the key intermediate in the multiple pathways is UDP-glucose, a substrate for cellulose synthesis. Our results showed that several CesA genes are expressed at higher levels during secondary cell wall synthesis than they are at earlier stages (Figure 6B). Ten*AtCESA* genes have been reported in *Arabidopsis*, and *AtCESA4, 7*, and *8* are specifically required to form the cellulose synthase complex (CSC) that is essential for secondary cell wall synthesis [40]–[42]. Similarly, three CESA isoforms have been identified during secondary cell wall synthesis in rice, maize, and *Populus* [10], [48], [49]. Also, many genes that participate in the degradation of poly- and oligo-saccharides were found to be up-regulated at 15 and 20 DPA, in order to produce more UDP-glucose for cellulose biosynthesis. Similarly, it has also been reported that during the secondary cell wall synthesis stage, certain metabolic pathways, including hydrolysis of fatty acids and non-cellulose poly- and oligo-saccharides, would be up-regulated [27]. Sucrose synthase (SuSy) has long been studied as a cytoplasmic enzyme in plant cells, where it serves to degrade sucrose and provide carbon for respiration and synthesis of cell wall polysaccharides and starch [50]. It has also been shown that genes associated with secondary cell wall biosynthesis are involved in sugar metabolism [51].

Multiple Mechanisms Affect Fiber Strength Development

Except for carbohydrate metabolism, recent research has shown that transcription factors also affect fiber development during secondary cell wall biosynthesis. Several NAC- and MYB-type transcription factors were up-regulated in the CSILs compared to TM-1 from 15 DPA to 25 DPA, and these included cotton homologs of *AtMYB2*, *AtMYB43*, and *AtNAC52*etc. (Figure 7A). The NAC-mediated transcriptional regulation of secondary wall biosynthesis is a conserved mechanism throughout vascular plants [44], [52]. *SND2*, a NAC transcription factor gene, regulates genes involved in secondary cell wall development in*Arabidopsis* fibers and increases fiber cell area in *Eucalyptus* [53]. A MYB75-associated protein complex is likely to be involved in modulating secondary cell wall biosynthesis in both the *Arabidopsis* inflorescence and stem [54]. It has also been found that the rice and maize MYB transcription factors, OsMYB46 and ZmMYB46, are functional orthologs of*Arabidopsis* MYB46/MYB83 and, when overexpressed in *Arabidopsis*, are able to activate the entire secondary wall biosynthetic program [55].

Several metabolic pathways were examined to determine the mechanism behind changes in fiber strength; these included cell wall, lipids, minor CHO metabolism, and two secondary metabolic pathways. Although results of the GO and KEGG analyses showed that CSIL-35431, CSIL-31010, and CSIL-35368 had similar patterns, fiber strength in these three lines were different. Our results support the hypothesis that different metabolic pathways can affect fiber strength, and the same pathway in the CSILs can be altered differentially at various times in development. DEGs in CSIL-31010 were up-regulated at 20 DPA, while the opposite was found for DEGs in CSIL-31134, especially those genes involved in cell wall modification. The expression levels of genes involved in flavonoid biosynthesis in the weak fiber line CSIL-35368 were changed dramatically at 15 DPA, but there were few genes changed at 25 DPA; this patter was the opposite of that in CSIL-35431 and CSIL-31010, lines with high quality fiber. We hypothesize that phenylpropanoid and flavonoid metabolism generally affected the fiber strength of CSIL-35368. Genes for phenylpropanoid and flavonoid biosynthesis showed significant enrichment and temporal differences in gene expression patterns which are associated with xylem formation [56]. It has been reported that expression levels of phenylpropanoid genes showed high correlations with specific fiber properties, supporting a role in determining fiber strength [57].

In conclusion, upland cotton CSILs carrying distinct *G. barbadense* chromosomal segments provide valuable material for research into fiber development. The *G. barbadense*chromosome segments resulted in different patterns of differentially expressed genes, and altered different metabolic pathways, mainly in carbohydrate metabolism. In addition, several transcription factor genes were found to be specifically up-regulated in the CSILs. Metabolic pathways involved in cell wall, lipid, phenylpropanoid, and flavonoid biosynthesis play a significant role during secondary cell wall formation, and are associated with the development of cotton fiber strength.

SUPPORTING INFORMATION

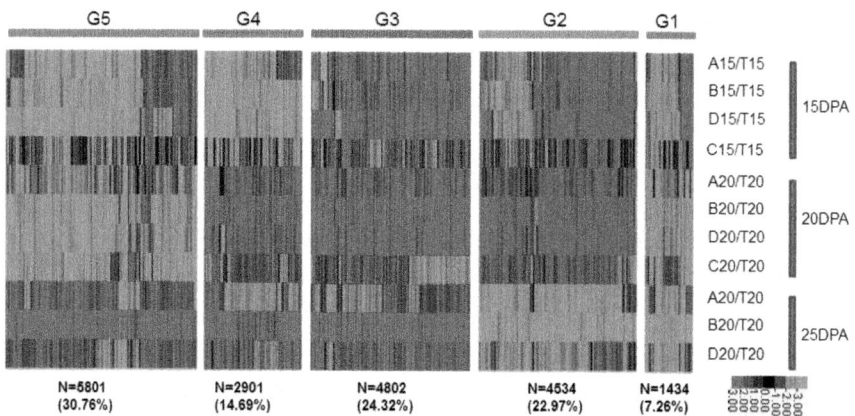

Figure S1. Heat map of the expression of DEGs between 4 CSILs at 15–25 DPA.

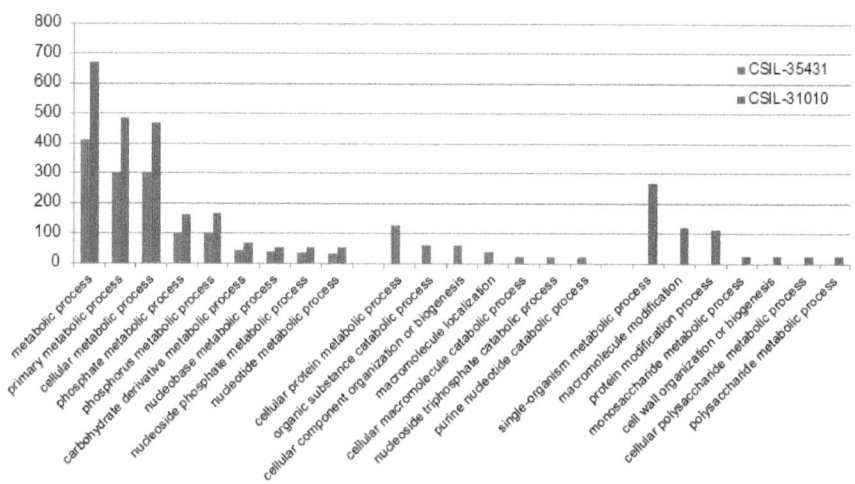

Figure S2. Enrichment analysis of common DEGs at 15DPA and 20DPA in CSIL-35431 and CSIL-31010.

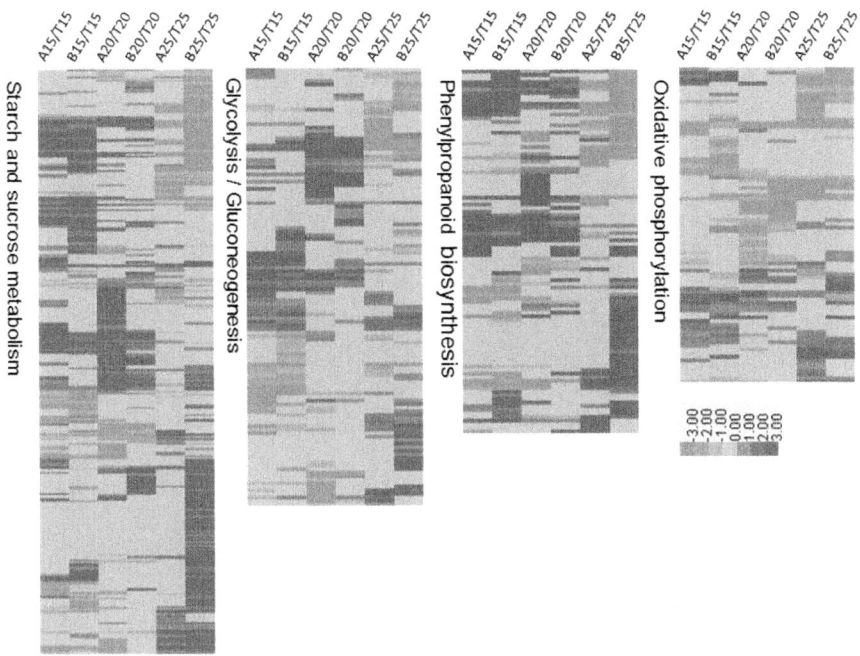

Figure S3. Heat map of DEGs participated in four metabolic pathways from 15 DPA to 25 DPA.

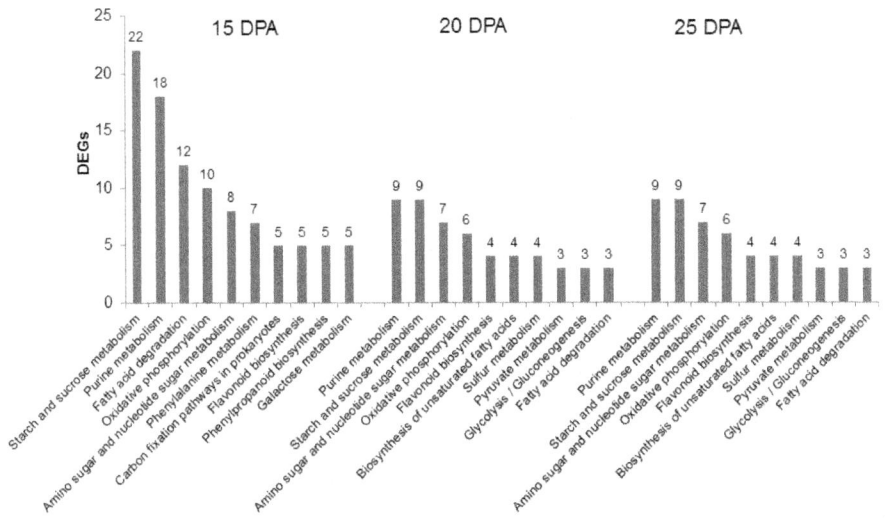

Figure S4. Pathway analysis of genes only down-regulated in CSIL-35368 from 15 DPA to 25 DPA.

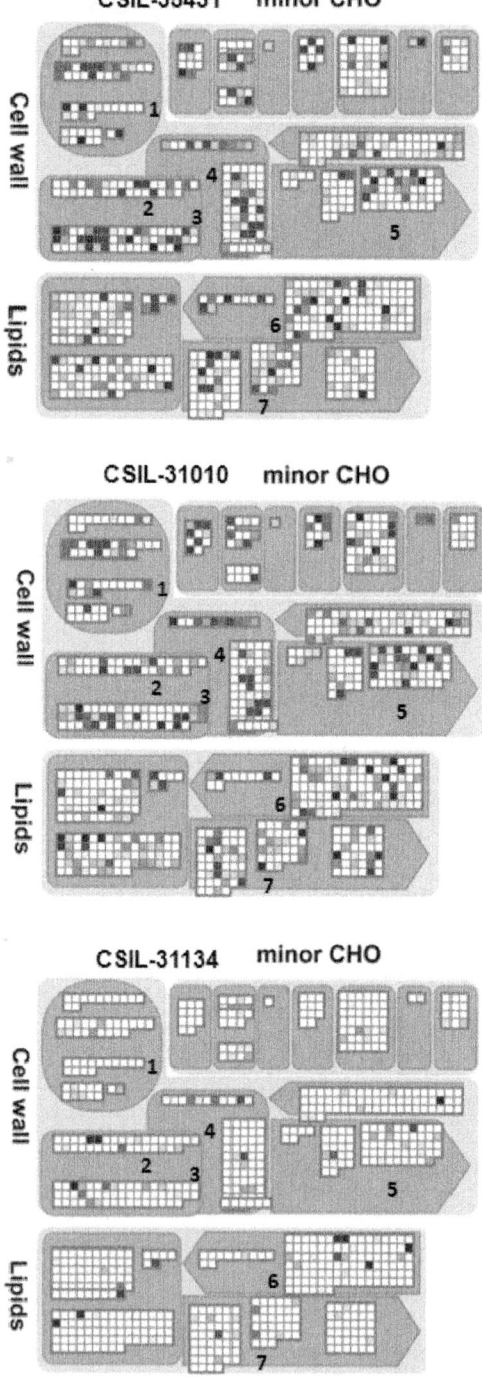

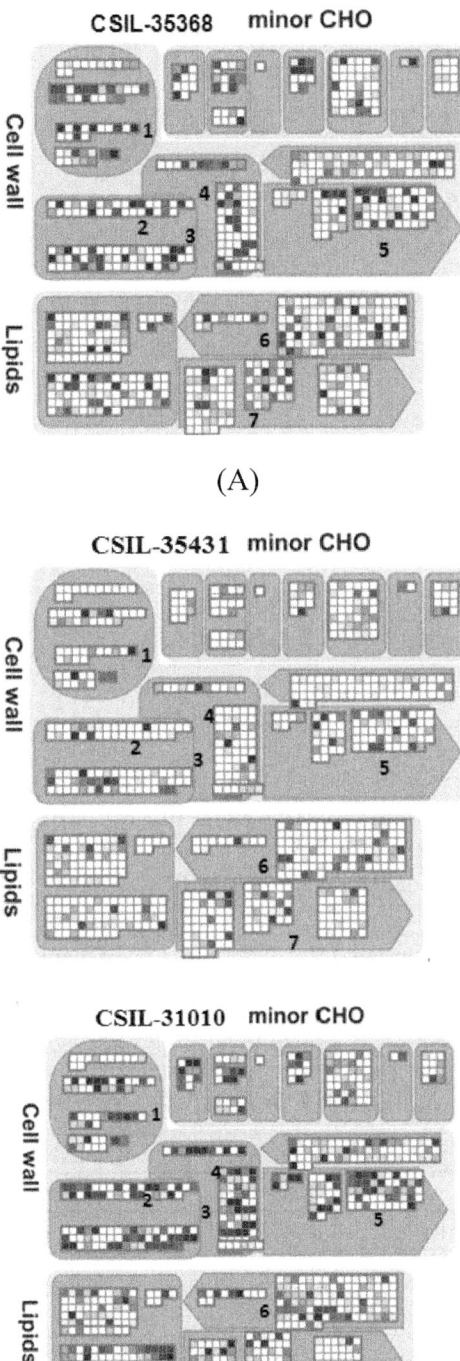

(A)

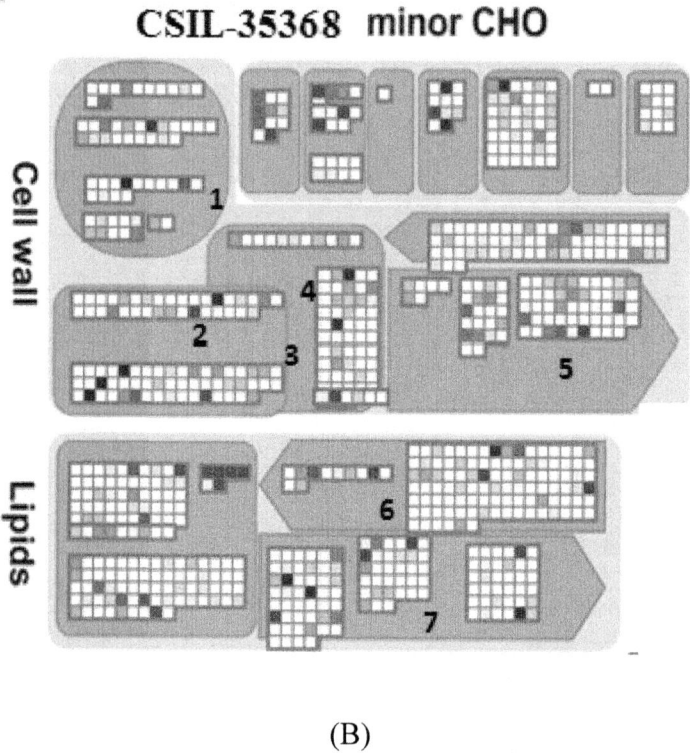

(B)

Figure S5. Metabolism analysis of DEGs in CSILs at 15 DPA and 25 DPA.

Table S1. Average fiber quality of 4 CSILs and TM-1.

CSILs	Chomosome segment	Fiber length (mm)	Fiber strength (cN/tex)	Mrc
CSIL-35431	A8(Chr.8) D10(Chr.20)	31.33±1.95**	35.10±2.12**	4.84±0.64
CSIL-31134	A8(Chr.8) D1(Chr.15)	30.63±1.14**	34.73±2.03**	4.59±1.09
CSIL-31010	D2(Chr.14)	30.01±1.05**	34.28±1.10**	5.05±0.94
TM-1		28.89±0.53	30.11±0.43	4.81±0.75
CSIL-35368	D11(Chr. 20)	27.67±1.18**	28.71±3.19	4.47±0.45
**: CSILs have significant difference with TM-1 at 1% level				

Table S2. Primer for quantitative RT-PCR.

Reference Gene	Forward Primer 5'-3'	Reverse Primer 5'-3'
Gorai.004G057400.1	GTGCTGAGAATGCTCTGGTT	GCCTTACTATTGTTGATG-TAATGAT
Gorai.009G009700.1	TGTCCATCTCTATCCATTCCTTAA	GAAGCTAATAGCACTGAC-CATAT
Gorai.011G037900.1	TGCTCGTTACGCCATAAGTC	TTCCACTCATTGCCTATT-GTTCA
Gorai.007G095000.1	ACCCATCAAAGAAATAGC-CATAGAC	AGGAAGCGAGTGGTGGAA
Gorai.009G078400.1	CGCACCCAATAATCCATTCTTA	CGATGTATCAAC-CAAGACTTTGG
Gorai.009G195400.1	GTAGCCATAGCCTGTAACTGATT	CGGGCATGCCTCAAA-GATACG
Gorai.011G116600.1	TAAGGACAGGCGAAGGAAGAA	CGAGAGTGAAGCAGTGTT-GATT
Gorai.009G211400.1	TTGCCGAGGAGAGTGAACTT	TGGAGTA-AACTTTGGGTGGGT
Gorai.009G276400.1	CTTCAACTCAAACAGCATCCT	ATACGGATCCTAAGT-CAGTGA
Gorai.010G051900.1	GTCGGAGTCGGTGCCGAGA	ATTGGTGGTGGTGGTGTTG
Gorai.002G073700.1	GGGAGTCTTGGTGGGCTTA	AAGTGTAGGCGGCATTGTG

Table S3. Categorization and abundance of tags.

CSILs	Raw Data	Distinct Tag	Clean Tag	Distinct Tag	Unambiguous Tag Mapping to Gene	Unambiguous Tag Mapping to Gene	Unambiguous Distinct Tag Mapping to Gene	Unambiguous Distinct Tag Mapping to Gene	Unambiguous Tag-mapped Genes	Unambiguous Tag-mapped Genes
35431-15	7461562	375999	7264426	182629	3961228	54.53%	53187	29.12%	21781	58.07%
35431-20	7389820	371748	7213791	199257	3411249	47.29%	48453	24.32%	22325	59.53%
35431-25	7298224	317147	7124575	148121	3463314	48.61%	41065	27.72%	20590	54.90%
TM-1-15	8374304	389162	8144920	169700	3983215	48.90%	45380	26.74%	21498	57.32%
TM-1-20	7231305	394430	7013147	177887	3206328	45.72%	40475	22.75%	20901	55.73%
TM-1-25	14267931	494345	14002534	251173	6829327	48.77%	56172	22.36%	22594	60.24%
35368-15	6938486	282172	6835546	182227	3268867	47.82%	44091	24.20%	21527	57.40%
35368-20	7165879	336637	7020147	194219	3429087	48.85%	51112	26.32%	22434	59.82%
35368-25	7414675	364911	7236602	190038	3469505	47.94%	47697	25.10%	22287	59.42%
31134-15	8710599	404122	8472498	175897	4055377	47.87%	43632	24.81%	21888	58.36%
31134-20	8265105	342671	8070435	158013	3964336	49.12%	42184	26.70%	20962	55.89%

31010-15	7471212	283130	7369012	182184	3671014	49.82%	46720	25.64%	21900	58.39%
31010-20	7007447	337525	6840101	172775	3381955	49.44%	45762	26.49%	22457	59.88%
31010-25	7203054	386302	7002766	188639	3552566	50.73%	48147	25.52%	22811	60.82%

15,15 DPA; 20, 20 DPA; 25, 25 DPA

Clean tags: tags after filtering dirty tags (low quality tags) from raw data.

Distinct tags: different kinds of tags.

Unambiguous tags: the clean tags after removing tags mapped to reference sequences from multiple genes.

Table S4. Enrichment analysis of gene ontologies in CSIL-35368 and CSIL-31010 at 15 DPA and 20 DPA.

CSILs	GO-ID	GO Ontology (Biologic process)
35368/TM-1 common DEGs at 15DPA and 20DPA	GO:0009809	lignin biosynthetic process
	GO:0010200	response to chitin
	GO:0055069	zinc ion homeostasis
	GO:0005985	sucrose metabolic process
	GO:0009834	secondary cell wall biogenesis
	GO:0046686	response to cadmium ion
	GO:0006200	ATP catabolic process
	GO:0009737	response to abscisic acid stimulus
	GO:0007155	cell adhesion
31134/TM-1 common DEGs at 15DPA and 20DPA	GO:0015977	carbon fixation
	GO:0042777	ATP synthesis proton transport
	GO:0060003	copper ion export
	GO:0019318	hexose metabolic process
	GO:0019761	glucosinolate biosynthetic process
	GO:0010252	auxin homeostasis
	GO:0006107	oxaloacetate metabolic process
	GO:0019318	hexose metabolic process
	GO:0016645	oxidoreductase activity

Table S5. Different expression level of 97 MYB transcription factors.

Gene NO.	Annotation from Tail10	35431-15/T15	31010-15/T15	35431-20/T20	31010-20/T20	35431-25/T25	31010-25/T25
Gorai.010G016800.1	AT-MYB44	1.6639	1.6660	2.3719	1.3344		
Gorai.009G215900.1	AT-MYB44			7.8806			
Gorai.007G312700.1	AT-MYB97	7.9529		11.7818	8.9573	-10.4945	-4.6585
Gorai.007G037100.1	AT-MYB96		-1.0492		1.2649		4.2777
Gorai.012G054500.1	AT-MYB96		-2.6621	-2.3626			
Gorai.008G192900.1	AT-MYB94	-1.4062	-1.3368	-1.1857			1.3236
Gorai.009G236900.1	AT-MYB93					-10.2625	-10.2625
Gorai.007G362200.1	ATMYB9	9.4834	8.4636	11.3167	9.8471	-4.1267	-10.5227
Gorai.007G348600.1	ATMYB9		8.9726	12.3967	10.7153	-10.6880	-10.6880
Gorai.013G083300.1	AT-MYB88						1.6653
Gorai.006G195700.1	AT-MYB85	-2.8122	-1.1596	-2.2030	-1.9336	2.1863	
Gorai.013G251100.1	AT-MYB83	3.6724	4.4733		1.7719		-13.0322
Gorai.004G172700.1	AT-MYB83		7.8506	-8.2261			
Gorai.004G276500.1	AT-MYB82					6.8111	
Gorai.006G254200.1	AT-MYB79						5.9539
Gorai.011G203900.1	AT-MYB78	2.2646			7.5703		
Gorai.005G068900.1	AT-MYB78			7.7931			2.9997
Gorai.006G080800.1	AT-MYB78						7.6433
Gorai.012G149000.1	AT-MYB73	-2.3107	-3.0086				2.4147
Gorai.010G023500.1	AT-MYB73	1.3119		3.9593	3.5027	-1.0340	
Gorai.007G055600.1	AT-MYB73	1.5588					3.5697

Gorai.009G208900.1	AT-MYB73	2.5760	1.4407	12.4513	9.8698	2.7684	3.5357
Gorai.005G234900.1	AT-MYB73	3.2182	1.7775	4.7444	3.7365	2.7822	
Gorai.009G276400.1	AT-MYB73	3.2822	2.4328	2.4046	1.9634	-2.6031	-1.2363
Gorai.007G218100.1	AT-MYB73	3.6542	1.8614	2.6741	2.4313		2.1775
Gorai.009G275100.1	AT-MYB73			8.1854			
Gorai.012G078200.1	AT-MYB70	8.9529	9.6458	1.6720			-2.0003
Gorai.004G130900.1	ATMYB7						7.2954
Gorai.001G020500.1	ATMYB7						7.1579
Gorai.007G227500.1	AT-MYB68	-1.6581	-9.1415				7.0059
Gorai.001G024300.1	AT-MYB68			7.7000			4.8823
Gorai.012G061800.1	AT-MYB66	-5.7176	-11.5006	-9.5342	-3.6644		9.4506
Gorai.004G196800.1	AT-MYB66	-2.2944	-1.8403	-2.6801	-1.3038	1.2572	5.2921
Gorai.010G096800.1	AT-MYB66	2.3026				2.5598	
Gorai.007G001800.1	AT-MYB66						2.9997
Gorai.012G011700.1	AT-MYB66			-8.0817			4.5847
Gorai.004G269400.1	AT-MYB63			7.3781			
Gorai.002G106300.1	AT-MYB63					-3.4402	-3.0003
Gorai.007G118200.1	AT-MYB61	-1.8896	-1.6247	-1.0407	-1.0764		1.3525
Gorai.001G148500.1	AT-MYB61	1.9420	2.9537			-1.4791	-3.9736
Gorai.004G147600.1	AT-MYB61			-1.4768		1.2469	
Gorai.013G196800.1	AT-MYB60	-3.2670	-3.8229	-3.5643		2.7297	6.4260
Gorai.007G252700.1	AT-MYB60						3.7001
Gorai.013G113000.1	AT-MYB60			7.4636			
Gorai.007G179300.1	ATMYB6	-1.0264	-1.9337	2.6963	2.0599	-1.5731	

Gorai.008G133700.1	AT-MYB55		-1.4871	-1.1369		-1.5187	
Gorai.007G002200.1	AT-MYB52	2.4870		3.4829	2.4420	-1.8997	-10.1030
Gorai.008G293100.1	AT-MYB52	8.8705	8.4636	3.6147			-8.2625
Gorai.009G251700.1	AT-MYB52			7.4935	7.4548	-3.9321	
Gorai.007G260600.1	ATMYB5	1.0038		1.4404			
Gorai.007G350500.1	ATMYB5						-1.4234
Gorai.008G126700.1	ATMYB5		-1.7555	1.2693		-1.1056	3.6709
Gorai.008G288900.1	ATMY-B4R1	2.4870					
Gorai.013G265300.1	AT-MYB48		2.1958			-2.1749	-3.1501
Gorai.003G136400.1	AT-MYB48					-7.5911	-7.5911
Gorai.006G129100.1	AT-MYB46	2.2805	2.6191	-1.8359		-1.0252	-9.7431
Gorai.001G205300.1	AT-MYB43	7.2424	7.2226	4.2072		-5.1647	-9.9758
Gorai.004G064100.1	AT-MYB43		3.3151			-4.3106	-9.1216
Gorai.008G060100.1	AT-MYB43			7.7000	7.9573		
Gorai.004G201900.1	AT-MYB42	8.0309	9.3779			-2.2069	-10.7729
Gorai.008G238800.1	AT-MYB42		8.7632				-7.5911
Gorai.007G074500.1	AT-MYB40			-2.4063	-10.8999	1.8863	-8.2286
Gorai.006G037500.1	AT-MYB40			2.4829		-7.8806	2.6695
Gorai.002G180300.1	ATMYB4	-8.5722	-8.5722				6.6433
Gorai.002G196700.1	ATMYB4	-1.7368				-1.3388	-1.5853
Gorai.007G192900.1	ATMYB4	-1.2735		-2.8398	-1.0319		
Gorai.005G162600.1	ATMYB4	-1.0494	-1.1648		1.0879		
Gorai.008G035700.1	ATMYB4			1.1273			-2.8281
Gorai.006G172800.1	ATMY-B3R-5	1.2906		1.5287			1.4880

Gorai.001G249400.1	ATMY-B3R-3			8.1854			
Gorai.009G301100.3	AT-MYB33	-1.2335	-2.2533			-3.0608	
Gorai.012G022000.1	AT-MYB33					-3.9321	1.9130
Gorai.001G239000.1	AT-MYB30						7.7428
Gorai.004G150800.1	ATMYB3	-7.6180		2.2323			
Gorai.006G192400.1	ATMYB3	-2.7418	-2.3022	-7.2932			9.8803
Gorai.009G288900.1	ATMYB3		1.5237	-1.3384			1.2011
Gorai.004G044000.1	ATMYB3						7.6433
Gorai.011G063900.1	AT-MYB26	-4.1925				-1.9321	1.2425
Gorai.009G142200.1	AT-MYB26	1.1839	1.3067				
Gorai.007G121000.1	ATMYB2	2.4662		4.5241	3.6605		5.1856
Gorai.004G146000.1	ATMYB2	8.4269	9.5445	11.3630			
Gorai.004G145800.1	ATMYB2	9.4834	11.5641	7.2401	5.0804	2.3374	
Gorai.008G277900.1	AT-MYB19	-3.0049	-9.5952				3.0701
Gorai.001G169700.1	AT-MYB17	-2.9779	-9.3459				2.8070
Gorai.004G157600.1	AT-MYB17						7.2954
Gorai.011G122800.1	AT-MYB16	-2.7290	-2.7015			-2.2079	
Gorai.013G088300.1	AT-MYB16						7.7428
Gorai.008G179600.1	AT-MYB16		3.3151				1.3533
Gorai.005G206800.1	AT-MYB15			7.7000			
Gorai.011G173900.1	AT-MYB15			8.6511			
Gorai.001G177100.1	AT-MYB15		-1.6621	4.8419	4.5596		2.9566
Gorai.012G132400.1	AT-MYB15			8.1854			
Gorai.011G021300.2	AT-MYB14	2.3874		1.5760			

Gorai.006G060300.1	AT-MYB116		-8.3728				9.1213
Gorai.007G346400.1	AT-MYB105	-8.0103					2.2627
Gorai.003G155400.1	AT-MYB103	2.5658	3.2282	-2.0357			-13.2135
Gorai.004G138300.1	AT-MYB103	4.3473	4.2035				-14.8496
15, 25DPA; 20, 20DPA; 25, 25DPA; T, TM-1. value=log2Ratio							

Table S6. Different expression level of 68 NAC transcription factors.

Gene NO.	Annotation from Tail10	35431-15/T15	31010-15/T15	35431-20/T20	31010-20/T20	35431-25/T25	31010-25/T25
Gorai.003G073300.1	ATNAC9-like	1.0776				-1.2476	1.2798
Gorai.008G261400.1	ATNAC9-like	1.9460	1.2992	1.9988	1.5716		2.6674
Gorai.001G150000.1	ATNAC9-like	2.2525		3.4187			
Gorai.007G017500.1	ATNAC9-like	2.8127		3.1723	2.7059		
Gorai.004G186700.1	ATNAC96			2.5760	-1.3683	-2.8446	-2.8197
Gorai.009G399900.1	ATNAC90						3.0872
Gorai.009G083500.1	ATNAC89			7.4935			
Gorai.007G043900.1	ATNAC83						2.5847
Gorai.005G195400.1	ATNAC83			8.3167			8.8803
Gorai.005G195300.1	ATNAC83			4.2373	2.6914		3.8410
Gorai.011G087700.1	ATNAC83		-3.3783			2.3671	6.6756
Gorai.001G274500.1	ATNAC83			8.3781			
Gorai.012G083700.1	ATNAC83			10.4935			
Gorai.012G125500.1	ATNAC83		-1.7173	3.7096	3.2731		4.7301
Gorai.007G065300.1	ATNAC82		-1.6621				-1.5325
Gorai.009G214000.2	ATNAC76		-1.5069	1.7248			1.5729
Gorai.009G028900.1	ATNAC75	5.9979	4.4671	2.0932	1.6210	-2.5141	-5.9487
Gorai.010G078900.1	ATNAC75			7.3781			
Gorai.004G221300.1	ATNAC74	-2.3851	-8.9755				8.2954
Gorai.007G147300.1	ATNAC74		-8.2618				11.6939

Gorai.001G178800.1	ATNAC74						11.9857
Gorai.008G155200.1	ATNAC73	11.1317	11.6640	1.5183	1.0742	-3.5418	-13.7105
Gorai.010G157600.2	ATNAC73					-2.0794	-6.4238
Gorai.008G227800.1	ATNAC71	1.6741	2.1452	2.0684			
Gorai.003G180600.1	ATNAC71			7.3781			
Gorai.004G149000.1	ATNAC69					-8.0457	
Gorai.006G203800.1	ATNAC62	4.3676	3.3615	5.2406	3.8203		-1.5792
Gorai.001G122800.1	ATNAC61	4.4505	2.8457	2.9228	1.8016		1.5205
Gorai.006G113000.1	ATNAC61	11.0959	10.0486	13.1897	11.3045	2.4984	
Gorai.006G060900.1	ATNAC58			5.5036	3.5804		
Gorai.010G012000.3	ATNAC53	3.7011	1.9526	1.8634			1.2292
Gorai.011G234200.1	ATNAC52	1.4546	1.4127	1.6244			
Gorai.009G309300.1	ATNAC52			1.8073	1.2991		
Gorai.012G007000.1	ATNAC52						2.1737
Gorai.007G267900.1	ATNAC47	-1.0674	-3.9708		1.4146	-4.4232	4.2911
Gorai.007G114500.1	ATNAC45		1.5478	2.3635		1.2254	
Gorai.006G034600.1	ATNAC44			2.3378			
Gorai.002G001900.1	ATNAC36	8.2424	7.7632	4.7142			
Gorai.005G257800.1	ATNAC36			4.6963	2.6990		
Gorai.008G038800.1	ATNAC36			3.9974	2.3162		-2.0998
Gorai.009G204700.1	ATNAC28	2.0263		3.5552	2.6851		4.5543
Gorai.009G309000.1	ATNAC2			1.7927			
Gorai.009G260000.1	ATNAC17	1.7885	1.5132	2.1034	1.3369		
Gorai.005G076200.1	ATNAC17			1.5302			1.6497
Gorai.007G038100.1	ATNAC100						-2.1158
Gorai.007G079900.1	ATNAC100		-2.6921		2.5172		6.1738
Gorai.008G194400.1	ATNAC100		-1.7059	2.4923		-3.7519	
Gorai.009G166300.1	ATNAC10	1.4507	1.5550				-7.0997
Gorai.009G354900.1	ATNAC1		-3.1027			-3.4346	-4.4097
Gorai.013G167100.1	ATNAC2	1.1408	1.1452	3.2667		-3.0493	-3.4769
Gorai.009G433100.1	ATNAC2	3.1110	2.3308	2.9001	2.1553	-3.2263	-5.1823
Gorai.002G073800.1	ATNAC2					-1.1030	

Gorai.005G013300.2	ATNAC14			3.9891		4.3671	
Gorai.003G114100.1	ATNAC7		3.2495	-3.6257			-8.3598
Gorai.003G077700.1	ATNAC43	1.4651	2.1013	-2.5183		1.5167	-2.2509
Gorai.008G259700.1	ATNAC43	2.0395	2.9696		1.7098		-9.6176
Gorai.002G073700.1	ATNAC72	2.9464		6.1881	3.5596		8.4798
Gorai.012G037600.1	ATNAC81	3.8558	4.1055	4.0000	2.8711	1.0013	2.4927
Gorai.007G112500.1	ATNAC43	5.5459	6.2397			-2.1291	-12.1100
Gorai.008G130300.1	ATNAC43	8.8274	10.5181		2.1829	-2.5727	-10.0842
Gorai.004G129200.1	ATNAC43	9.8054	7.2226			-2.5295	-10.8641
Gorai.010G051900.1	ATNAC2			8.4370			4.5233
Gorai.009G433200.1	ATNAC72		-3.8288	4.1131			5.6435
Gorai.009G186000.1	ATNAC31						9.2622
Gorai.002G113300.1	ATNAC31			8.3781			
Gorai.005G088800.1	ATNAC2			10.4513			
Gorai.008G186300.2	ATNAC8						1.8332
Gorai.001G150800.1	ATNAC43			2.5838			
15, 25DPA; 20, 20DPA; 25, 25DPA; T, TM-1. value=log2Ratio							

AUTHOR CONTRIBUTIONS

Conceived and designed the experiments: TZ. Performed the experiments: LF RT SW XL PW. Analyzed the data: LF JC. Wrote the paper: LF TZ.

REFERENCES

1. Basara AS, Malik CP (1984) Development of cotton fiber. Inter Rev Cyto 65–113.

2. Haigler TA, Jernstedt JA (1999) Molecular genetics of developing cotton fibers. In: AM Basra (Ed), Cotton Fibers. Hawthorne Press, New York, 231–267.

3. Kim HJ, Triplett BA (2001) Cotton fiber growth in planta and in vitro. Models for plant cell elongation and cell wall biogenesis. Plant Physiol 127: 1361–1366. doi: 10.1104/pp.010724

4. Lee JJ, Hassan OS, Gao W, Wei NE, Kohel RJ, et al. (2006) Developmental and gene expression analyses of a cotton naked seed mutant. Planta 223: 418–432. doi: 10.1007/s00425-005-0098-7

5. Lee JJ, Woodward AW, Chen ZJ (2007) Gene expression changes and early events in cotton fibre development. Ann Bot 100: 1391–1401.

6. Wilkins TA, Arpat AB (2005) The cotton fiber transcriptome. Physiol Plant 124: 295–300. doi: 10.1111/j.1399-3054.2005.00514.x

7. Meinert MC, Delmer DP (1977) Changes in biochemical composition of the cell wall of the cotton fiber during development. Plant Physiol 59: 1088–1097. doi: 10.1104/pp.59.6.1088

8. Bolton JJ, Soliman KM, Wilkins TA, Jenkins JN (2009) Aberrant Expression of Critical Genes during Secondary Cell Wall Biogenesis in a Cotton Mutant, Ligon Lintless-1 (Li-1). Comp Funct Genom 659301. doi: 10.1155/2009/659301

9. Richmond TA, Somerville CR (2000) The cellulose synthase superfamily. Plant Physiol 124: 495–498. doi: 10.1104/pp.124.2.495

10. Tanaka K, Murata K, Yamazaki M, Onosato K, Miyao A, et al. (2003) Three distinct rice cellulose synthase catalytic subunit genes required for cellulose synthesis in the secondary wall. Plant Physiol 133: 73–83. doi: 10.1104/pp.103.022442

11. Paterson AH, Wendel JF, Gundlach H, Guo H, Jenkins J, et al. (2012) Repeated polyploidization of Gossypium genomes and the evolution of spinnable cotton fibres. Nature 492: 423–427. doi: 10.1038/nature11798

12. Persson S, Wei H, Milne J, Page GP, Somerville CR (2005) Identification of genes required for cellulose synthesis by regression analysis of public microarray data sets. Proc Natl Acad Sci USA 102: 8633–8638. doi: 10.1073/pnas.0503392102

13. Ruan YL, Chourey PS, Delmer DP, Perez-Grau L (1997) The Differential Expression of Sucrose Synthase in Relation to Diverse Patterns of Carbon Partitioning in Developing Cotton Seed. Plant Physiol 115: 375–385.

14. Brill E, van Thournout M, White RG, Llewellyn D, Campbell PM, et al. (2011) A novel isoform of sucrose synthase is targeted to the cell wall during secondary cell wall synthesis in cotton fiber. Plant Physiol 157: 40–54. doi: 10.1104/pp.111.178574

15. Potikha TS, Collins CC, Johnson DI, Delmer DP, Levine A (1999) The involvement of hydrogen peroxide in the differentiation of secondary walls in cotton fibers. Plant Physiol 119: 849–858. doi: 10.1104/pp.119.3.849

16. Yang YM, Xu CN, Wang BM, Jia JZ (2001) Effects of plant growth regulators on secondary wall thickening of cotton fibres. Plant Growth Regul 35: 233–237.

17. Arpat AB, Waugh M, Sullivan JP, Gonzales M, Frisch D, et al. (2004) Functional genomics of cell elongation in developing cotton fibers. Plant Mol Biol 54: 911–929. doi: 10.1007/s11103-004-0392-y

18. Wang QQ, Liu F, Chen XS, Ma XJ, Zeng HQ, et al. (2010) Transcriptome profiling of early developing cotton fiber by deep-sequencing reveals significantly differential expression of genes in a fuzzless/lintless mutant. Genomics 96: 369–376. doi: 10.1016/j.ygeno.2010.08.009

19. Padmalatha KV, Dhandapani G, Kanakachari M, Kumar S, Dass A, et al. (2012) Genome-wide transcriptomic analysis of cotton under drought stress reveal significant down-regulation of genes and pathways involved in fibre elongation and up-regulation of defense responsive genes. Plant Mol Biol 78: 223–246. doi: 10.1007/s11103-011-9857-y

20. Chaudhary B, Hovav R, Rapp R, Verma N, Udall JA, et al. (2008) Global analysis of gene expression in cotton fibers from wild and domesticated *Gossypium barbadense*. Evol Dev 10: 567–582. doi: 10.1111/j.1525-142x.2008.00272.x

21. Hovav R, Udall JA, Chaudhary B, Hovav E, Flagel L, et al. (2008) The evolution of spinnable cotton fiber entailed prolonged development and a novel metabolism. PLoS Genet 4: e25. doi: 10.1371/journal.pgen.0040025

22. Hovav R, Udall JA, Hovav E, Rapp R, Flagel L, et al. (2008) A majority of cotton genes are expressed in single-celled fiber. Planta 227: 319–329. doi: 10.1007/s00425-007-0619-7

23. Ji SJ, Lu YC, Feng JX, Wei G, Li J, et al. (2003) Isolation and analyses of genes preferentially expressed during early cotton fiber development by subtractive PCR and cDNA array. Nucleic Acids Res 31: 2534–2543. doi: 10.1093/nar/gkg358

24. Udall JA, Flagel LE, Cheung F, Woodward AW, Hovav R, et al. (2007) Spotted cotton oligonucleotide microarrays for gene expression analysis. BMC Genomics 8: 81. doi: 10.1186/1471-2164-8-81

25. Qin YM, Hu CY, Pang Y, Kastaniotis AJ, Hiltunen JK, et al. (2007) Saturated very-long-chain fatty acids promote cotton fiber and *Arabidopsis* cell elongation by activating ethylene biosynthesis. Plant Cell 19: 3692–3704. doi: 10.1105/tpc.107.054437

26. Shi YH, Zhu SW, Mao XZ, Feng JX, Qin YM, et al. (2006) Transcriptome profiling, molecular biological, and physiological studies reveal a major

role for ethylene in cotton fiber cell elongation. Plant Cell 18: 651–664. doi: 10.1105/tpc.105.040303

27. Gou JY, Wang LJ, Chen SP, Hu WL, Chen XY (2007) Gene expression and metabolite profiles of cotton fiber during cell elongation and secondary cell wall synthesis. Cell Res 17: 422–434. doi: 10.1038/sj.cr.7310150

28. Eshed Y, Zamir D (1995) An introgression line population of *Lycopersicon pennellii* in the cultivated tomato enables the identification and fine mapping of yield-associated QTL. Genetics 141: 1147–1162.

29. Liu S, Zhou R, Dong Y, Li P, Jia J (2006) Development, utilization of introgression lines using a synthetic wheat as donor. Theor Appl Genet 112: 1360–1373. doi: 10.1007/s00122-006-0238-x

30. Takai T, Nonoue Y, Yamamoto SI, Yamanouchi U, Matsubara K, et al. (2007) Development of chromosome segment substitution lines derived from backcross between indica donor rice cultivar *'Nona bokra'* and japonica recipient cultivar *'Koshihikari'*. Breeding Sci 57: 257–261. doi: 10.1270/jsbbs.57.257

31. Wang P, Ding YZ, Lu QX, Guo WZ, Zhang TZ (2008) Development of *Gossypium barbadense* chromosome segment substitution lines in the genetic standard line TM-1 of *Gossypium hirsutum*. Chi Sci Bull 53: 1512–1517. doi: 10.1007/s11434-008-0220-x

32. Kohel R, Richmond T, Lewis C (1970) Texas marker-1. Description of a genetic standard for *Gossypium hirsutum* L. Crop Sci 10: 670–671. doi: 10.2135/cropsci1970.0011183x001000060019x

33. Pan J, Zhang T, Kuai B (1994) Studies on the inheritance of resistance to *Verticillium* dahliae in cotton. J Nanj Agric Univ 17.

34. Yang C, Guo W, Li G, Gao F, Lin S, et al. (2008) QTLs mapping for *Verticillium* wilt resistance at seedling and maturity stages in *Gossypium barbadense* L. Plant Sci 174: 290–298. doi: 10.1016/j.plantsci.2007.11.016

35. Wang P, Zhu Y, Song X, Cao Z, Ding Y, et al. (2012) Inheritance of long staple fiber quality traits of *Gossypium barbadense* in *G. hirsutum* background using CSILs. Theor Appl Genet 124: 1415–1428. doi: 10.1007/s00122-012-1797-7

36. Jiang JX, Zhang TZ (2003) Extraction of total RNA in cotton tissues with CTAB-acidic phenolic method. Cott Sci 15: 166–167.

37. Benjamini Y, Yekutieli D (2001) The control of the false discovery rate in multiple testing under dependency. Ann Stat 1165–1188. doi: 10.1214/aos/1013699998

38. Herrero J, Valencia A, Dopazo J (2001) A hierarchical unsupervised growing neural network for clustering gene expression patterns. Bioinformatics 17 (2) 126–136. doi: 10.1093/bioinformatics/17.2.126

39. Kanehisa M, Araki M, Goto S, Hattori M, Hirakawa M, et al. (2008) KEGG for linking genomes to life and the environment. Nucleic Acids Res 36: D480–D484. doi: 10.1093/nar/gkm882

40. Taylor NG, Howells RM, Huttly AK, Vickers K, Turner SR (2003) Interactions among three distinct CesA proteins essential for cellulose synthesis. Proc Natl Acad Sci USA 100: 1450–1455. doi: 10.1073/pnas.0337628100

41. Taylor NG, Laurie S, Turner SR (2000) Multiple cellulose synthase catalytic subunits are required for cellulose synthesis in *Arabidopsis*. Plant Cell 12: 2529–2540. doi: 10.1105/tpc.12.12.2529

42. Taylor NG, Scheible WR, Cutler S, Somerville CR, Turner SR (1999) The irregular xylem3 locus of Arabidopsis encodes a cellulose synthase required for secondary cell wall synthesis. Plant Cell 11: 769–780. doi: 10.2307/3870813

43. Olsen AN, Ernst HA, Leggio LL, Skriver K (2005) NAC transcription factors: structurally distinct, functionally diverse. Trends Plant Sci 10: 79–87. doi: 10.1016/j.tplants.2004.12.010

44. Zhong R, Lee C, Ye ZH (2010) Functional characterization of poplar wood-associated NAC domain transcription factors. Plant Physiol 152: 1044–1055. doi: 10.1104/pp.109.148270

45. Zhong R, Lee C, Zhou J, McCarthy RL, Ye ZH (2008) A battery of transcription factors involved in the regulation of secondary cell wall biosynthesis in *Arabidopsis*. Plant Cell 20: 2763–2782. doi: 10.1105/tpc.108.061325

46. Delmer DP, Amor Y (1995) Cellulose biosynthesis. Plant Cell 7: 987–1000. doi: 10.1105/tpc.7.7.987

47. Delmer DP, Haigler CH (2002) The regulation of metabolic flux to cellulose, a major sink for carbon in plants. Metab Eng 4: 22–28. doi: 10.1006/mben.2001.0206

48. Appenzeller L, Doblin M, Barreiro R, Wang HY, Niu XM, et al. (2004) Cellulose synthesis in maize: isolation and expression analysis of the cellulose synthase (CesA) gene family. Cellulose 11: 287–299. doi: 10.1023/b:cell.0000046417.84715.27

49. Song DL, Shen JH, Li LG (2010) Characterization of cellulose synthase complexes in *Populus* xylem differentiation. New Phytol 187: 777–790. doi: 10.1111/j.1469-8137.2010.03315.x

50. Amor Y, Haigler CH, Johnson S, Wainscott M, Delmer DP (1995) A membrane-associated form of sucrose synthase and its potential role in synthesis of cellulose and callose in plants. Proc Natl Acad Sci USA 92: 9353–9357. doi: 10.1073/pnas.92.20.9353

51. Hinchliffe DJ, Meredith WR, Yeater KM, Kim HJ, Woodward AW, et al. (2010) Near-isogenic cotton germplasm lines that differ in fiber-bundle strength have temporal differences in fiber gene expression patterns as revealed by comparative high-throughput profiling. Theor Appl Genet 120: 1347–1366. doi: 10.1007/s00122-010-1260-6

52. Zhong R, Lee C, Ye ZH (2010) Evolutionary conservation of the transcriptional network regulating secondary cell wall biosynthesis. Trends Plant Sci 15: 625–632. doi: 10.1016/j.tplants.2010.08.007

53. Hussey SG, Mizrachi E, Spokevicius AV, Bossinger G, Berger DK, et al. (2011) SND2, a NAC transcription factor gene, regulates genes involved in secondary cell wall development in *Arabidopsis* fibres and increases fibre cell area in Eucalyptus. BMC Plant Biol 11: 173. doi: 10.1186/1471-2229-11-173

54. Bhargava A, Ahad A, Wang S, Mansfield SD, Haughn GW, et al. (2013) The interacting MYB75 and KNAT7 transcription factors modulate secondary cell wall deposition both in stems and seed coat in *Arabidopsis*. Planta 237: 1199–1211. doi: 10.1007/s00425-012-1821-9

55. Zhong R, Lee C, McCarthy RL, Reeves CK, Jones EG, et al. (2011) Transcriptional activation of secondary wall biosynthesis by rice and maize NAC and MYB transcription factors. Plant Cell Physiol 52: 1856–1871. doi: 10.1093/pcp/pcr123

56. Brown DM, Zeef LA, Ellis J, Goodacre R, Turner SR (2005) Identification of novel genes in *Arabidopsis* involved in secondary cell wall formation using expression profiling and reverse genetics. Plant Cell 17: 2281–2295. doi: 10.1105/tpc.105.031542

57. Al-Ghazi Y, Bourot S, Arioli T, Dennis ES, Llewellyn DJ (2009) Transcript profiling during fiber development identifies pathways in secondary metabolism and cell wall structure that may contribute to cotton fiber quality. Plant Cell Physiol 50: 1364–1381. doi: 10.1093/pcp/pcp084

CITATION

CHAPTER 1

Teresa Cuellar, Eric Belhassen, Begoña Fernández-Calvín, Juan Orellana, and Jose L Bella (1996). Chromosomal differentiation in *Helianthus annuus* var.*macrocarpus*: heterochromatin characterization and rDNA location, *Heredity* (1996) 76, 586–591; doi:10.1038/hdy.1996.84

CHAPTER 3

Nagaki K, Tanaka K, Yamaji N, Kobayashi H and Murata M (2015). Sunflower centromeres consist of a centromere-specific LINE and a chromosome-specific tandem repeat. *Front. Plant Sci.* 6:912. doi: 10.3389/fpls.2015.00912

CHAPTER 4

Zamariola L, Tiang CL, De Storme N, Pawlowski W and Geelen D (2014) Chromosome segregation in plant meiosis. *Front. Plant Sci.* **5**:279. doi: 10.3389/fpls.2014.00279

CHAPTER 5

Cristina Maria Pinto de Paula and Vânia Helena Techio (2014). Immunolocalization of chromosome-associated proteins in plants – principles and applications. Botanical Studies 2014 55:63. Doi:10.1186/s40529-014-0063-5

CHAPTER 6

Katsuyuki Ichitani, Satoru Taura, Takahiro Tezuka, Yuuya Okiyama, and Tsutomu Kuboyama (2011). Chromosomal Location of *HWA1* and *HWA2*, Complementary Hybrid Weakness Genes in Rice, Rice (2011) 4: 29–38. DOI 10.1007/s12284-011-9062-2

CHAPTER 7

Koo D-H, Sehgal SK, Friebe B, Gill BS (2015) Structure and Stability of Telocentric Chromosomes in Wheat. PLoS ONE 10(9): e0137747. doi:10.1371/journal.pone.0137747

CHAPTER 8

Genc Y, Taylor J, Rongala J, Oldach K (2014) A Major Locus for Chloride Accumulation on Chromosome 5A in Bread Wheat. PLoS ONE 9(6): e98845. doi:10.1371/journal.pone.0098845

CHAPTER 9

Fang L, Tian R, Chen J, Wang S, Li X, Wang P, et al. (2014) Transcriptomic Analysis of Fiber Strength in Upland Cotton Chromosome Introgression Lines Carrying Different *Gossypium barbadense* Chromosomal Segments. PLoS ONE 9(4): e94642. doi:10.1371/journal.pone.0094642

INDEX